M000234917

THE ECOLOGICAL VISION

THE ECOLOGICAL VISION

Reflections on the American Condition

Peter F. Drucker

Transaction Publishers
New Brunswick (U.S.A.) and London (U.K.)

First paperback printing 2000
Copyright © 1993 by Peter F. Drucker

All rights reserved under International and Pan-American Copyright Conventions. No part of this book may be reproduced or transmitted in any form or by any means, electronic or mechanical, including photocopy, recording, or any information storage and retrieval system, without prior permission in writing from the publisher. All inquiries should be addressed to Transaction Publishers, Rutgers—The State University, 35 Berrue Circle, Piscataway, New Jersey 08854-8042.

This book is printed on acid-free paper that meets the American National Standard for Permanence of Paper for Printed Library Materials.

Library of Congress Catalog Number: 92-3967
ISBN: 1-56000-061-9 (cloth) 0-7658-0725-4 (paper)
Printed in the United States of America

Library of Congress Cataloging-in-Publication Data

Drucker, Peter Ferdinand, 1909-
    The ecological vision: reflections on the American condition/Peter F. Drucker.
    p.   cm.
    ISBN: 1-56000-061-9 (cloth:alk. paper) 0-7658-0725-4 (pbk.:alk. paper)
    1. United States—Social conditions. 2. Human ecology—United States. 3. Man—Influence on nature—United States. 4. Community. I. Title.
HN57.D76   1992
304.2'0973   92-3967
                                                                  CIP

Contents

Part Five   Work, Tools, and Society

Part Six   The Information-Based Society

Part Seven   Japan as Society and Civilization

Part Eight   Why Society Is Not Enough

Introduction

The thirty-one essays in this volume were written over a period of more than forty years—the earliest, chapter 10, "Keynes: Economics as a Magical System" in 1946; the most recent, "Afterword: Reflections of a Social Ecologist," especially for this volume in 1992. They are reprinted here as they first appeared, without change except occasionally in their title. These essays range over a wide array of disciplines and subject matter. Yet they all have in common that they are "Essays In Social Ecology" and deal with the man-made environment. They all, one way or another, deal with the interaction between individual and society or with the interaction between individual and community. And they try to look upon the economy, upon technology, upon art, as dimensions of social experience and as expressions of social values. None of these essays are essays in theory—the ones that come closest are chapter 11, "Management's New Role," and chapter 22, "Information, Communications, and Understanding"—the former attempting to present a theory of management as the specific social function of a society of organizations; the latter attempting to present information as both logic and perception, that is, as both communications and meaning. Neither essay, however, comes close to being "pure" theory. But even the essays that come closest to being "practical" and discuss "what to do," have a theorectical foundation or, at least, a theoretical dimension. And all were written to be easy and, I hope, enjoyable reading.

PETER F. DRUCKER

Part One

American Experiences

Introduction to Part One

The first three essays in this section: "The American Genius is Political," "Calhoun's Pluralism," and "Henry Ford: The Last Populist," are the only chapters I actually finished for a planned book tentatively entitled, "American Experiences," one of many planned books I never wrote or finished. The book was going to have some twenty chapters. Each, except the first one, "The American Genius is Political," was going to be built around a representative person such as Calhoun or Henry Ford. And each chapter was going to represent and discuss a uniquely American trait, value, or concept which, though often held unconsciously, shapes the way Americans see society, economy, government, and politics. The second chapter, for instance, was going to discuss Jonathan Edwards, the last of the great Puritan divines of Colonial America and America's greatest metaphysician—but also the father of the uniquely American concept of the relationship between State and Church. Whereas Edwards' contemporaries in eighteenth-century Enlightenment Europe worked to separate State and Church in order to protect the State from a power-hungry, bigoted Church, Edwards argued for a separation of the two in order to protect religion from government and politics. This explains why there has been no anticlericalism in America and why, alone among developed countries, America, the most laical of states is the most religious society. And the concluding chapter of the book was tentatively entitled "Lincoln: What We Believe In," and was to discuss the meaning and import of the fact that America, alone of all countries, has a politician for its great public saint.

The book was never finished; it fell victim to my growing interest in, and concern with, the new social organization of the increasingly pluralist society of organizations—first, of course, with the business enterprise—and with the new social function of this society, the management of institutions. But I never lost my interest in the subject. The two concluding essays in this section, though written many years later, also deal with American Experiences, that is, with the sociology of American society and American politics.

1

The American Genius Is Political

1

The genius of the American people is political. The one truly saintly figure this country has produced, the one name that symbolizes the "fulfilled life" to most Americans, the one man whose very life was dedicated toward directing human activity onto a higher goal, was a politician: Abraham Lincoln.

The phrase which since early days expresses the essence of their own society to Americans is a political promise: the "equal chance of every American boy to become President." One has only to translate the slogan—for instance, into a promise of "equal opportunity for every boy to become Prime Minister"—to see by contrast that it is uniquely American, and this not because the promise of equal opportunities in itself would be absurd, but because the political sphere is the meaningful sphere of social values only in this country.

The American nation itself has been formed out of a multitude of diverse national traditions not by imposing on the newcomer a uniform religion, uniform customs, a new culture—not even by imposing on him the American language. It has been formed by imposing on him a common political creed. What makes the immigrant into an American is an affirmation of abstract political principles, the oath of citizenship in which he promises to uphold "the republican form of government."

Above all, the meaning of this country—what it stands for for Americans themselves as well as for the world—is a political meaning.

First published in *Perspectives*, 1953.

It is a form of government, a social order, and an economic system that are equally in people's minds when they praise America and when they condemn her. And when Americans sing of their own country, they sing of her—in the words of our most popular anthem—as the "sweet land of liberty," which would hardly occur as a definition of one's country and as an avowal of one's identification with it even to the most ardent of European liberals.

We have to go back all the way to the Rome of Augustus with its concept of "Latinitas" to find a society that so completely understands itself in political terms as does the United States. Yet Latinitas was wishful thinking and never became reality, whereas the political meaning of the United States furnishes its essence: its ideal personality, its promises, its power of integration and assimilation. That the American genius is political is therefore the major key to the understanding of America, of its history, and of its meaning.

2

In the political sphere lie the American ideas and institutions which are peculiar to this country and which give it its distinction.

There is, first, the peculiarly American symbiosis of secular state and religious society which is the cornerstone of the American commonwealth. The United States today is both the oldest and the most thoroughgoing secular state. It is also at the same time the only society in the West in which belief in a supernatural God is taken for granted and in which the traditional religious bodies, the churches, continue to discharge, unchallenged, many important community functions. And this coincidence of secular state and religious society is not simply accidental. In the American mind the two serve as each other's main support, if not as each other's precondition. Everywhere else the secular state arose out of a revolt against religion. In this country the secular state owes its existence largely to the demand of the leaders of dominant, indeed of established, religious creeds that civil power and religious society be strictly separated for the greater good of religion and church.

In the political sphere also lies the concept of "constitutionalism" which is the chief organizing principle of American society. Constitutionalism is much more than respect for law; that is something, indeed, for which the American is not renowned. Rather, constitutionalism is a view of the nature and function of abstract principles and of their relationship to social action. It is a belief that power, to be beneficial, must be subjected to general and unchangeable rules. It is an assertion that ends and means cannot be meaningfully separated or considered apart from each other. It is a belief that the validity of actions can be determined by rational criteria. It is, in other words, a political ethics. The Constitution of the Republic is only one application of this fundamental belief. Constitutionalism has been the organizing force in every one of our major institutions. Today, for instance, we see it at work in the industrial sphere fashioning a "common law" in labor-management relations and reinforcing such principles as "federalism" and "legitimate succession" for management inside business enterprise.

Peculiarly American also is the political view of education. This is shown in the insistence of Americans that education, on all levels, be equally accessible to all, if not, indeed, obligatory on all. It is shown in the naïve but general belief that the level of general schooling is a fairly reliable index of civic competence and that any increase in it is a step toward a better and fuller citizenship. Most characteristically, the political view of education is expressed in the definition of the educational goal—the feature that makes understanding the American school system so difficult for the foreigner. American education rejects alike the traditionally European concept of the "educated individual" and the "trained robot" of modern totalitarianism. To both it opposes the demand that the school has to educate responsible self-governing citizens who, in Lincoln's words, "do not want to be masters because they do not want to be slaves." And because the American school is both free from control by the central government and a central political institution, it is bound to be the subject of violent political dispute whenever this country examines the premises on which its society and government rest, as, for instance, during the early days of the New Deal and again today.

Finally, there is the American political party—which has but the name in common with parties in the rest of the Western world. The major purpose of the American party is not to express principles, but to provide at all times a functioning and a legitimate government. It does not express a political philosophy to the realization of which governmental power is to be directed. It expresses the need of the body politic for a strong, a national, and a unifying government. For these reasons no American party has survived unless it succeeded in appealing to all classes and conditions, that is, unless it succeeded in being truly nationwide—which explains why both parties always are moderate in their actions. No party has survived unless it could appeal at once to men on the extreme right and to men on the extreme left. This, in turn, explains why both parties always appeal to the extremists in their campaign oratory. And no party has survived unless it succeeded in sinking differences of interests and principles in the unifying appeal of a common American creed.

3

The real achievement of American history in the first half of this century also lay in the political sphere. It is an achievement so great, so recent, and still so much in a state of becoming as to be almost beyond our vision: the creation of a new industrial society with new institutions—and of an industrial society that promises to be a stable, a free, and a moral society.

We ourselves in this country are prone to discuss our industrial achievements in economic, if not in purely technical, terms. But every single "productivity team" which came to this country from Europe in the forties and fifties to study the causes of American productivity soon saw that economic and technological factors explain less than nothing about the American economy. The key to an understanding lies in the social beliefs and the new social and political institutions that we have developed since the turn of the century. The clue is to be found in our development of the corporation as an organization of joint human efforts to a common goal, or in our new social organ, management, the function of which must be defined in sociopolitical terms, that is, as the organization and leadership of people.

More important even than any single one of these new institutions—more important even than such new ideas as productivity considered as a social responsibility of business, or the concept of the market as something that management through its own actions creates and expands—is the new American theorem as to the relationship between business and society—a theorem as opposed to nineteenth-century laissez-faire thinking as it is to nineteenth- and twentieth-century socialism. Business activity is no longer seen, as it was in laissez-faire theory, as something separate and distinct, without direct relationship to social and political goals. But it is also not seen, as in all socialist creeds, as something which, if it is to be in the public interest, must be controlled and perhaps suppressed by the government. Business activity is seen as a necessarily private activity which, for its own good and its own justification, has to strive for the common good and the stated ends of society. Business enterprise is thus seen as local and autonomous self-government which, by serving the ends of society, serves its own self-interests and guarantees its own survival. This, the most decisive contribution America has been making to the Western world these last fifty years, is unquestionably a theorem of political science.

4

In the political sphere, once more, lies that peculiarly American form of behavior: voluntary group action. Perhaps nothing sets this country as much apart from the rest of the Western world as its almost instinctive reliance on voluntary, and often spontaneous, group action for the most important social purposes.

We hear a great deal of "American individualism" these days. And there is certainly a fundamental belief in the individual, his strength, his integrity and self-reliance, his worth in the American tradition. But this "individualism" is much less peculiar to this country—and much less general—than its (usually overlooked) "collectivism." Only it is not the collectivism of organized governmental action from above. It is a collectivism of voluntary group action from below.

This shows itself in the way in which people in this country tackle their social and community problems. If the teenagers in a small town get out of control, the local service club, Rotary or Lions, will build a clubhouse for them and get them off the streets. If the hospital needs a new wing, the Woman's Club will get the money for it. American schools are run as much by a voluntary group organization—the Parent-Teacher Association—as they are by the Superintendent of Schools. And when we suddenly have to draft millions of Americans into uniform, we dump the social and community problems that this creates into the lap of voluntary committees and organizations. Even the drafting of young men into the armed forces is left to ordinary citizens who volunteer to serve without compensation on the draft boards of their communities.

Alone among Western countries, the United States knows no "Ministry of Education." But, then, no other country has anything to compare with the power and influence that private groups—e.g., the great foundations—wield over higher education. And yet we take this completely as a matter of course. Every American knows instinctively that this country is actually ruled by thousands of purely voluntary, purely private, mostly local groups. He takes it for granted that it is the easiest thing in the world to get such a group going to meet any local needs whatsoever and that despite its private character it will be responsive to public opinion.

It is not "competition" that characterizes American life, as some social scientists want us to believe. It is the symbiosis between competition and cooperation organized in and through voluntary, private groups. And while the roots of this behavior reach deep into the past— in the brotherhoods of the members of the small religious sects and in their interdependence, as well as in the neighborliness of the frontier— it is just as much at work in our newest institutions. No other feature of our business system is so much remarked upon by visiting management and labor delegates from abroad as the close, highly organized cooperation on policy matters, on technology, on business problems, and so forth that obtains between the most bitter business rivals; or the close, constant cooperation on day-to-day problems that obtains between managers and union officials in a given plant.

5

Americans themselves tend to take it for granted that American genius is political. If they ask at all what explains this, their answer is likely to be: the frontier. Certainly the human explosion which carried settlement and civilization in less than one century from the Atlantic across a wild, hostile, and uncharted continent to the shores of the Pacific is the greatest achievement of the American people and their deepest experience. And yet the frontier and the course of settlement were themselves already results of the American political genius rather than its origins. It was the ability to organize spontaneously for group action, it was the idea of "constitutionalism," that made orderly commonwealths out of the immigrant trains moving over the plains or mountains; it was the ability to adapt, as if instinctively, inherited patterns of social, political, and economic organization to the new and unexpected conditions of the frontier that alone made settlement of the country possible. Above all, the rapid settlement of the continent rested on the country's ability to make, so to speak overnight, Americans out of millions of immigrants from all over Europe. This presupposed the absorption of the alien into American citizenship and his integration into American political institutions.

The formative influence on America's political spirit is something deeper than climate, geography, or even the experiences of our history. It consists in our fundamental beliefs regarding the nature of man and the nature of the universe. To say that the genius of this country is political is to say that America, from earliest colonial times, has refused to split the world of ideas from the world of matter, the world of reason from the world of the senses. It is to say that she has alike refused to grant more than a temporary visitor's visa to the philosophies of Descartes, Hume, or the German Idealists as well as that of Marx—indeed, to the entire body of post-Cartesian philosophy on which modern European thought is founded. For every one of these philosophies denies meaning, if not existential reality, to the sphere of politics.

To the politician, matter can never be unreal, can never be an illusion, can never be base. His task is to use matter for constructive purposes. Nor can he ever consider ideas illusory or unreal. Without them he would have neither direction nor the ability to move toward his goal. The politician must always oppose to the "either-or" of the monistic philosopher—whether idealist or materialist, realist or nominalist, rationalist or intuitionist—the philosophical dualism of "both." He can neither, with the European humanist, divorce "sordid" politics from "pure" intellectual and artistic culture, nor can he, with the European materialist, see in politics simply a rationalization of material conditions or an automatic product of material forces. He can be neither "starry-eyed idealist" and "uplifter" nor the "money-mad chaser of the almighty dollar." He must always combine elements of both extremes in the attempt to find balance and harmony. For to the politician ideas and matter are the two poles of a single world, opposed to each other but at the same time inseparable, mutually interdependent, and each other's complement.

America has insisted—at least since Jonathan Edwards first started philosophizing on these shores two hundred and fifty years ago—that ideas and matter, reason and experience, logic and intuition, must always be held equally real and equally valid. And it is this philosophical world view that alone explains the central position of the political sphere in the American tradition.

In this view politics becomes at once a responsibility of man; it becomes the chief moral duty in human life. Politics becomes a respectable, in fact a creative, sphere of action—creative in the aesthetic meaning of the word—"the endless adventure," as one eighteenth-century politician called it; and creative in the spiritual meaning of the word, as partaking of charity through which man is moved from the pursuit of his own, empty self toward his real mission of making the creation glorify the creator. Politics as the creative, the meaningful, the moral, the responsible, the charitable sphere—that lies at the center of what America believes, what she has achieved, and what she stands for. The great men to whose memory she always turns in the hours of darkness and confusion have been those few political leaders who realized to the full the opportunity and the obligation of politics.

At bottom, the American philosophical position—the position that believes in politics as the human activity through which matter is made to serve spirit—stands on a religious foundation. It is not incorrect to call it "Christian." But actually it is the great contribution of Israel. For it is in the Old Testament that the Lord looked upon his material creation and "saw that it was very good." Yet the creation is nothing without the Spirit that created it. And it is man's specific task, his own mission and purpose, to make manifest the creator in and through the creation—to make matter express spirit.

6

The challenge that faces this country today is again primarily a challenge of politics. There is, first, the challenge of the Welfare State, of maintaining a free society and a free government under the threat of an indefinite "cold war" under the constant pressure of high taxation, the dangers of sneaking inflation, the tensions and worries of a world neither at peace nor at war.

There is the even greater challenge of the development of a foreign policy that is appropriate to America's power and to her responsibilities. This is doubly great a challenge since foreign policy is the one department of political action in which this country has failed so far. Ours is the third time that the United States has won military victory in a great war only to lose the peace almost immediately.

How we will meet these challenges will decide whether the American experiment has, after all, been nothing more than an episode in the suicidal Western civilization, or whether America will be capable of giving the West political inspiration leading it toward an age of freedom, stable peace, and prosperity. And the outcome of the test will depend not primarily on America's economic wealth or her military strategy, but on her political leadership, her political wisdom, her political maturity. What is on trial today is America's political genius.

2

Calhoun's Pluralism

The American party system has been under attack almost continuously since it took definite form in the time of Andrew Jackson. The criticism has always been directed at the same point: America's political pluralism, the distinctively American organization of government by compromise of interests, pressure groups, and sections. And the aim of the critics, from Thaddeus Stevens to Henry Wallace, has always been to substitute for this "unprincipled" pluralism a government based, as in Europe, on "ideologies" and "principles." But never before—at least not since the Civil War years—has the crisis been as acute as in this last decade; for the political problems which dominate our national life today—foreign policy and industrial policy—are precisely the problems which interest and pressure-group compromise is least equipped to handle. And while the crisis symptoms—a left-wing third party and the threatened split-off of the southern wing—are more alarming in the Democratic party, the Republicans are hardly much better off. The 1940 boom for the "idealist" Willkie and the continued inability to attract a substantial portion of the labor vote are definite signs that the Republican party, too, is under severe *ideological* pressure.

Yet, there is almost no understanding of the problem—precisely because there is so little understanding of the basic principles of American pluralism. Of course, every politician in this country must be able instinctively to work in terms of sectional and interest compro-

First published in *The Review of Politics*, 1948.

mise, and the voter takes it for granted. But there is practically no awareness of the fact that organization on the basis of sectional and interest compromise is both the distinctly American form of political organization and the cornerstone of practically all major political institutions of the modern United States. As acute an observer as Winston Churchill apparently did not understand that Congress works on a basis entirely different from that of Britain's Parliament, neither do nine out of ten Americans and 999 out of 1,000 teachers of those courses in "Civics." There is even less understanding that sectional and interest-group pluralism is not just the venal expediency of that stock villain of American folklore, the "politician," but that it in itself is a basic ideology, a basic principle—and the one which is the very foundation of our free society and government.

1

To find an adequate analysis of the principle of government by sectional and interest compromise, we have to go back more than a hundred years to John C. Calhoun and to his two political treatises[1] published after his death in 1852. Absurd, you will say, for it is practically an axiom of American history that Calhoun's political theories, subtle, even profound though they may have been, were reduced to absurdity and irrelevance by the Civil War. Yet, this "axiom" is nothing but a partisan vote of the Reconstruction period. Of course, the specific occasion for which Calhoun formulated his theories, the slavery issue, has been decided; and for the constitutional veto power of the states over national legislation, by means of which Calhoun proposed to formalize the principle of sectional and interest compromise, was substituted in actual practice the much more powerful and much more elastic but extraconstitutional and extralegal veto power of sections, interests, and pressure groups in Congress and within the parties.[2] But *his basic principle itself: that every major interest in the country, whether regional, economic, or religious, is to possess a veto power on political decisions directly affecting it*, the principle which Calhoun called, rather obscurely, *"the rule of concurrent majority,"* has become the organizing principle of American politics. And it is precisely this principle that is under fire today.

What makes Calhoun so important as the major key to the understanding of American politics is not just that he saw the importance in American political life of sectional and interest pluralism; other major analysts of our government, de Tocqueville, for instance, or Bryce or Wilson, saw that, too. But Calhoun, perhaps alone, saw in it more than a rule of expediency, imposed by the country's size and justifiable by results, if at all. He saw in it a basic principle of free government.

> Without this [the rule of concurrent majority based on interests rather than on principles] there can be . . . no constitution. The assertion is true in reference to all constitutional governments, be their forms what they may: It is indeed, the negative power which makes the constitution—and the positive which makes the government. The one is the power of acting—and the other the power of preventing or arresting action. The two, combined, make constitutional government.
>
> . . . it follows, necessarily, that where the numerical majority has the sole control of the government, there can be no constitution . . . and hence, the numerical, unmixed with the concurrent majority, necessarily forms, in all cases, absolute government.
>
> . . . The principle by which they (governments) are upheld and preserved . . . in constitutional governments is *compromise;*—and in absolute governments is *force.* . . .[3]

And however much the American people may complain in words about the "unprincipled" nature of their political system, by their actions they have always shown that they, too, believe that without sectional and interest compromises there can be no constitutional government. If this is not grasped, American government and politics must appear not only as cheap to the point of venality; they must appear as utterly irrational and unpredictable.

2

Sectional and interest pluralism has molded all American political institutions. It is the method—entirely unofficial and extraconstitutional—through which the organs of government are made to function, through which leaders are selected, policies developed, men and groups organized for the conquest and management of political power. In particular it is the explanation for the most distinctive features of the American political system: the way in which Congress operates, the way in which major government departments are set up and run, the qualifications for "eligibility" as a candidate for elective office, and the American party structure.

To all foreign observers of Congress two things have always remained mysterious: the distinction between the official party label and the "blocs" which cut across party lines; and the power and function of the congressional committees. And most Americans, though less amazed by the phenomena, are equally baffled.

The "blocs"—the "Farm Bloc," the "Friends of Labor in the Senate," the "Business Groups," and so on—are simply the expression of the basic tenet of sectional and interest pluralism that major interests have a veto power on legislation directly affecting them. For this reason they must cut across party lines—that is, lines expressing the numerical rather than the "concurrent" majority. And because these blocks have *(a)* only a negative veto and *(b)* that only on measures directly affecting them, they cannot in themselves be permanent groupings replacing the parties. They must be loosely organized; and one and the same member of Congress must at different times vote with different blocs. The strength of the blocs does not rest on their numbers, but on the basic mores of American politics which grant every major interest group a limited self-determination—as expressed graphically in the near sanctity of a senatorial filibuster. The power of the Farm Bloc, for instance, does not rest on the numerical strength of the rural vote—a minority vote even in the Senate with its disproportionate representation of the thinly populated agricultural states—but on its "strategic" strength, that is, on its being the spokesman for a recognized major interest.

Subordination of a major interest is possible, but only in a "temporary emergency." Most of the New Deal measures were, palpably, neither temporary nor emergency measures; yet their sponsors had to present them, and convincingly, as "temporary emergency measures" because they could be enacted only by overriding the extraconstitutional veto of the business interest.

Once the excuse of the "temporary emergency" had fully lost its plausibility, that major interest could no longer be voted down; and the policy collapsed. By 1946, for instance, labor troubles could be resolved only on a basis acceptable to both labor and employer: higher wages *and* higher prices. (Even if a numerical majority had been available to legislate against either party—and the business group could probably still have been voted down in the late forties—the solution had to be acceptable to both parties.)

The principle of sectional and interest compromise leads directly to the congressional committee system—a system to which there is no parallel anywhere in the world. Congress, especially the House, has largely abdicated to its committees because only in the quiet and secrecy of a committee room can sectional compromise be worked out. The discussion on the floor as well as the recorded vote is far too public and therefore largely for the folks back home. But a committee's business is to arrive at an agreement between all major sectional interests affected; which explains the importance of getting a bill before the "right" committee. In any but an American legislature the position of each member, once a bill is introduced, is fixed by the stand of his party, which, in turn, is decided on grounds that have little to do with the measure itself, but are rather dictated by the balance of power in the government and by party programs. Hence it makes usually little difference which committee discusses a bill or whether it goes before a committee at all. In the United States, however, a bill's assignment to a specific committee decides which interest groups are to be recognized as affected by the measure and therefore entitled to a part in writing it ("who is to have standing before the committee"), for each committee represents a specific constellation of interests. In many cases this first decision therefore decides the fate of a proposed measure, especially as the compromise worked out by the committee is generally accepted once it reaches the floor, especially in the House.

It is not only Congress but every individual member of Congress himself who is expected to operate according to the "rule of concurrent majority." He is considered both a representative of the American people and responsible to the national interest and a delegate of his constituents and responsible to their particular interests. Wherever the immediate interests of his constituents are not in question, he is to be a statesman; wherever their conscience or their pocketbooks are affected, he is to be a business agent. This is in sharp contrast to the theory on which any parliamentary government is based—a theory developed a full two hundred years ago in Edmund Burke's famous speech to the voters at Bristol—according to which a member of Parliament represents the commonwealth rather than his constituents. Hence in all parliamentary countries, the representative can be a stranger to his constituency—in the extreme, as it was practiced in Weimar Germany, there is one long national list of candidates who run in all constituencies—whereas the congressman in this country must be a resident of his constituency. And while an American senator considers it a compliment and an asset to be called "Cotton Ed Smith," the Speaker of the House of Commons not so long ago severely reprimanded a member for calling another member—an official of the miners' union—a "representative of the coal miners."

The principle of sectional and interest pluralism also explains why this is the only nation where Cabinet members are charged by law with the representation of special interests—labor, agriculture, commerce. In every other country an agency of the government—any agency of the government—is solemnly sworn to guard the public interest against "the interests." In this country the concept of a government department as the representative of a special interest group is carried down to smaller agencies and even to divisions and branches of a department. This showed itself, for example, during World War II in such fights as that between OPA, representing the consumer, and the War Production Board, representing the producer; or within WPB between the Procurement branches, speaking for the war industries, and the Civilian Requirements Branch, speaking for the industries producing for the "home front."

The mystery of "eligibility"—the criterion which decides who will make a promising candidate for public office—which has baffled so many foreign and American observers, Bryce, for instance—also traces back to the "rule of the concurrent majority." Eligibility simply means that a candidate must not be unacceptable to any major interest, religious, or regional group in the electorate; it is primarily a negative qualification. Eligibility operates on all levels and applies to all elective offices. It has been brilliantly analyzed in "Boss" Flynn's *You're the Boss.* His classical example is the selection of Harry Truman as Democratic vice-presidential candidate in 1944. Truman was "eligible" rather than Wallace, Byrnes, or Douglas precisely because he was unknown; because he was neither easterner nor westerner nor southerner, because he was neither New Deal nor Conservative, and so on; in short, because he had no one trait strong enough to offend anybody anywhere.

But the central institution based on sectional pluralism is the American party. Completely extraconstitutional, the wonder and the despair of every foreign observer who cannot fit it into any of his concepts of political life, the American party (rather than the states) has become the instrument to realize Calhoun's "rule of the concurrent majority."

In stark contrast to the parties of Europe, the American party has no program and no purpose except to organize divergent groups for the common pursuit and conquest of power. Its unity is one of action, not of beliefs. Its only rule is to attract—or at least not to repel—the largest possible number of groups. It must, by definition, by acceptable equally to the right and the left, the rich and the poor, the farmer and the worker, the Protestant and the Catholic, the native and the foreign-born. It must be able to rally Mr. Rankin of Mississippi and Mr. Marcantonio of New York—or Senator Flanders and Colonel McCormick—behind the same Presidential candidate and the same "platform."

As soon as it cannot appeal at least to a minority in every major group (as soon, in other words, as it provokes the veto of one section, interest, or class) a party is in danger of disintegration. Whenever a party loses its ability to fuse sectional pressures and class interests into one national policy—both parties just before the Civil War, the Republican party before its reorganization by Mark Hanna, both parties again today—the party system (and with it the American political system altogether) is in crisis.

Consequently, it is not that Calhoun was repudiated by the Civil War, which is the key to the understanding of American politics, but that he has become triumphant since.

The apparent victors, the "Radical Republicans," Thaddeus Stevens, Seward, Chief Justice Chase, were out to destroy not only slavery and states' rights but the "rule of the concurrent majority" itself. And the early Republican party—before the Civil War and in the Reconstruction period—was, indeed, determined to substitute principle for interest as the lodestar of American political life. But in the end it was the political thought of convinced pluralists such as Abraham Lincoln and Andrew Johnson, rather than the ideologies of the Free Soilers and Abolitionists, which molded the Republican party. And ever since, the major developments of American politics have been based on Calhoun's principle. To this the United States owes the strength as well as the weaknesses of its political system.

3

The weaknesses of sectional and interest compromise are far more obvious than its virtues; they have been hammered home for a hundred years. Francis Lieber, who brought the dominant German political theories of the early nineteenth century to this country, attacked pluralism in Calhoun's own state of South Carolina a century ago. Twenty years later Walter Bagehot contrasted, impressively, General Grant's impotent administration with those of Gladstone and Disraeli to show the superiority of ideological party organization. The most thorough and most uncompromising criticism came from Woodrow Wilson; and every single one of the Professor's points was amply borne out by his later experience as President. Time has not made these weaknesses any less dangerous.

There is, first of all, the inability of a political system based on the "rule of the concurrent majority" to resolve conflicts of principles. All a pluralist system can do is to deny that "ideological" conflicts (as they are called nowadays) do exist. Those conflicts, a pluralist must assert, are fundamentally either struggles for naked power or friction between interest groups which could be solved if only the quarreling parties sat down around a conference table. Perhaps, the most perfect, because most naïve, expression of this belief remains the late General Patton's remark that the Nazis were, after all, not so very different from Republicans or Democrats. (Calhoun, while less naïve, was as unable to understand the reality of "ideological" conflict in and around the slavery problem.)

In nine cases out of ten the refusal to acknowledge the existence of ideological conflict is beneficial. It prevents fights for power, or clashes of interests, from flaring into religious wars where irreconcilable principles collide (a catastrophe against which Europe's ideological politics have almost no defense). It promotes compromise where compromise is possible. But in a genuine clash of principles—and, whatever the pluralists say, there *are* such clashes—the "rule of concurrent majority" breaks down; it did, in Calhoun's generation, before the profound reality of the slavery issue. A legitimate ideological conflict is actually aggravated by the pluralists' refusal to accept its reality: the compromisers who thought the slavery issue could be settled by the meeting of good intentions, or by the payment of money, may have done more than the Abolitionists to make the Civil War inevitable.

A weakness of sectional and interest pluralism just as serious is that it amounts to a principle of inaction. The popular assertion, "It's better to make the wrong decision than to do nothing at all," is, of course, fallacious; but no nation, however unlimited its resources, can have a very effective policy if its government is based on a principle that orders it to do nothing important except unanimously. Moreover, pluralism increases exorbitantly the weight of well-organized small interest groups, especially when they lobby *against* a decision. Congress can far too easily be high-pressured into emasculating a bill by the expedient of omitting its pertinent provisions; only with much greater difficulty can Congress be moved to positive action. This explains, to a large extent, the eclipse of Congress during the last hundred years, both in popular respect and in its actual momentum as policymaking organ of government. Congress, which the founding fathers had intended to be the central organ of government—a role which it fulfilled up to Andrew Jackson—became the compound representative of sections and interests and, consequently, progressively incapable of national leadership.

Pluralism gives full weight—more than full weight—to sections and interests; but who is to represent the national welfare? Ever since the days of Calhoun, the advocates of pluralism have tried to dodge this question by contending that the national interest is equal to the sum of all particular interests and that it therefore does not need a special organ of representation. But this most specious argument is contradicted by the most elementary observation. In practice, pluralism tends to resolve sectional and class conflicts at the expense of the national interest which is represented by nobody in particular, by no section and no organization.

These weaknesses had already become painfully obvious while Calhoun was alive and active—during the decade after Andrew Jackson, the first president of pluralism. Within a few years after Calhoun's death, the inability of the new system to comprehend and to resolve an ideological conflict—ultimately its inability to represent and to guard the national interest—had brought catastrophe. For a hundred years and more, American political thought has therefore revolved around attempts to counteract if not to overcome these weaknesses.

Three major developments of American constitutional life were the result: the growth of the functions and powers of the president and his emergence as a "leader" rather than as the executive agent of the Congress; the rise of the Supreme Court, with its "rule of law," to the position of arbiter of policy; the development of a unifying ideology—the "American Creed."

Of these the most important—and the least noticed—is the American Creed. In fact, I know of no writer of major importance since de Tocqueville who has given much attention to it. Yet even the term "un-American" cannot be translated successfully into any other language, least of all into "English" English. In no other country could the identity of the nation with a certain set of ideas be assumed—at least not under a free government. This unique cohesion on principles shows, for instance, in the refusal of the American voter to accept socialists and communists as "normal" parties, simply because both groups refuse to accept the assumption of a common American ideology. It shows, for another example, in the indigenous structure of the American labor movement with its emphasis on interest pressure rather than on a political philosophy. And this is also the only Western country where "Civics" are taught in schools—the only democratic country which believes that a correct social philosophy could or should be part of public education.

In Europe, a universal creed would be considered incompatible with a free society. Before the advent of totalitarianism, no European country had ever known anything comparable to the flag salute of the American school child.[4] For in Europe all political activity is based on ideological factions; consequently, to introduce a uniform ideology in a European country is to stamp out *all* opposition. In the United States ideological homogeneity is the very basis of political diversity. It makes possible the almost unlimited freedom of interest groups, religious groups, pressure groups, and the like; and in this way it is the very fundament of free government. (It also explains why the preservation of civil liberties has been so much more important a problem in this country—as compared to England or France, for instance.) The assumption of ideological unity gives the United States the minimum of cohesion without which its political system simply could not have worked.

4

But is even the "American dream" enough to make a system based on the "rule of the concurrent majority" work today? Can pluralism handle the two major problems of American politics—the formulation of a foreign policy, and the political organization of an industrial society—any more successfully than it could handle the slavery issue? Or is the American political system as much in crisis as it was in the last years of Calhoun's life, and for pretty much the same reasons?

A foreign policy can never be evolved by adding particular interests—regional, economic, or racial—or by compromising among them; it must supersede them. If Calhoun's contention that the national interest will automatically be served by serving the interests of the parts is wrong anywhere, it is provably wrong in the field of foreign affairs.

A foreign policy and a party system seem to be compatible only if the parties are organized on programmatic grounds, that is, on principles. For if not based on general principles, a foreign policy will become a series of improvisations without rhyme or reason. In a free society, in which parties compete for votes and power, the formulation of a foreign policy may thus force the parties into ideological attitudes which will sooner or later be reflected in their domestic policies, too.

This was clearly realized in the early years of the Republic, when foreign policy was vital to a new nation, clinging precariously to a long seaboard without hinterland, engaged in a radical experiment with new political institutions, surrounded by the Great Powers of that time, England, France, and Spain, all of them actually or potentially hostile. This awareness of foreign policy largely explains why the party system of the founding fathers—especially of Hamilton—was an ideological one; it also explains why the one positive foreign-policy concept this country developed during the nineteenth century, the Monroe Doctrine, was formulated by the last two politically active survivors of the founding generation, Monroe and John Quincy Adams. No matter how little Calhoun himself realized it, his doctrine would have been impossible without the French Revolution and the Napoleonic Wars which, during the most critical period of American integration, kept our potential European enemies busy. By 1820, the country had become too strong, had taken in too much territory, to be easily attacked; and it was still not strong enough, and far too much absorbed in the development of its own interior, to play a part in international affairs. Hence Calhoun, and all America with him, could push foreign policy out of their minds—so much so that this is the only country in which it is possible to write a comprehensive work on an important historical period without as much as a mention of foreign affairs, as Arthur M. Schlesinger, Jr., managed to do in his *Age of Jackson*.

But today foreign policy is again as vital for the survival of the nation as it ever was during the administrations of Washington and Jefferson. And it has to be a foreign *policy*, that is, a making of decisions; hence neither "isolationism" nor "internationalism" will do. (For "internationalism"—the search for formulas which will provide automatic decisions, even in advance—is also a refusal to have a foreign policy; it may well have done this country, and the world, as much harm as "isolationism"—perhaps more.) To survive as the strongest of the Great Powers, the United States might even have to accept permanently the supremacy of foreign policies over domestic affairs, however much this may go against basic American convictions, and indeed against the American grain. But no foreign policy can be evolved by the compromise of sectional interests or economic pressures; yet neither party, as today constituted, could develop a foreign policy based on definite principles.

The other great national need is to resolve the political problems of an industrial society. An industrial society is by nature ultrapluralistic, because it develops class and interest groups that are far stronger, and far more tightly organized, than any interest group in a preindustrial age. A few big corporations, a few big unions, may be the actually decisive forces in an industrial society. And these groups can put decisive pressure on society: they can throttle social and economic life.

The problem does not lie in "asocial behavior" of this or that group, but in the nature of industrial society, which bears much closer resemblance to feudalism than to the trading nineteenth century. Its political problems are very similar to those which feudalism had to solve—and failed to solve. It is in perpetual danger of disintegration into virtually autonomous fiefs, principalities, "free cities," "robber baronies," and "exempt bishoprics"—the authority and the interest of the nation trampled underfoot, autonomous groups uniting to control the central power in their own interest or disregarding government in fighting each other in the civil conflict of class warfare. And the alternative to such a collapse into anarchy or civil war—the suppression of classes and interest groups by an all-powerful government—is hardly more attractive.

An industrial society cannot function without an organ able to superimpose the national interest on economic or class interests. More than a mere arbiter is needed. The establishment of the "rules of civilized industrial warfare," as was done by both the Wagner Act and the Taft-Hartley Act, tries to avoid the need for policies by equalizing the strength of the conflicting sections; but that can lead only to deadlock, to collusion against the national interest, or, worse still, to the attempt to make the national authority serve the interest of one side against the other. In other words, an industrial society cannot fully accept Calhoun's assumption that the national good will evolve from the satisfaction of particular interests. An industrial society without national policy will become both anarchic and despotic.

Small wonder that there has been increasing demand for a radical change which would substitute ideological parties and programmatic policies for the pluralist parties and the "rule of the concurrent majority" of the American tradition. Henry Wallace's Third Party Movement, while the most publicized, may well be the least significant development; for third parties are, after all, nothing new in our political history. But for the first time in a hundred years there is a flood of books—and by serious students of American government—advocating radical constitutional reform. However much Senator Fulbright, Henry Hazlitt, and Thomas Finletter disagree on details, they are one in demanding the elimination—or at least the limitation—of the "rule of the concurrent majority" and its replacement by an ideological system functioning along parliamentary lines. More significant even may be Walter Reuther's new unionism with its blend of traditional pressure tactics and working-class, that is, ideological, programs and aims.

5

Yet all these critics and reformers not only fail to ask themselves whether an ideological system of politics would really be any better equipped to cope with the great problems of today—and neither the foreign nor the industrial policy of England, that most successful of all ideologically organized countries, looks any too successful right now; the critics also never stop to consider the unique strength of our traditional system.

Our traditional system makes sure that there is always a legitimate government in the country; and to provide such a government is the first job of any political system—a duty which a great many of the political systems known to man have never discharged.

It minimizes conflicts by utilizing, rather than suppressing, conflicting forces. It makes it almost impossible for the major parties to become entirely irresponsible: neither party can afford to draw strength from the kind of demagogic opposition, without a governmental responsibility, which perpetually nurtures fascist and communist parties abroad. Hence, while the two national parties are willing to embrace any movement or any group in the country that commands sufficient following, they in turn force every group to bring its demands and programs into agreement with the beliefs, traditions, and prejudices of the people.

Above all, our system of sectional and interest compromise is one of the only two ways known to man in which a free government and a free society can survive—and the only one at all adapted to the conditions of American life and acceptable to the American people.

The central problem in a free government is that of factions, as we have known since Plato and Aristotle. Logically, a free government and factions are incompatible. But whatever the cause—vanity and pride, lust for power, virtue or wickedness, greed or the desire to help others—factionalism is inherent in human nature and in human society. For two thousand years the best minds in politics have tried to devise a factionless society—through education (Plato), through elimination of property (Thomas More), through concentration on the life of the spirit outside of worldly ambition (the political tradition of Lutheranism). The last great attempt to save freedom by abolishing faction was Rousseau's. But to create the factionless free society is as hopeless as to set up perpetual motion. From Plato to Rousseau, political thought had ended up by demanding that factions be suppressed, that is, that freedom, to be preserved, be abolished.

The Anglo-American political tradition alone has succeeded in breaking out of this vicious circle. Going back to Hooker and Locke, building on the rich tradition of free government in the cities of the late Middle Ages, Anglo-American political realism discovered that if factions cannot be suppressed, they must be utilized to make a free government both freer and stronger. This one basic concept distinguishes Anglo-American political theory and practice from continental European politics and accounts for the singular success of free and popular governments in both countries. Elsewhere in the Western world the choice has always been between extreme factionalism, which makes government impotent if not impossible and inevitably leads to civil war, and autocracy, which justifies the suppression of liberty with the need for effective and orderly government. Nineteenth-century France with its six revolutions, or near revolutions, stands for one, the totalitarian governments of our time for the other alternative of continental politics.

But—and this is the real discovery on which the Anglo-American achievement rests—factions can be used constructively only if they are encompassed in a frame of unity. A free government on the basis of sectional interest groups is possible only when there is no ideological split in the country. This is the American solution. Another conceivable solution is to channel the driving forces, the vectors of society, into ideological factions which obtain their cohesion from a program for the whole of society and from a creed. But that presupposes an unquestioned ruling class with a common outlook on life, with uniform mores and traditional, if not inherent, economic security. Given that sort of ruling class, the antagonist in an ideological system can be expected to be a "loyal opposition," that is, to accept the rules of the game and to see himself as a partner rather than as a potential challenger to civil war. But a ruling class accepted by the people as a whole, and feeling itself responsible to the people as a whole, cannot be created by fiat or overnight. In antiquity only Rome, in modern times only England, achieved it. On the Continent, all attempts to create a genuine ruling class have failed dismally.

In this country, the ruling-class solution was envisaged by Alexander Hamilton and seemed close to realization under the presidents of the "Virginia Dynasty." Hamilton arrived at his concept with inescapable consistency; for he was absorbed by the search for a foreign policy and for the proper organization of an industrial society—precisely the two problems which, as we have seen, pluralism is least equipped to resolve. But even if Hamilton had not made the fatal mistake of identifying wealth with rulership, the American people could not have accepted his thesis. A ruling class was incompatible with mass immigration and with the explosive territorial expansion of nineteenth-century America. It was even more incompatible with the American concept of equality. And there is no reason to believe that contemporary America is any more willing to accept Hamilton's concept, Mr. James Burnham's idea of the managerial elite notwithstanding. This country as a free country has no alternative, it seems, to the "rule of the concurrent majority," no alternative to sectional pluralism as the device through which factions can be made politically effective.

It will be very difficult, indeed, to resolve the problems of foreign and of industrial policy on the pluralist basis and within the interest-group system, though not provably more difficult than these problems would be on another, ideological, basis. It will be all the harder as the two problems are closely interrelated; for the effectiveness of any American foreign policy depends, in the last analysis, on our ability to show the world a successful and working model of an industrial society. But if we succeed at all, it will be with the traditional system, horse trading, logrolling, and politicking all included. An old saying has it that this country lives simultaneously in a world of Jeffersonian beliefs and in one of Hamiltonian realities. Out of these two, Calhoun's concept of "the rule of the concurrent majority" alone can make one viable whole. The need for a formulated foreign policy and for a national policy of industrial order is real, but not more so than the need for a real understanding of this fundamental American fact: the pluralism of sectional and interest compromise is the warp of America's political fabric; it cannot be plucked out without unraveling the whole.

Notes

1. *A Disquisition on Government* and *A Discourse on the Constitution and Government of the United States.*
2. Calhoun's extreme legalism, his belief that everything had to be spelled out in the written Constitution—a belief he shared with his generation—is one of the major reasons why the importance of his thesis has not been generally recognized. Indeed, it is of the very essence of the concept of "concurrent majority" that it cannot be made official and legal in an effective government—the express veto such as the U.N. Charter gives to the Great Powers makes government impossible.

3. Quotations from *A Disquisition on Government* (Columbia, S.C., 1852), 35–37.
4. The perhaps most profound discussion of the American ideological cohesion can be found in the two decisions of the Supreme Court on the compulsory flag salute, and in the two dissents therefrom, which deserve high rating among American state papers.

3

Henry Ford: The Last Populist

1

Henry Ford's hold on America's imagination—indeed, on the imagination of the world's masses—was not due to his fabulous financial success. And it can only partly be explained by the overwhelming impact of the automobile on our way of life. For Henry Ford was less the symbol and embodiment of new wealth and of the automobile age than the symbol and embodiment of our new industrial mass-production civilization.

He perfectly represented its success in technology and economics; he also perfectly represented its political failure so far—its failure to build an industrial order, an industrial society. The central problem of our age is defined in the contrast between the functional grandeur of the River Rouge plant, with its spotless mechanical perfection, and the formlessness and tension of the social jungle that is Detroit. And the two together comprise Henry Ford's legacy.

Both his success and his failure can be traced to his being thoroughly representative of that most native and most dominant of all American traditions, the one which in populism found its major political expression. Indeed, Henry Ford was both the last populist and perhaps the greatest one. He owed all his basic convictions to Bryan: pacifism, isolationism, hatred of monopoly and of "Wall Street" and of "international bankers," firm belief in a sinister international conspiracy, and so forth. He also made true the great dream of the political

First published in *Harper's Magazine*, 1947, shortly after Ford's death.

crusaders of 1896: that industrial production might be made to serve the common man. This dream had obsessed the American people since Brook Farm and Robert Owen's New Harmony, half a century before Bryan.

The populists had believed that a Jeffersonian millennium would result automatically from eliminating "monopoly" and the "money power" and the "satanic mills" of crude industrialism—as these terms were understood in the nineteenth century. Ford fulfilled the dream. He succeeded without benefit of monopoly; he defied the big bankers; he gave his factories a clean and airy efficiency which would have delighted nineteenth-century reformers. But in fulfilling the dream he dispelled it. And in the place of the old enemies which he vanquished, we have today, in the industrial system which Ford did so much to develop, new problems to face: the long-term depression, and the political and social problems of industrial citizenship in the big plant. Henry Ford's solution of the industrial problems with which the nineteenth century had wrestled unsuccessfully constituted his success, his achievement. His inability to solve the problems of the new industrial system, his inability to see even that there were such problems, was the measure of his final and tragic failure.

It may seem paradoxical to interpret Henry Ford's importance in terms of a concept—especially a political concept such as populism. He himself had nothing but contempt for concepts and ideas and prided himself on being neither a theoretician nor a politician, but a "practical man." And the main criticism which has been leveled against him and against everything he stood for—the criticism embodied in, for instance, Charlie Chaplin's *Modern Times*—has been that he made mechanical perfection an end in itself. But even his contribution to technology was not really a technical but a conceptual one—superb production man and engineer though he was. For he invented nothing, no new technique, no new machine, not even a new gadget. What he supplied was *the idea of mass production itself:* organization of men, machines, and materials into one productive whole.

In economics, too, Ford discovered no new facts; the data showing the effect of volume production on costs had all been collected and analyzed. But Ford was the first manufacturer to understand that these data disproved the traditional theory that restricted production and a high profit margin—that is, monopoly—provided the most profitable form of industrial production. He demonstrated that one could raise wages, cut prices, produce in tremendous volume, and still make millions.

Above all, Ford himself regarded his technical and economic achievements primarily as means to a social end. He had a definite political and social philosophy to which he adhered to the point of doctrinaire absurdity. Concern with the social effects of his actions determined every one of his steps and decisions throughout his entire life. It underlay the break with his early partners who wanted to produce a luxury car for the rich rather than follow Ford's harebrained idea of a cheap utility car for the masses. It motivated the radical wage policy of the early Ford who in 1914 fixed his minimum wage at the then utopian figure of $5.00 a day for unskilled labor. It showed in Ford's lifelong militant pacifism, of which the tragicomic Peace Ship episode of 1915–1916 was only one manifestation. It showed in his isolationism, in his hostility to Wall Street, and in the raucous pamphleteering of the Dearborn *Independent* in the twenties. This social philosophy explains the millions he poured into "chemurgy" or into utopian village communities of self-sufficient, sturdy, yeoman farmers. It was responsible for his belief in decentralization and for his nostalgic attempt to recreate the atmosphere of an earlier and simpler America in a museum community—right next door to the River Rouge plant.

It might almost be said that Henry Ford's lifework, despite these moves of his, brought about the opposite kind of world from the one he hoped for and believed in. Thus Ford, the pacifist, built up one of the world's greatest armament plants and helped to make possible the mechanized warfare of our age. Ford, the isolationist, more than any other man has made it impossible for this country to stay out of international politics and international wars: for he made this country the most powerful industrial nation on earth. Ford, the agrarian decentralist, left as his life's work the River Rouge plant, the most highly centralized and most completely mechanized concentration of industrial power in the world. The enemy of finance capital and bank credit, he made installment buying a national habit. An orthodox Jeffersonian, he has come to stand for the extreme application of the assembly-line principle, with its subordination of the individual to the machine. And the very workers at the Ford Motor Company, whose mass production was to give economic security and full industrial citizenship to all, organized before his death the most class-conscious union in America—and a Communist-dominated local at that.

Yet it would be wrong to argue from the failure of Ford's social ideas that they never were anything but "eccentric hobbies," as the obituaries rather condescendingly called them. The tragic irony with which his every move turned against him in the end does not alter the fact that his was the first, and so far the only, systematic attempt to solve the social and political problems of an industrial civilization. There is also little doubt that Ford himself believed—certainly until 1941 when the Ford workers voted for the CIO, and perhaps even afterward—that he had actually found the answer for which the American people had been searching for almost a century: the realization of the Jeffersonian society of independent equals through industrial technology and economic abundance.

Nor was he alone in this appraisal of the meaning of his work. It was shared by the American people as a whole in the years immediately following World War I; witness Wilson's urging in 1918 that Ford run for the Senate and the powerful "Ford for President" boom of 1923. The view was also held abroad, especially in the Europe of the early twenties and in Lenin's Russia—perhaps even more generally there than here. Indeed, it was the performance of Henry Ford's America which in 1918 and 1919 gave substance to Wilson's promise of the millennium of peace, democracy, and abundance and which established America's moral and political authority in those years. And the Ford spell remained potent long after Wilson's promise had faded under the cold light of the international realities of the 1920s.

The world of Ford's death, the world after World War II, was at least as much under the spell of Frankin D. Roosevelt's name as an earlier generation had been under that of Wilson. But Henry Ford in 1946 no longer symbolized an America that had successfully solved the basic social problems of an industrial world. He stood instead for the lack of a solution. And that surely accounted in large measure for the difference between 1919 and 1947 in the acceptance and the effectiveness of America's moral and economic leadership.

2

Henry Ford took the conveyor belt and the assembly line from the meat-packing industry, where they had been in general use as early as

1880. The interchangeability of precision-made parts was an even older principle; it went back to the rifle plant which Eli Whitney built in Bridgeport for the War of 1812. The idea of breaking down a skilled job into the constituent elementary motions, so that it could be performed by unskilled men working in series, had been thoroughly explored—by Taylor among others—and had been widely used in American industry twenty years before Ford came on the scene, as, for example, by Singer Sewing Machine and National Cash Register. Yet we associate all these principles with Henry Ford, and rightly so. For each of them had been employed only as an auxiliary to the traditional manufacturing process. It was Ford who first combined them and evolved out of them consciously and deliberately a new concept of industrial production, a new technology. It is this new concept of mass production which in scarcely more than one generation gave us a new industrial civilization.

To Ford the importance of this new principle lay in its impact on society as the means for producing an abundance of cheap goods with the minimum of human effort and toil. Mass production itself, however, he considered as something purely technical, as simply a new method of organizing *mechanical* forces. Ford disciples, heirs, and imitators, the engineers and production men who still run our big industries, are certainly as convinced as their master that mass production is a mechanical technique; many use it as if it were a mere gadget. And Charlie Chaplin took the same view when, in *Modern Times,* he caricatured our modern industrial civilization.

But if mass production were, indeed, only a technique, and primarily mechanical—if it were different in degree but not in kind from pulley, lever, or wheel—it could be applied only to mechanical tasks similar to the ones for which it was first developed. But long before the World War II, mass production principles were used for such jobs as the sorting and filling of orders in a mail-order house and the diagnosis of patients in the Mayo Clinic. Henry Luce even used it successfully to organize writers—traditionally regarded as extreme individualists—for the mass production of interchangeable "formula-writing." During World War II we applied mass-production principles to thousands of new products and processes and to such problems as the selection and training of men in the armed services. In all these uses the mechanisms of the assembly line are purely subordinate, if, indeed, applied at all. In other words, mass production is not fundamentally a mechanical principle, but *a principle of social organization*. It does not coordinate machines or the flow of parts; it organizes men and their work.

Ford's importance lies precisely in the fact that his principle of mass production substitutes the coordination of human beings for the coordination of inanimate parts and of mechanical forces on which industry was originally based. When we talk of the Industrial Revolution, we think at once of Watt's steam engine. It is true that there was a lot more to the Industrial Revolution than new machines; but the steam engine is a good symbol for it because the essence of early industry was the new organization of mechanical forces. Mass production is based, however, on the organization of human beings and of human work—something radically different from anything that was developed in the early days of industry. Indeed, it has brought about a new Industrial Revolution. The assembly line is a symbol for a new principle of social organization, a new relationship between men who work together in a common task, if not for a common purpose.

On what basis does this mass-production principle organize men? What kind of society does it either assume or create? It assumes or creates a society in which things are produced by the cooperation of individuals, not by a single individual. By himself the individual in modern mass-production industry is completely unproductive and ineffectual. But the organized group produces more, better, and more effectively than any individual or any number of individuals working by themselves ever could. In this society the whole—the organized group—is clearly not only more than, but different from, the sum of its parts.

Proof of this is what happens when a man loses his place in the organized group or his access to the productive organism; when, in other words, he becomes unemployed. Under modern mass-production conditions, the man who has lost his job is not just out of luck economically; in fact, in a rich country such as ours, the direct economic consequences of unemployment can be minimized almost to the vanishing point. But he is incapable of producing anything, of being effective in society; in short, he is incapable of being a citizen; he is cast out. For he derives his productiveness, his function in the community, his citizenship—at least his effective rather than purely formal citizenship—from his position in the group effort in the team, in the productive organism.

It is this social effect of unemployment, incidentally, rather than the economic effect, that makes it the major catastrophe it is. That unemployment endangers people's standards of living is, of course, bad enough; but that it endangers their citizenship and self-respect is its real threat and explains our panicky fear of the "next depression."

In the society of the modern mass-production plant everyone derives his effectiveness from his position in an organized group effort. From this follow some important consequences. One is that such a society needs a government, a direction, a management responsible to no one special-interest group, to no one individual, but to the overall purpose, the overall maintenance and strengthening of the whole without which no individual, no special-interest group, could be effective. It also follows that in such a society there must be rank: a difference of authority and prestige based on the differentiation of functions. But at the same time, in such a society no one individual is less important or more important than another. For while no one individual is irreplaceable—only the organized relationship between individuals is irreplaceable and essential—every single operation, every single function, is equally necessary; the whole order would collapse, the entire productive machine would come to a stop, were one to take out one function, one job—just as the whole chain becomes useless if one takes out one link. That is why, in such a society, there should be simultaneously an inequality of subordination and command, based on the differentiation of functions, and a basic equality, based on membership and citizenship.

This is by no means a new type of social organization; on the contrary, it is a very old one. It was described in the old Roman fable retold in Shakespeare's *Coriolanus* which likened society to the human body, none of whose organs—neither feet, nor hands, nor heart, nor stomach—could exist or work by itself, while yet the body could not properly work without any of them. It was expressed in the medieval metaphors of the order of the spheres and of the chain of being. And even as a practical way of organizing men for economic production, the mass-production principle is not new. Indeed, the first thorough applications of mass production and the assembly line were not in the Ford plant in Detroit, but hundreds of years earlier and thousands of miles away, in the workshops of the medieval stonemasons who built the great cathedrals. In short, mass-production society, of which the assembly line is the symbol, is a hierarchical one.

This shows clearly when we analyze what popularly passes for a clear explanation of the essence of mass production: the saying that it replaces skilled by unskilled labor. That is pure nonsense, if taken literally. Of course, in mass production manual skill is eliminated by breaking up each operation into the component simple operations, with each worker performing only one unskilled operation or a series of such. But this presupposes a fantastic skill in analyzing and breaking up the operation. The skill that is taken out of the manual operation has to be put back again further up the line, in the form of much greater knowledge, much more careful planning for the job; for there is such a thing as a law of the preservation of skill. And in addition mass production needs a new skill: that of organizing and leading the human team. Actually "unskilled" mass production needs proportionately more and more highly skilled men than "skilled" production. The skills themselves have changed from those of the craftsman to those of engineer, draftsman, and foreman; but the number of trained and skilled men in American industry has been growing twice as fast since 1910 as that of unskilled and semiskilled men.

Above all, the cooperation and coordination which are needed to make possible the elimination of manual skill presuppose an extraordinarily high level of social skill and social understanding, of experience in working together. The difficulties that our war plants had with new labor showed that very graphically. And, contrary to popular belief, it is no more difficult to export the old methods of industrial production to a new industrial country, even though those methods require considerable manual skill on the part of the worker, than it is to export mass-production techniques where no manual but a great deal of social skill is required.

What we mean when we say that mass production is based on unskilled labor is simply that the individual becomes effective and productive only through his contribution to the whole, and not if viewed separately. While no individual does the job, each one is necessary to get the job done. And the job, the end product of cooperative effort, is more skilled than anything the most skilled person could have produced by himself. As in every hierarchical society, there is no answer in the mass-production plant to the question of who does the job, but there is also no answer to the question of who does not do the job. For everybody has a part in it.

There were a good many industries even in the forties which did not use

the mass-production principle. Among them are some of the most efficient ones, for instance, the modern cotton mills (in which one worker
may manage a great many looms) and a good many of our chemical
industries (in which one worker may perform a number of different
functions). Nevertheless, the mass-production industries were representative of American industry as a whole because they expressed in the
purest form the essence of industrial production, i.e., a principle of
social organization. The real Industrial Revolution of our day—the one
which Henry Ford led and symbolized—was not a technological one,
was not based on this or that machine, this or that technique, but on the
hierarchical coordination of human efforts which mass production
realizes in its purest form.

3

It is understandable that Henry Ford's disciples and imitators failed
to see the political and social implications of mass production until
they were confronted by them in the form of an aggressive union
movement, and very often not even then. For most of these men were
really concerned with technical problems only and really believed in
mechanical efficiency as an end in itself. But Henry Ford's own
blindness cannot be so simply explained as due to a lack of social or
political concern—not even as due to a lack of social or political
imagination. The real explanation is that Ford was concerned exclusively with the solution of the *social and political problems of the pre-
Ford, the pre-mass-production industrial civilization.* And because his
answers really did solve these problems, or at least the more important
of them, it never entered his mind to subject this answer of his in turn
to a social and political analysis. His gaze was firmly fixed on the industrial reality of his own youth—the industrial reality against which
populism had revolted in vain. He never even saw what he himself had
called into being. As a high official of his own company once said,
"What Mr. Ford really sees when he looks at River Rouge is the
machine shop in which he started in 1879."

Though Henry Ford may never have heard of Brook Farm, of Robert
Owen's New Harmony, or of any of the scores of utopian communities
that had dotted the Midwest not so many years before his birth in 1863,
they were his intellectual ancestors. He took up where they had left off;
and he succeeded where they had failed. Colonel McCormick's Chi-

cago *Tribune* called him an "anarchist" in the red-hunting days of
1919 when the term meant more or less what "communist" would
mean today in the same paper. But in spite of the obvious absurdity of
the charge, the jury awarded Ford only six cents in damages when he
sued for libel; for he was undeniably a radical. He turned into a stand-
patter after 1932, when his life's work had shown itself a failure in its
inability to produce the stable and happy society of which he had
dreamed. But the Henry Ford of the earlier Model T days was an
iconoclast attacking in the name of morality and science the established order of J. P. Morgan and of Mark Hanna's Republican party.

The utopias of the 1830s and 1840s were in themselves the reaction
to a failure: the abortive attempt during Jackson's administration to
bring back to America the lost innocence of the Jeffersonian society of
self-sufficient independent farmers. The utopians no longer hoped to
be able to do away with the modern division of labor or even with
industry. On the contrary, they promised to obtain for mankind the full
benefits of industrial productivity, but without its having to pay the
price of subjecting itself to the "money power" or to "monopoly" or
of having to work in the "satanic mills" of Blake's great and bitter
poem. These were to be eliminated by a blend of pious sentiment,
community regulations, and social science.

Of all the utopians only the Mormons survived—and they only by
flight from the land of the Gentiles. But though they failed, Brook
Farm, New Zion, New Harmony, and all the other attempts at the
American industrial Jerusalem left a very deep imprint on the consciousness of the American people. Neither Fourier, whose ideas
fathered Brook Farm, nor Robert Owen, was an American. Yet it is
possible, indeed probable, that the mixture of earnest, semireligious
sentiment and trust in a "scientific" principle which is so typical of the
American "reformer" or "radical" has its roots in much older and
deeper layers in our history than the utopias. But it is certain that the
utopias determined the specific form which American radicalism was
to take for a whole century. They provided the targets, the battle cries,
and the weapons for populism, for Wilson's New Freedom, and even
for much of the early New Deal (such as the "scientific" gold magic
of 1933). They fathered Henry George, Bellamy, and the antitrust
laws. They molded the beliefs and the hopes of America's inland
empire in the Midwest. But they remained a futile gesture of revolt
until Henry Ford came along.

Today we know that in depression and unemployment we have as serious an economic problem as "monopoly" and the "money power" ever were. We see very clearly that mass production creates as many new social and political questions as it answers. Today we realize that as a *final* solution to the problems of an industrial civilization, Henry Ford's solution is a failure.

But Ford's mass production was not aimed at these new dangers, but at the traditional devils of American radicalism. And these it actually did exorcise. Ford succeeded in showing that industrial production can be production *for* the masses—instead of production for the benefit of monopolist or banker. Indeed, he showed that the most profitable production is production for the masses. He proved that industrial production could give the workers increasing purchasing power to buy industrial products and to live on a middle-class standard; that was the meaning of his revolutionary $5.00-a-day minimum wage.

Finally—and to him most importantly—he proved that, properly analyzed and handled, industrial production would free the workers from arduous toil. Under modern mass-production conditions, the worker is confined to one routine operation requiring neither skill nor brawn nor mental effort. This fact would not have appeared to Henry Ford as a fatal defect, but as a supreme achievement; for it meant that—in contrast to the tradition of the "satanic mills"—the worker's skill, intelligence, and strength would be fully available for his community life as an independent Jeffersonian citizen outside of the plant and after working hours.

At Brook Farm, too, the "real life" was supposed to come in the "communion of the spirits" in the evening after the day's work had been done; but the day's work took so much time and effort that the "real life" could be lived only by neglecting the work. Mass production cuts both time and energy required for the day's work so as to give the worker plenty of scope for this "real life." No wonder that Ford— the Ford of 1919—thought he had built the "New Jerusalem" on a permanent foundation of steel, concrete, and four-lane highways.

4

It was Ford's personal tragedy to live long enough to see his utopia crumble. He was forced to abandon his basic economic principle—the principle of the cheapest possible production of the most utilitarian commodity. First he scrapped the Model T. That was in 1927. Then, five years later, he abandoned the Model A and adopted the annual model change which substitutes the appeal of prestige and fashion for the appeal of cheapness and utility. When he did this he became just another automobile manufacturer. Even so, his share in the market dropped from nearly half in 1925 to less than 20 percent in 1940. Even more decisively proven was his failure to give the worker industrial citizenship; in 1941 the Ford workers voted to join the CIO almost three to one.

Up to the hour when the results were announced, the old man is said to have firmly believed that "his" workers would never vote for a union. All along he had fought off realization of his defeat by pretending to himself that his downfall was being caused by sinister conspiracies rather than by faults in the structure of the community which he had built. This tendency to look for personal devils—itself a legacy from the utopians—had shown itself quite early in the tirades of the Dearborn *Independent* against international bankers, Wall Street, and the Jews during the 1920s. It became the basis on which he fought the unions all through the thirties. It also probably explains why Harry Bennett, starting as the plant's chief cop, rose to be the most powerful individual in the Ford organization of the thirties and the only one who really seemed to enjoy the old man's confidence. But the union victory, followed shortly by the unionization of the foremen, must have hit Henry Ford as a repudiation of all he had thought he had achieved, and had achieved primarily for the workers. The last years of the old man must have been very bitter ones, indeed.

The lesson of Ford's ultimate failure is that we cannot hope to solve the problems of the mass-production society by technological devices or by changing the economics of distribution. These were the two approaches on which all nineteenth-century thought had relied, whether orthodox or rebel. Henry Ford went as far along these lines as it is possible to go.

For the time being, the political results of Ford's achievement were extraordinary. It took the wind out of the sails of the socialist critique of capitalist society. In this country it brought about the change from the fiery political action of Eugene Debs to the politically impotent moralism of Norman Thomas; in continental Europe it converted social democracy from a millennial fighting creed into a respectable but timid bureaucracy. Even more telling was the reaction of Communist Russia to Ford. In the twenties the Russians had to add to the messianic hopes of Karl Marx the promise of achieving eventually in a socialist society what Ford had already achieved in a capitalist one: a chance for the worker to drive to the plant in his own car and to work in collar and tie, and without getting calluses on his hands. And until 1929, as every meeting of the Third International affirmed, the communists were completely convinced that Ford's America had actually solved the basic problems of capitalism and had restored it to ascendancy all the world over. Not until the Great Depression were the communist leaders able to revitalize their creed by making it appear to do what it cannot do: to solve, by the sheer force of the police state, the new, the post-Ford problems of industrial society as they appeared after 1929.

As we in America confront these problems, the economic ones will not be the most difficult. Indeed, the chief economic problem of our time—the prevention of depressions—should be solvable by basically mechanical means: by adapting our employment, fiscal, and budgeting practices to the time span of industrial production—that is, to the business cycle. Much more baffling, and more basic, is the political and social problem with which twentieth-century industrialism confronts us: the problem of developing order and citizenship within the plant, of building a free, self-governing industrial society.

The fact that Henry Ford, after his superb success, failed so signally—that there is today such a grim contrast between his social utopia and our social reality—emphasizes the magnitude of the political task before us. But however treacherous the social jungle of post-Ford, mass-production society, however great the danger that it will fester into civil war and tyranny, the twentieth-century evils which Henry Ford left to us may well be less formidable than the nineteenth-century evils which he vanquished.

4

IBM's Watson: Vision for Tomorrow

Everybody knows that Thomas Watson, Sr. (1874–1956), built IBM into a big computer company and was a business leader. But "everybody" is wrong. Thomas Watson, Sr., did not build the IBM we now know; his son, Tom Watson, Jr., did this after he moved into the company's top management in 1946, only thirty-two years old (though the father stayed on until 1956 as a very active chairman). The father's IBM was never more than medium-size; as late as 1938 it had only $35 million in sales. Of course, it did not make or sell computers; its mainstays were punch-card machines and time clocks.

Instead of being a business leader, Thomas Watson, Sr., did not attain personal success and recognition until he was well past sixty. Twice in the 1930s he personally was on the verge of bankruptcy. What saved him and spurred IBM sales during the Depression were two New Deal laws: the Social Security Act in 1935 and the Wage-Hours Act of 1937–38. They mandated records of wages paid, hours worked, and overtime earned by employees, in a form in which the employer could not tamper with the records. Overnight they created markets for the tabulating machines and time clocks that Thomas Watson, Sr., had been trying for long years to sell with only moderate success.

In 1939, when I was working in New York as a correspondent for a group of British papers, I wanted to write a story on Watson and IBM. I had become interested in the company because of the big pavilion it

First published in *Esquire* Magazine, 1983

had put up at the New York World's Fair and thought a story on so small a frog behaving like a big shot might be amusing. "Forget it," my editor wrote back. "We are not interested in a story on an unsuccessful company which as far as anyone can tell is never going to amount to much." And Watson was then already sixty-five years old.

But Thomas Watson, Sr., was something far more important than a successful businessman who built a big company. He was the seer and, very largely, the maker of what we now call postindustrial society and one of the great social innovators in American history. Fifty years ago he had the vision of "data processing" and of "information." In the midst of the Depression and half broke himself, he financed the research that produced both the theoretical foundations for the computer and the first highly advanced model of it.

Fifty years ago or more he also invented and put into practice what in this country is now known, studied, and imitated as Japanese management—needless to say, he did so without the slightest knowledge of Japan and without owing anything to the then-nonexistent Japanese practices.

In fact, Watson's problem was that he was far ahead of his time, in his vision as well as in his practices.

I first met Watson in the early 1930s. He had been elected president of the American section of the International Chamber of Commerce—a job nobody else had wanted—and I, then a very young reporter, was sent to interview him. But the interview made absolutely no sense to my editor and never appeared. Of course, Watson delivered himself of the platitudes expected from the Chamber of Commerce, about free trade, about international cooperation, and about "the role of the businessman as an ambassador of peace." But then he started to talk of what clearly was his real interest—things he called *data* and *information*. But what he meant by those terms—it was not at all what anyone else at the time meant by them—he could not explain. Indeed, newspapers usually put him down as a crank.

I doubt that Watson himself *understood*. He had a vision—he *saw*. But unlike most seers, Watson *acted* on his vision.

Some kind of computer would have come without him. There were a good many people at the time working at fast calculating machines, especially under the impetus of World War II, with its need for navigational equipment for fast-moving airplanes, for firing mechanisms for long-range cannon shooting at invisible targets way beyond the horizon, for aerial bombs, and for antiaircraft guns. But without Watson it could have been a very different computer, a "calculating machine" rather than an "information processor." Without Watson and his vision the computer would have emerged as a "tool" rather than as a "technology."

Watson did not invent a single piece of hardware. He had no technical education, indeed little formal education altogether, having gone to work as a salesman of sewing machines, pianos, and organs in his native upstate New York when barely eighteen. He also had no technical talent, or much grasp of mathematics or of theory altogether. He was definitely not an inventor in the mold of Edison or his contemporary Charles Kettering. The computers that he sponsored, financed, and pushed for—one finished in 1943, another four years later, in 1947—contributed little by way of hardware and engineering solutions even to IBM's own early commercial computer in the 1950s.

But Watson saw and understood the computer fifteen years before the term was even coined. He knew right away that it had to be radically different from a high-speed calculator. And he did not rest until his engineers had built the first operational true "computer." Watson specified very early—in the late 1930s at the latest—what computer people now call the *architecture* of a computer: capacity to store data; a memory, and random access to it; ability to receive instructions and to change them, to be programmed, and to express logic in a computer language. Watson's 1947 Selective Sequence Electronic Calculator (SSEC) was much more powerful and far more flexible than any machine then in existence or being planned. Its 12,500 vacuum tubes and 21,400 relays could, for instance, solve partial differential equations. It was, above all, the first—and for several years the only—machine to combine electronic computation with stored programs, with its own computer language, and with the capacity to handle its instructions as data: that is, to change, to correct, and to update them on the basis of new information. These are, however, precisely the features that distinguish a computer from a calculator.

When IBM embarked on designing a computer to be sold in large quantities, it applied hardware largely developed by others—mainly university labs at MIT, Princeton, Rochester, and Pennsylvania. But it was all designed to Watson's original specifications. This explains why the 650, IBM's first successful commercial machine, immediately became the industry leader and the industry standard when it was introduced in 1953; why it sold eighteen hundred units during its first five years on the market—twice as many as the best market research had predicted total computer sales worldwide to come to in the entire twentieth century; and why it gave IBM the world leadership in computers the company still holds.

The early *technical* history of the computer is quite murky. And what part Watson and IBM, or anyone else, played in it is remarkably controversial. But there is no doubt at all that Watson played a key role in the *conceptual* history of the computer. There are many others who were important, if not central, in the engineering of the computer. But Watson created the computer age.

The story begins at 1933 when IBM, on Watson's orders, designed a high-speed calculator for Columbia University's scientists and built it out of standard tabulating-machine parts. Four years later, in 1937, Howard Aiken, a Harvard mathematician, suggested that IBM try to link up several existing machines—designed originally for bookkeeping purposes—to create an even faster calculator for the time-consuming and tedious computations needed by astronomers. The work, Aiken estimated, would take a few months and would cost around $100,000. Instead, the work took six years and the cost was well over half a million dollars. For instead of the high-speed calculator Aiken had in mind, Watson went ahead and produced a computer.

First Watson specified an all-electronic machine; all earlier designs, including IBM's own machines, were electromechanical with levers and gears doing the switching. Of course, *electronic* then meant using vacuum tubes—the transistor was well into the future. And the resulting machines were therefore big and clumsy, tended to overheat, and used a great deal of electricity. Still, no one at the time even knew how to use electronic devices as switches—the key, of course, to computer technology. Even the needed theoretical work had yet to be done. For Watson and IBM to decide to desert the well-known and thoroughly familiar electromechanical technology for the totally unexplored electronics was a leap into the dark, but it was the decision that made the computer possible.

Watson's second specification was equally innovative. His project envisaged what no engineer at the time even dreamed of: a computer memory. Howard Aiken in his research proposal had been concerned with rapid calculation; he wanted a "numbers-cruncher." His idea was to build something that would be able to do very quickly what traditional adding machines and slide rules did slowly. IBM added to this the capacity to store data. And this meant that IBM's projected machine would be able to process information. It would have a data bank—a term that, of course, did not exist at the time—and would be able to refer back to it. Thus it could memorize and analyze. It could also—and this is crucial—correct and update its own instructions on the basis of new information.

And this then meant—another one of IBM's then quite visionary specifications—that IBM's machine was going to be capable of being *programmed,* that is, of being used for any information capable of being expressed in a logical notation. All the designs then being worked on were for single-purpose machines. The best known of these, called ENIAC, constructed at the University of Pennsylvania during the World War II years for Army Ordnance and completed in 1946, was, for instance, designed to do high-speed calculations for fast-firing guns. But it could do little more, as it had neither memory nor programming capacity. IBM, too, had specific applications in mind—especially calculations for astronomical tables. But, perhaps because Watson knew firsthand of the data-processing needs of a host of institutions—public libraries, the Census Bureau, banks, insurance companies, and the like—his specifications called for a machine capable of being programmed for all kinds of data. IBM's machine was a multipurpose machine from the start. This, then, also meant that when IBM finally had a computer that could be manufactured and sold—in the early 1950s—it was both willing and able to supply the users whose demand actually created the computer industry, but of whom no one had even thought in the early stages of computer design: business people who wanted computers for such mundane, unscientific purposes as payroll and inventory. Watson's insistence on memory and program thus explains in large measure why there is a computer industry.

Because of Watson's specifications of memory and program, the IBM research project of 1937 helped create computer science. The "analytical machine" that Watson tried to develop needed both computer theory and computer language—again, of course, terms that did not exist then. It needed, of course, first-class engineering, as did all the other calculating machines then designed. But it also created, or at least stimulated, the first computer scientists.

This research produced a prototype in late 1943—called the Automatic Sequence Controlled Calculator (ASCC)—and Howard Aiken did do astronomical calculations on it, in early 1944. But it was far too advanced for the manufacturing technology of its time. In fact, only recently have the materials and the technology become available to manufacture the kind of computer for which Watson organized the research in 1937. Watson then donated the prototype, together with money to run and maintain it, to Harvard and immediately commissioned the work on the next design—the work out of which the SSEC evolved several years later. Its first public demonstration, on January 27, 1948, in New York City—at which it calculated all past, present, and future positions of the moon—ushered in the computer age. Even now, thirty-five years later, the computer that popular cartoons show, with its blinking lights and whirling wheels, is the SSEC as it was displayed at this first public demonstration. And looking at Watson's specifications in retrospect, this is precisely what must have been in his mind's eye all along.

As significant as Watson the computer seer is Watson the social innovator. He was even further ahead in his social vision than he was in his vision of data and information and had perhaps even more of an impact—and was equally (or more) misunderstood by his contemporaries.

Watson became a national figure in 1940, when *Fortune* published a violent, mudslinging attack on him. It insinuated that Watson was the American Führer and tried altogether to make him the symbol of everything repulsive in American capitalism. And it also fixed the popular perception of Thomas Watson as a reactionary and as a paternalist, a perception that still lingers on. But today it is clear that Watson's real offense was to be far ahead of his time.

Rereading the story in 1983, forty-three years later, I saw right away that what had most irked the *Fortune* writer in 1940 was Watson's ban on liquor in IBM's employee clubs and at IBM parties. He was indeed passionately opposed to alcohol—I once heard that his father or one of his uncles had been an alcoholic. But whatever Watson's reasons, it is perhaps not such a bad idea to keep work and liquor apart.

Then, of course, there was Watson's regimentation of his sales force, that is, his demand that IBM's salesmen wear dark suits and white shirts. But what Watson tried to do, as he had explained to me in my interviews with him a year before the *Fortune* story, was to instill self-respect in salesmen and respect for them among the public. When Watson began, the salesman was a "drummer," disreputable and a smelly crook, who spent his evenings in the small town's whorehouse unless he spent it debauching the farmer's innocent daughter in the hayloft. Watson, who had been a drummer himself, had deeply suffered from the contempt for his calling and was determined that his associates have self-respect and be respected. "I want my IBM salesmen," he had said in his 1939 talks with me, "to be people to whom their wives and their children can look up. I don't want their mothers to feel that they have to apologize for them or have to dissimulate when they are being asked what their son is doing" (and there was a clear implication, I thought then, that he was speaking of his own mother).

Watson's heinous crime, however, and the one the *Fortune* writer could not forgive him for, was that he believed in a worker who took responsibility for his job, was proud of it, and loved it. He believed in a worker who saw his interests as identical to those of the company. He wanted, above all, a worker who used his own mind and his own experience to improve his own job, the product, the process, and the quality. Watson's crime—and it was indeed seen as a crime in the late 1930s, when a vulgar Marxism infested even fairly conservative members of the intelligentsia—was to make workers enjoy their work and thus accept the system rather than feel exploited and become revolutionary proletarians.

The way Watson did this was, first, to give his employees what we now call lifetime employment. As in Japan today, there was no contractual obligation on the part of Watson's IBM not to lay off people. But there was a moral commitment. In the first Depression years, 1931 and 1932, IBM did indeed lay off a few people, but Watson immediately stopped it. His refusal later, in the darkest Depression years of 1933 and 1934, to lay off workers could have bankrupted IBM. And employment security for workers has been IBM practice for fifty years now.

In the mid-1930s Watson abolished "foremen." They became

"managers," whose job it was to assist workers, to make sure that workers had the tools and the information they needed, and to help them when they found themselves in trouble. Responsibility for the job itself was placed firmly on the work group.

The worker, Watson argued, knew far better than anyone else how to improve productivity and quality. And so around 1935 he invented what we now know as quality circles and credit to the Japanese. The industrial engineer was to be a "resource" to the work group and its "consultant," rather than the "expert" who dictated to it. Watson also laid down that the individual worker should have the biggest job possible rather than the smallest one, and as far back as 1920 he initiated what we now call job enrichment. (And after the old man's death his son, in 1958, put into effect his father's old idea and put *all* employees on a monthly salary, abolishing both the hourly wage and the distinction between blue-collar and white-collar employees.)

Finally, every employee at IBM had the right to go directly to the company's chief executive officer, that is, to Watson, to complain, to suggest improvements, and to be heard—and this is still IBM practice.

Much earlier Watson had invented what the Japanese now call continuous learning. He began in the early 1920s with his salespeople, who were called into headquarters again and again for training sessions to enable them to "do better what they are already doing well." Later Watson extended continuous learning to the blue-collar employees in the plant, and in the 1940s he hired Dwayne Orton, a college president from San Francisco, to be the company's director of education.

These practices very largely explain IBM's success. They exorcised the great bugaboo of American business, the employees' alleged resistance to change. They enabled the company in the 1950s and 1960s to grow as fast as any business had ever grown before—and perhaps faster—without much internal turbulence, without horrendous labor turnover, and without industrial strife. And because Watson's salesmen had all along been used to learning new things, middle-aged punch-card salesmen without experience in computers or in big-company management could overnight become effective executives in a big high-technology company.

But forty years ago Watson's policies and practices were very strange indeed. They either made Watson look like a crank, which is what most people believed, or they made him into the sinister force the

Fortune writer saw. Today we realize that Watson was only forty or fifty years ahead of his time. Watson was actually one year older than Alfred Sloan. But whereas Sloan in the 1920s created modern management by building General Motors, and with it the modern "big corporation," Watson ten years later and quite independently created the "plant community" that we know to be the successor to Sloan's "big business enterprise" of the 1920s. He created in the 1930s the social organization and the work community of the postindustrial society.

The first ones to see this, by the way, were the Japanese. Again and again I have been laughed at in Japan when I talk about Japan's management embodying Japanese values. "Don't you realize," my Japanese friends ask, "that we are simply adapting what IBM has done all along?" And when I ask how come, they always say, "When we started to rebuild Japan in the 1950s, we looked around for the most successful company we could find—it's IBM, isn't it?"

Watson also anticipated the multinational. And this, too, no one in 1940 could understand.

Watson very early established foreign branches of IBM, one in France, one in Japan, and so on. IBM actually had little foreign business at the time, and Watson knew perfectly well that he could have handled whatever there was through exporting directly. But he also realized that the IBM he envisaged for the future, the IBM of what we now call data processing and information handling, would have to be a multinational company. And so he created, more than fifty years ago, the structure he knew IBM would have to have and would have to know how to manage were it ever to become the company he dreamed of. He also saw that the central problem of the multinational is to combine autonomy of the local affiliate with unity of direction and purpose of the entire company. This is in large measure the reason that Watson deliberately created what to most people seemed such extreme paternalism—today we would call it the IBM culture. Way back in 1939 he tried to explain it to me. "Foreign affiliates," he said,

must be run by natives of their own country and not by expatriates. They must be accepted as members of their own society. And they must be able to attract the best people their own country produces. Yet they have to have the same objectives, the same values, the same vision of the world as the parent company. And this means that their management people and their professional people share a common view of the company, of the products, and of their own direction and purpose.

Most other multinationals still struggle with the problem Watson foresaw—and solved—fifty years ago.

Watson was an autocrat, of course. Visionaries usually are. Because visionaries cannot explain to the rest of us what they see, they have to depend on command.

Watson was a generous person but a demanding boss and not one bit permissive. But he demanded the right things: dedication and performance to high standards. He was irascible, vain, opinionated, a publicity hound, and an incorrigible name-dropper. And as he grew older he became interestingly addicted to being grossly flattered. But he was also intensely loyal and quite willing to admit that he had been wrong and to apologize. The people who worked for him feared him—but almost no one left.

Of course, he did not fit into the intellectual climate of the New York of the late 1930s. He was already sixty-five at that time and his roots were in the small rural towns of the 1880s, with their church socials and their service-club luncheons, with communal singing and pledges of allegiance to the flag—all the things that the "smart set" of the time had learned to sneer at as the marks of the "booboisie." And what was worse and made Watson a menace was that his employees shared Watson's values, rather than those of the intellectuals, and loved to work for him.

Watson was an extraordinarily complex man who fit into no category. He was among the first American businessmen to use modern design and modern graphics for the company logo and company stationery, products, and offices. But he did so mainly because it was the cheapest way to make his company stand out—and he had no advertising money. To get attention for IBM he used the slogan THINK, which he had invented at NCR. His salesmen used THINK notebooks by the thousands—again, it was the only advertising he could afford. And although he himself didn't have much of a sense of humor, he made up a good many of the THINK jokes and saw to it that they got wide circulation—again, to get attention for his company. Finally he hit upon the brilliant idea of building a huge pavilion at the 1939 New York World's Fair as the only way to make millions of visitors from all over the world aware of his then still small and practically unknown company and, as he told me at the time, "at the lowest cost per thousand viewers anyone ever had to pay."

Watson's popular image is that of the hidebound conservative, and he lent credence to this view by wearing the stiff Herbert Hoover collar long after it had gone out of fashion. But, almost alone among the businessmen of his time, he was an ardent New Dealer and a staunch supporter of Franklin D. Roosevelt. Indeed, FDR offered him major posts in his administration: first secretary of commerce, and then ambassador to Great Britain (when Watson said no, the job went to Joseph Kennedy instead). And at the very end of his life Watson urged President Eisenhower to return to New Deal principles and was upset over Ike's insistence on reducing the government's role in economy and society.

Watson had a most unusual personal history, perhaps the most unusual one in the annals of American business. He started out dirt poor as a teenage drummer—not much more than a peddler. He rose to become the star salesman for the then-new National Cash Register Company of Dayton, Ohio, which was the first business-machine company in history. In about fifteen years he had become the company's sales manager. But then the company was indicted for antitrust violations, with which, we now know, Watson had little or nothing to do.

The trial was a sensation, for National Cash had been *the* growth company of the early 1900s. Watson drew a stiff fine and a one-year jail sentence. This conviction was set aside two years later, in 1915, by the Court of Appeals, and the government then dropped the case. But Watson had lost his job; both his career and his reputation were in tatters. It was an experience that marked him for life, an experience that he never forgot.

Already forty, he then had to begin all over again as general manager for a small and until then quite unsuccessful company that owned the punch-card patents. It was not until he was past sixty, in the mid-1930s, that he succeeded in establishing this company, by then renamed IBM, on a firm footing.

Watson—and this, I think, makes the man singularly interesting—was a uniquely American type and one the establishment has never understood throughout our history. He was possessed of a towering intellect but was totally nonintellectual. He belongs to the same type as Abraham Lincoln, who similarly offended the establishment of his day, the cosmopolitan, polished, learned Bostonians who considered themselves so superior to the yokel president in whose cabinet they were forced to serve. He was also similar in many ways to America's first great novelist, James Fenimore Cooper, who was also sneered at, despised, and pilloried by the intellectuals and the establishment of his time, such as the New England Transcendentalists. They could not make sense of the powerful, tragic, and prophetic vision in which Cooper foretold the end of the American Dream in such books as *The Pioneers* and *The Prairie*.

This American type is totally native and owes nothing to Europe—which is one reason that the intellectuals of his time do not know what to do with him. Typically the men of this type have a gift for words—and Watson fully shared it. But they are not men of ideas. They are men of vision. What makes them important is that, like Watson, they act on their vision.

5

The Myth of American Uniformity

1

"How can you Americans stand all this uniformity?" Every one of the dozens of visitors from all over Europe who, during these past few years, have discussed their American impressions with me, has asked this question in one form or another. Yet what makes every single one—businessman, clergyman, or scientist; teacher, lawyer, or journalist; labor leader or civil servant—come to me for information is bewilderment, if not shock, at the incomprehensible and boundless diversity of this country.

"But *somebody* must lay out the standard curriculum for the liberal arts college. If the federal or the state governments do not do it, who does?"

"In what grade does the American high-school student start Latin? How many hours a week are given to it? And what works of Shakespeare are normally read in the American high schools?"

"It can't really be true that there is no one labor union policy on industrial engineering. I am told that some unions actually insist on a time and motion study of each job, some unions acquiesce in it, and others refuse to allow any industrial engineers. But surely no union movement could possibly operate pulling in opposite directions on a matter as important as this?"

"Please explain to us what American managements mean when they talk of 'decentralization.' Wouldn't this mean that different units of a

First published in *Harper's Magazine,* 1952

company would do things differently, adopt different policies, follow different ideas? And how could any management allow that and still keep its authority and control?''

The going gets really tough when the talk returns to political institutions or to the churches. That it makes all the difference in the world what congressional committee a pending bill is assigned to, will upset even the urbane visitor—if indeed he believes it. And among the most frustrating hours of my life was the evening I spend with a Belgian Jesuit who insisted that there must be one simple principle that decides when and where agencies of the Catholic Church in this country work together with other faiths, and when not. The only comfort was that he obviously had got no more satisfaction from his American brethren in the order than from me.

Yet it is quite clearly not in diversity that the visitors see the essence of America. They are baffled by it, shocked by it, sometimes frightened by it. But they don't really believe in it. Their real convictions about this country come out in the inevitable question: ''But don't you find it trying to live in so uniform a country?''

It is not only the casual visitor, spending a few weeks here, who believes in ''American uniformity'' despite all he sees and hears. The belief survives extended exposure to the realities of American life.

A few months ago a well-known English anthropologist, reviewing an exhibition of American paintings for a most respectable London Sunday paper, explained the ''mediocrity of American painting'' by a reference to ''the uniformity of the American landscape—all prairie and desert.'' One might remind the reviewer that nothing is more startling to the immigrant who comes to America to live than the tremendous variety of the landscape and the violence of contrasts in the American climate, soils, geology, fauna, and flora. Or one might reduce the argument to its full absurdity by asking which of these sons of Kansas, for example, is the typically uniform prairie product— William Allen White, Earl Browder, or General Eisenhower? But the essential fact is not that the argument is nonsense. It is that Geoffrey Gorer, the anthropologist, knows this country well, and that the newspaper that printed his nonsense is unusually knowledgeable about things American on the whole. Yet though they know all about New England or Virginia or Minnesota or Oregon, though they probably also know about the artists who paint in the desert of Cape Cod—or is it a prairie?—they immediately think of "uniformity" when something American needs explanation.

Or take the case of the young Danish lawyer who came to see me just before sailing back home. He was going to stay just a few minutes as he had only one question to ask. In the end he stayed almost the whole day—yet left with it still unanswered. His question? In one plant of the American company where he worked for seven months as a trainee, he found that output standards for the workers were set by a joint management-union committee. In another plant of the same company, located just a few miles away and organized by the same local of the same union, output standards, he found, were considered strictly a "management prerogative," with union action confined to formal protests against management decisions. He was sure he must have been mistaken in his observation; at the least, management and union must both be eager to have a uniform policy, whereas both seemed perfectly happy with the existing "disorder." It could not convince him that this was a fairly common situation all over the country. He left, certain that our labor relations must be uniform, if not for the whole country then at least for an industry, let alone for one company or one union.

And the "productivity teams" that have come over from Europe to study American methods these past few years insist in their reports that one of the basic reasons for the greater productivity of American industry is the "standardization" of the individual manufacturer on a very small number of models or lines. Yet most of the productivity reports themselves contain figures which show the exact opposite to be true: the typical American automobile manufacturer (even the smaller one), the typical shoe manufacturer, or the typical foundry turns out more models than its European counterpart. The people who write these reports seem quite unconscious of the contradiction.

Clearly, "American uniformity" is an axiom for the European, before and beyond any experience. It is indeed the one thing the European *knows* he knows about this country. There are today plenty of people in Europe who know that not all Americans are millionaires—though there are still far too few who know from firsthand experience how high the standard of living of the American worker really is. There are even some Europeans who have come to suspect that race relations in this country are quite a bit more complex than in the books of Richard Wright. But it is a very rare European indeed—if he exists at all—who does not *know* that America is "uniform."

2

How is this dogma to be explained? The standard answer is that there is an outward sameness, a uniformity to all things material in this country. I have been skeptical of this answer ever since, a few years back, a magnificently accoutered cowboy, complete from white ten-gallon hat to woolly chaps and silver spurs, complained to me bitterly about the "uniformity" of the American costume, and contrasted it with the picturesque leather pants, white knee stockings, and green suspenders of the Austrian students among whom he had spent a few months before the war. The "cowboy" was an earnest young social worker riding the range in the great cow center of Chicago. And he delivered himself of his plaint on the way from a lecture on psychology to one on urban community problems during a YMCA conference. His excuse for his dress—had he felt the need for any—would have been that he went folk and square dancing both in the morning, before the day's lectures, and in the evening. But the Austrian students—I did not have the heart to tell him—wear their leather pants because they possess at best one good suit and have to go easy on it.

Altogether there is as little diversity in Europe's outward material appearance as there is in this country. People all through Europe, right through the Iron Curtain, dress pretty much alike. And when they don't—surely even the quaintest Sunday costume of a Slovak maiden can hardly rival the colors of a Californian going out on the golf course, or the ties a Midwestern salesman wears on his rounds.

Our towns and cities, ugly maybe, are not as much uniform as they are nineteenth- and twentieth-century towns and cities. Even in Europe it is primarily the old cities which look different. At least I know nothing in this country to rival the bleak monotony of sooty brick and broken chimney pots of the railroad ride into London, or the pea-in-the-pod uniformity of the famous Dutch housing developments with their endless rows of identically neat bungalows. And even the sun-drenched limbo—frowzy palms and peeling stucco—of the middle-class sections of Los Angeles offers occasional variety and architectural surprise compared to the numbing grayness through which one drives from the airport into the city that to most Americans stands as the symbol of European diversity: Paris.

When it comes to manufactured goods there is actually more diversity in this country than Europe has ever known. The variety of goods carried by our stores is the first thing that impresses any visitor from abroad. Nor is this a postwar phenomenon. As far back as 1938 one of the leading department store chains in England studied the Sears Roebuck catalogue and concluded that in every single category of goods the American mail-order business, for all its "standardization," offered a wider range of goods in far more models than any European retailer could obtain from European manufacturers, let alone afford to carry.

But, you may say, when the European talks about "American uniformity" he is not thinking of the material and outward aspects of American life, but of American culture and society. And here the dogma of American uniformity becomes totally incomprehensible. For it is in the nonmaterial realms—in religion, political institutions, education, business life, even in entertainment—that the diversity of this country most deeply confuses the visitor from the other side.

Well-informed Europeans have heard that this country's political life is founded on pluralism and that our religious organization knows no rules—though they seldom seem to realize that these facts alone deny the legend of American uniformity. Most of them, however, believe that this country has uniformity in education. A hand-picked team of British educators, scientists, and industrialists who recently studied the relationship between industry and the universities in this country, obviously took uniformity for granted—though every single fact in their own report contradicted the assumption. Actually the diversity in politics and religion here is as nothing compared to the riot that prevails in education—the matrix of society.

There are colleges which look with distrust upon any book written later than 1300, are pained if they have to teach anything that is at all tainted with "usefulness," and occasionally even dream of going back to teaching in Latin. There are other colleges—giving the same B.A. and enjoying the same acceptance by the general public—in which a student can earn a degree through courses in night-club etiquette, horseback riding, and fashion drawing. And in at least one South-western university you can now get a Ph.D. in square and folk dancing; last summer I was shown with great pride the first accepted doctor's thesis, a formidable tome of 652 pages, mostly footnotes, on "the Left-Turn Hopsa Step in Lithuanian Polkas." Greater still is the diversity among different kinds and types of colleges, and among private schools, church schools, and state schools, let alone such phenomena, totally incomprehensible to any visitor from abroad, as the half-private, half-state university, or the church-supported but nonsectarian college. Some of the larger "liberal arts" colleges have flourishing engineering schools of their own, and one large engineering school, Carnegie Tech in Pittsburgh, also runs a first-rate art and music school. And how can one explain to a European, accustomed to a Ministry of Education, the role of the private foundations, such as Rockefeller or Carnegie, and their power?

It is scarcely possible to talk about trends in American higher education, so mixed are the currents. Many engineering schools for instance have lately broadened their curricula to include more and more of the arts and humanities; but Columbia University—itself preponderantly dedicated to the liberal arts—has just announced plans for the most highly specialized Engineering Center. The University of Chicago has for some years now been admitting freshmen after only two years of high school, with the avowed aim of making the undergraduate college an intermediate rather than a "higher" institution; whereas other well-known schools, in order to make "higher education" really "high," increasingly prefer men in their early twenties who have spent a few years at work after leaving high school.

Nor is the situation any more uniform in secondary schools. Within thirty miles of New York City there are public schools so progressive as to live up to the caricatures in the funny papers of the twenties, and others so conservative as to justify every word of the progressives' indictment of the traditional schools. I have taught college freshmen from public schools who had learned more mathematics than most college curricula offer, and others who had come from schools with an equal reputation where mathematics, beyond long division, was an "elective" and was taken only by children planning a career in science or medicine. There was one proper Bostonian who could remember only one American president, Benjamin Franklin; the main educational dish served at his very proper-Bostonian school had been a rich stew called "civics" which contained odd pieces of almost anything except the history of his own country. And I have taught other freshmen whom school had given a sound knowledge of historical facts and even the thrill of history. Yet every one of these high schools is unmistakably and characteristically "American."

Even less compatible with the myth of American uniformity is the reality of American literature—and if education is the matrix of society, literature is its truest reflection. How "uniform" for instance are the American writers who emerged in the literary explosion of the nineteen-twenties: Sinclair Lewis, Hemingway, Willa Cather, John Dos Passos, Wolfe, Faulkner, T. S. Eliot, Robert Frost, Carl Sandburg? A more diverse lot, in style, mood, and subject matter, could hardly be imagined; and the diversity becomes the greater the more names are added: Sherwood Anderson, for instance, or Ellen Glasgow, Dashiell Hammett or H. L. Mencken, Scott Fitzgerald, Ezra Pound, Eugene O'Neill, or e.e. cummings. And just how "uniform" are the three magazine successes of the same decade, the *Reader's Digest,* *Time,* and the *New Yorker?* The educated European knows American literature and avidly reads our magazines; a good many American writers may indeed be better known in Europe than here. Yet if there is one thing he is sure of it is the "standardization" of the American mind.

Writers and journalists, it may be said, are noncomformists to begin with. Well, what about entertainment in America? Our radio stands perhaps first among the targets of the European critics of "American uniformity." If everything they are saying about it were true, it still would not account for the twenty-five to thirty-five stations—one in almost every major city—which offer "serious" music eight to twelve hours a day. None of the "serious" or "high-brow" programs of European radio systems draws enough of an audience to exist without heavy subsidies. But a great many of the two or three dozen "serious" radio stations in this country manage to operate at a profit, though they are supported only by advertisers who are unlikely to be interested in anything but a listening audience large enough to justify the investment.

One station, WABF in New York, is even running an entire Sunday of music without any advertising, financed solely by voluntary contributions from its listening audience.

And how is one to explain that young people weaned on "Good Night, Irene," the "Hit Parade," and the "Lone Ranger" rush head-long into chamber music, as listeners and, increasingly, as players, as soon as they reach college? Walking across the campus on a fine spring evening one hears "long-haired" music, from Buxtehude to Bartok, streaming out of every other open window. Symphony orchestras are appearing in small towns as well as in large cities. And instead of being subsidized, these American orchestras are supported as they are formed, by voluntary community action; the *Wall Street Journal* reports that more money is spent on symphonies than on basketball.

But what should completely destroy the European's concept of American uniformity is the diversity existing in industry. The fact that in this country business and industry are part of the country's culture is to the European the most remarkable thing about the United States—to the point where he greatly overstates the extent to which America has become a "business society." The European businessmen and labor leaders who have been touring this country under the auspices of The Marshall Plan these last four years all report as their central finding the experiments in new techniques, new products or processes, in account-ing or in labor relations, in organization structure or in foreman training, carried on in almost every company they visit. No two, they find, do the same things. In the words of one of these teams, "Every American company feels that it has to do something different to stay in the race." Many teams feel that American competition is too "ex-treme"; and when pressed to explain why they use this term, they talk about the demands on managerial imagination and worker adaptability made by the need always to do something different and something new. The same sort of variety is found in our labor relations. In Europe relations between management and union tend to be rigidly and uni-formly molded by a central association of industries negotiating with a central federation of trade unions; the individual employer or the individual union local pays dues but is otherwise inert. Not so here.

Some of the members of these teams even feel that we carry diversity too far to be efficient. "I have seen a dozen plans for management development in as many companies," one of the senior men in British industry told me. "Every one—Standard Oil, Ford, Johnson & Johnson, Du Pont, General Electric, the Telephone Com-pany—has a plan of its own, a staff of its own, a philosophy of its own. That just makes no sense. Why don't you fellows get together, appoint a committee, and have them work out the *one best plan* which everybody could use and which could be run centrally by a few top-flight people?"

It is not even true that within the American plant there is more "standardization" than elsewhere, as we have all come to believe. The figures tell a different story. In this country of mass production a larger proportion of the workers in manufacturing industries are what the census calls "skilled workers and foremen" than in any other country on which we have data. And we have an even greater number of men, proportionately, in executive, technical, and managerial positions. In other words American productive strength lies in higher capital investment per worker and better management; better planning, better layout, better scheduling, better personnel relations, better marketing— all of which mean more skilled and more trained people rather than more unskilled repetitive work.

3

I am not discussing here the *quality* of American culture, whether it be crude, shallow, vulgar, commercialized, materialist, or, as the Marxists maintain, full of "bourgeois idealism." My concern here is solely with the prevailing European conviction of American uniformity. And that conviction is an obvious absurdity. Nor could it be anything else considering the pragmatic bent of the American people and their deeply engrained habit of voluntary and local community action and community organization.

Indeed any serious student of America has to raise the question whether there is not *too much* diversity in this country. There is the danger that diversity will degenerate into aimless multiplicity—difference for difference's sake. Jefferson, de Toqueville, and Henry Adams, as well as recent critics of American education such as Robert Hutchins, have seen in this the major danger facing American society and culture.

There is actually more uniformity in European countries, both materially and culturally, than in the United States. It may no longer be true that the French minister of education knows at every hour exactly what line of what page of what book is being read in every French school. But still, in education, in religious life, in political life, in business as well as in its cultural ideals, European countries tend to have at most a few "types," a few molds in which everything is formed. What then can the European possibly mean when he talks of the "uniformity of America"?

He himself, as a visitor, unconsciously furnishes the answer in the way he sorts out his American experiences, in the questions he asks, in the answers he understands and those he doesn't. When he thinks of "diversity" he tends to think of the contrast between the ways in which social and economic classes live. He is used to seeing a definite and clear-cut upper-class civilization and culture dominating. And that indeed he does not find in this country. Therefore the bewildering differences in American life appear to him meaningless—mere oddities.

I still remember how the sage of our neighborhood in suburban Vienna, the wife of the market gardener across the street, explained the "Great War," the war of 1914–18, when I was a small boy of ten or eleven: "The war had to come because you couldn't tell maids from their ladies by their dress any more." Frau Kiner's explanation of history differed from that offered during the nineteen-twenties by Europe's learned sociologists, whether of the Right or of the Marxist persuasion, mainly by being brief and simple. They all assumed that there must be a distinct upper-class way of life, an upper-class architecture, upper-class dress, upper-class goods in an upper-class market—and contrasted with it the "folk culture" of the peasantry or the equally distinct ways of life of the middle class and working class. Indeed that eminently sane, that notoriously Americophile magazine, the London *Economist,* echoed Frau Kiner only a few months ago when it reported with apparent amazement that "to the best of their ability—and their ability is great—the [American] manufacturers make clothes for the lower-income groups that look just as smart as those they make for the more fortunate"—and explained this perverse attempt to make maids look like their ladies as the result of the "egalitarian obsession" of this country.

The class-given differentiation in Europe is even more pronounced in the nonmaterial, the cultural spheres. One example is the tremendous importance of the "right speech" in practically every European country; for the "right speech" is upper-class speech. Another example is the extent to which European educational systems are based on the education of a ruling class. The Renaissance Courtier, the Educated Man of the Humanists, the Christian Gentleman of nineteenth-century England—the ideal types which embodied the three basic educational concepts of modern Europe—were all in origin and intent ruling-class types. The rising middle class not only did not overthrow the class concept of education, it emphasized it as a symbol of its own emergence into the ruling group. Similarly, in Occupied Germany the working-class leaders—to the chagrin as well as the complete bewilderment of American educational advisers—have shown no enthusiasm for the plan to convert the traditional *Gymnasium* into an American high school. To deprive these schools of their ruling-class character would actually deprive them of social meaning for working-class children.

Europe has even succeeded in turning diversities and differences that were not social in their origin into class distinctions. One of the best examples of this is the way in which the "gentry" and its retainers became identified with the Church of England while the "tradesman" went to "Chapel"—a distinction that held till very recent times and is not quite gone yet.

Thus the European myth of American uniformity tells us less about America than about Europe. For it is based, in the last analysis, on Frau Kiner's belief that a class structure of society is the only genuine moral order.

That today the theme of "American uniformity" is played on above all by communist propaganda is thus no accident. For the "proletariat" of communist ideology is indeed a "master class." It is a reaffirmation of the European ruling-class concept and of its ruling-class way of life in an extreme form—only turned upside down. On this rests to a considerable extent the attraction of communism for European intellectuals. There is an old Slav peasant proverb: "There will always be barons for there must always be peasants." All Vishinsky would have to do to change it into an orthodox Soviet proverb would be to change "barons" to "proletarian commissars." And Frau Kiner's philosophy of history he would not have to change at all.

4

But Frau Kiner's statement could never have been made in this country, not even by a sociology professor in a three-volume tome. Whether the United States really has no ruling class—and therefore no classes at all—or whether, as the Marxists assert, the classes are only camouflaged in this country, one thing is certain: this country knows no distinct upper-class or lower-class "way of life." It knows only different ways of making a living.

Indeed there has been only one genuine ruling-class way of life in this country since its beginning: that of the plantation aristocracy in the Old South between 1760 and 1860. When the *nouveaux riches* in the period between the Civil War and the First World War made the attempt to set themselves up as "Society" they failed miserably. They could not even develop an upper-class American architecture—and of all the arts architecture is the mirror of the way of life. The tycoons had to be contented with imitation French châteaux, Italian Renaissance palaces, and Tudor manors—the white elephants which their servant-less grandchildren are now frantically turning over to monasteries, hospitals, or schools. (It is not entirely an accident, perhaps, that the people most eager to live today in the baronial halls of yesterday's capitalists seem to be Soviet delegates.) To find an upper-class way of life the tycoons had to gate-crash the Scottish grouse moors, the Cowes Regatta, or the Kaiser's maneuvers in Kiel. In this country it was difficult indeed to lead a ruling-class life.

The closest we come today in this country to anything that might be called an "upperclass way of life" is to be found in the top hierarchy of the big business corporations. The way people in some of these companies talk about the "twelfth floor" or the "front office" faintly echoes Frau Kiner's concept of the "ladies." At work the big business executive has indeed some of the trappings of a distinct style of living in the ceremonial of receptionist, secretary, and big office, in his expense account, in the autographed picture of the "big boss" on the wall, the unlisted telephone, and so forth. But only at work. As soon as he leaves the office the "big shot" becomes simply another business-man, anonymous and indistinguishable from millions of others. And he is quite likely to live, like the president of our largest corporation, in an eight-room house in a pleasant and comfortable but not particularly swank suburb.

In fact, it does not even make too much sense to talk of this country as a "middle-class" society. A middle class has to have a class on either side to be in the middle. There is more than a grain of truth in the remark made jokingly by one of my European visitors, an Italian student of American literature: "If there were such a thing as a working-class literature, *Babbitt* and *Arrowsmith* would be its models."

5

Any European who has perchance read thus far, will growl that if Europe's mental picture of "American uniformity" is absurd, America's mental picture of "European class society" is absurder. And he is right. In fact, the one myth is the reverse of the other. To the American, for instance, "class society" means a society without social mobility. But Frau Kiner was anything but respectable lower-middle class knowing its place. She was a successful social climber who had fought her way up from a sharecropper's shanty and a job as scullery maid at fourteen—and had pushed her man up with herself. And in those years after the first world war she was capping her social triumph by marrying off her beautiful and well-dowered daughters to "gentlemen"—elderly and moth-eaten, but undeniably "gentlemen."

Nor is a society in which an Eliza Doolittle can jump from slum waif to "great lady" just by learning upper-class speech a society without social mobility. (Indeed there is no better sign of America's failure to understand Europe's "class society" than our tendency to play "Pygmalion" as a farce and as a take-off on upper-class snobbery—whereas it is as much crusading pamphlet as comedy of manners, the only snobs in it being the class-conscious cockneys.) Altogether there has been tremendous social mobility in any Western or Central European country whenever there was great economic expansion: in Britain between the Napoleonic Wars and 1860, in Germany a generation later, in Bohemia—perhaps the most startling example—between 1870 and 1900. The central difference between America and Europe may well be in the *meaning* rather than in the *extent* of social mobility. When the boss's son is made a vice president in this country the publicity release is likely to stress that his first job was pushing a broom. But when a former broom-pusher, born in the Glasgow slums, gets to be managing director in a British company the official announcement is likely to hint gently at descent from Robert Bruce.

I must break off here. Another European visitor has just come in for a chat, a young French philosopher, fresh from a six-month tour of American universities. I anticipate a pleasant and informative afternoon; the letter with which he introduced himself was interesting and intelligent. "The thing that impressed me most," he wrote, "is that no university I visited tries to develop a 'school of philosophy.' On the contrary each tries to stress different views and different schools in its faculty—the exact opposite from what we would normally do." Yet I know that sooner or later in the course of the afternoon he will ask me, "Mr. Drucker, don't you find it very trying to live in so mechanically uniform a country?"

Part Two

Economics as a Social Dimension

Introduction to Part Two

There is only one point on which the economists and I are in agreement: I am NOT an economist. It is not that I don't know enough economics—and if I didn't it would be easy enough to remedy the deficiency. And I actually sat at the feet of both of this century's great economists, John Maynard Keynes and Joseph Schumpeter. But I do not accept the basic premise on which economics as a discipline is based and without which it cannot be sustained. I do not accept that the economic sphere is an independent sphere, let alone that it is the dominant one (and even less that it is the ONLY sphere as so many economists maintain, especially the true believers of the Austrian School). It is surely an important sphere. And as Bertold Brecht said, "first comes the belly and then morality"—and filling the belly is what economics is all about in the main. I am not only willing, I insist that in all political and social decisions the economic costs are calculated and taken into account. To talk of "benefits" only—as so much of the social legislation of the post-World War II period did—I consider irresponsible and bound to lead to disaster. And I believe in free markets, having seen far too much of the alternative.

But still, for me the economic sphere is one sphere rather than *the* sphere. Economic considerations are restraints rather than overriding determinants. Economic wants and economic satisfactions are important but not absolutes. Above all, economic activities, economic institutions, economic rationality are means to noneconomic, that is, human or social ends, rather than ends in themselves. And this means that I do not see economics as an autonomous "science." In short, it means that I am not an economist—something I have known since, in 1934 as a young economist in a London merchant bank, I sat in the Keynes seminar in Cambridge. I suddenly realized that Keynes and all the brilliant economics students in the room were interested in the behavior of commodities while I was interested in the behavior of people.

But precisely for this reason I am deeply interested in economics as a social dimension. I am deeply interested in economics as the symbol around which social and political issues can be organized—which has largely been its role in American history. I am deeply interested in economics as an expression of social and political views and values, which is what the second essay in this part, "The Poverty of Economic Theory," discusses. I am deeply interested in economics as expression of social views and social values which is what the third essay in this part, "The Delusion of Profit," is about. And I am deeply interested in the key personalities of economics, the great economists, such as Keynes and Schumpeter (chapter 9).

A word of explanation about chapter 10, "Keynes: Economics as a Magical System." When it was first published right after Keynes' death in 1946, it became immensely popular, was reprinted in anthologies and college text books, and was used in college courses. Then it disappeared from view—Keynesian economics became the accepted wisdom in the United States and the way this essay viewed it came to be seen as irrelevant. Whether this was a blessing or not, is debatable. It may not be sheer accident that the two countries that never adopted Keynesian economics, Japan and West Germany, have been the two countries that have done spectacularly well in the forty years since World War II whereas the two countries of the Keynesian persuasion, Great Britain and the United States, have steadily been losing ground and are in danger of becoming noncompetitive. The main point, however, is that we now, in the 1990s, are reaching the point where the analysis of my 1946 essay will be tested. The Keynesian prescription—to create prosperity and productivity through government spending—has clearly reached its end in both the United States and the United Kingdom. In these countries government spending to stimulate consumption can only aggravate stagnation and inflation. At the same time radical deflation—the medicine the World Bank and the International Monetary Fund administer to sick economies—has such grave side effects and is of such doubtful efficacy that it would never be permitted if it had to be approved by an economic

"Food and Drug Administration." Will the United States and the United Kingdom now turn to the policies the West Germans and the Japanese have been using—the policies of promoting savings and capital investment—which this essay advocated more than forty years ago?

6

The Economic Basis of American Politics

Why is there not only one single American among the "great economists"? From its earliest days this country has had more economists than any other country. It has led in the development of the tools of economic analysis. Economists are everywhere—in government, in business, in the universities, in the labor unions. In no other country, indeed, is a knowledge of economics considered part of ordinary education; in the United States, however, we have for many years been trying to combat "economic illiteracy" in the secondary schools. And certainly there is no country in which popular interest in economics is greater and in which economic issues are more prominent.

Yet we conspicuously lack the Great Economist, the economist who changes our ideas about economics and gives us new approaches to the interpretation and direction of economic events.

Or, rather: the great economists America has produced are not known as such, are not recognized as such. Alexander Hamilton and Henry Clay certainly deserve being considered very great economists: Hamilton, at the very dawn of systematic economics, created a basic theory of economic development which has not been much improved since; Henry Clay's "American System" is the font and origin of all welfare economics. Yet their very names are rarely mentioned in histories of economic thought, whereas the German Friedrich List, who repeated what Clay had taught him, usually occupies a prominent place in these books.

First published in *The Public Interest*, 1968.

Of course, neither Hamilton nor Clay really wanted to be known as an economist. Their own ambitions were elsewhere: Hamilton on becoming the commanding general of a victorious United States Army, Clay on being elected president. To both men, their economics were totally incidental to their politics and a tool thereof. For both Hamilton and Clay, economic policy was clearly a means to a political end. And when their economic views are discussed, they are correctly treated as part of their political theories and political strategies.

The explanation for this state of affairs would seem to be this: Economics is too important to American politics to be left to the American economist. Economics has a political role to fulfill that transcends its own subject matter. For well over two hundred years it has been the unifying impulse in this country's political process. Since colonial days, "economic interests" have been used systematically to create political forces and political alignments and, above all, to unite regional and sectional groups behind one leader and one program. The names of these economic interests have changed; but whether we speak of the "manufacturing interest," the "farming interest," or the "silver bloc," the idea itself has not changed. Similarly, for well over two centuries, economic issues—such as the tariff, the currency, or free soil—have been used to overcome and neutralize ideological cleavages and conflicts that otherwise might have torn apart the nation. During all of our history, fundamental rifts in the country have been bridged by polarizing politics on economic issues; these are issues on which a compromise, distributing dollars and cents, is always possible. The classical example is, of course, the compromise over the "Tariff of Abomination" between the South and Andrew Jackson that postponed the conflict over slavery for thirty years. All along it has been good American political manners to talk dollars and cents when we really mean political decisions. The way in which Robert McNamara, as secretary of defense in both the Kennedy and Johnson administrations, used budgetary control radically to alter strategic concepts and military organization is another good example.

Perhaps most revealing is the way in which we have used the economic sphere to think through and work out basic issues of the relation of government to society. Big business is far more powerful in France or Germany, for instance, than it is in the United States. But only in the United States has the relationship between "government and business" come to be considered the key issue for a fundamental discussion of the role and power of government in society. Indeed, our public discussions for a century now have led many a naïve foreign observer to conclude that in the United States there are no noteworthy social institutions other than business institutions.

The Great Themes of American History

The result of this peculiar role of economic issues and economic controversy in the American political process is most paradoxical. On the one hand, economics in this country appears to be far more prominent and far more important in the political life than in any other country. American history seems to be dominated, at first sight, by economic conflict. Indeed, it is possible to overlook the fact that the great themes of American history have all along been moral and constitutional: slavery, the industrial versus the agrarian society, and federalism in the nineteenth century; racial equality, the role and function of the central government, and American's place in an international society in this century. In sharp contrast to countries whose politics have an ideological organization and pattern, such as all European countries, these great themes are barely mentioned in day-to-day, year-to-year American politics, where the slogans are primarily economic. It is, therefore, only too easy to mistake the appearance for reality. It is thus possible to argue, as a whole school of historians did, that the Civil War was an "unnecessary conflict" and could have been avoided by paying a few hundred million dollars to the slaveowners. It is possible, as Arthur M. Schlesinger, Jr., did in his brilliant *Age of Jackson,* to overlook completely that the central theme and the crowning achievement of the Jackson administration was to establish the sovereignty of the national government over all regional or sectional interests; instead, Mr. Schlesinger made economic and class interests paramount. It is even possible, as Charles Beard did in a long life as a historian, to see the whole of our history as determined by, and subordinated to, economic interest. But Beard lived long enough to find out that any attempt to predict the course of American history and American political behavior from economics is bound to misfire.

Every American politician must, indeed, know how to use economic measures for political ends. If he aspires to a national role, he must be a master of finding and creating economic alignments that unify diverse groups across the nation. Even John C. Calhoun, the most nearly "metaphysical" of our political thinkers, spent the last two decades of his life in an abortive attempt to bridge the moral gulf of slavery between South and West by means of their common economic interest as farmers.

So it is that, despite its appearance of centrality, economics in the American experience is actually a subordinate means to predominantly noneconomic ends. Our values are not economic values, nor is our economics autonomous. Politics even decides what economic issues are allowed to appear on the stage of American history. For, to be "available" as a political vehicle an economic issue must fit into our political logic. It must mobilize national energies and must unify large masses across regions for joint political action and for the conquest of the central political power that is the presidency.

This explains the absence of the Great Economist. Such a man must assume the autonomy of the economic sphere in human life. He must assume the reality of economic values. If he is interested at all in politics (as few of them were), he must treat politics as a handmaiden of economics and as a tool for achieving economic purposes. These assumptions make no sense to the American experience. What flourishes in this country, therefore, is economic technicians of skill and renown, economic analysts, and expert fashioners of economic tools. We have political economists and economic politicians galore. But the climate is most uncongenial to the Great Economist. Such a man must assume an autonomous economic reality, of which the political issues are merely a reflection. In our American experience, however, economics is the conventional shorthand and the lingua franca for issues and decisions which are not economic, but political and moral. One might, indeed, formulate a basic rule of American politics to read: *If at all possible, express a political issue and design a political alignment as an economic issue and an economic alignment.*

The insight that economic interests can be used as the hinge of politics is commonly traced to the famous No. 10 of the *Federalist Papers* in which James Madison (following Harrington's *Oceana* and John Locke) concluded that power follows property. But, when Madison wrote, American political life had already, for a century, been habitually organized around economic issues and in economic alignments. Madison only codified what had been fairly general American experience during the colonial period.

Colonial legislatures had, indeed, no alternative if they wanted to be effective at all. The matters that now occupy a legislature—public order and law, the administration of justice or education—may have been of very great interest to them, but were normally not within their reach. They had, perforce, to be left to the individual local communities, the towns and counties. In colonial America, distances were too great, population too sparsely settled, for any central authority to be effective. If the local community would not look after its own internal affairs, nobody else could. The major burning issue for the colonial legislature was relations with the mother country. And those turned on economic problems and economic questions: taxes and tariffs, coinage and credit, and so on. It was in these matters, above all, that the colony's Royal Governor was interested; for in eighteenth-century theory and practice, an overseas colony was an economic asset. Of course, such recurrent disturbances as Indian risings or the endemic war with the French occupied a good deal of the time and attention of the colonial politician. But his main job was to represent the colonist before and against the economic power represented and exerted by the Governor. His very *raison d'être* was economic. And only by identifying and organizing economic interests could the eighteenth-century colonial politician create unity in the electorate he represented—an electorate which, as the century wore on, became increasingly diversified in its other characteristics (religious beliefs, ethnic origins, and so on). There never was much need to do for the eighteenth-century politician, in the state houses of Boston, Philadelphia, and Williamsburg, what Sir Lewis Namier had to do in our century for his English counterpart: to seek out and identify his economic affiliations and interests. That was the one fact about the politician in colonial America that was always clear, evident, and known to everybody.

But to Madison—and to all the brilliant politicians who, in the first quarter-century of United States history, established the political conventions and the political processes for the infant country—should be given credit for one fundamental insight: Economic interests could be used to *unify*. They could be used to overcome the pernicious "factions," the cleavage of society into ideological camps divided by their basic beliefs, which the founding fathers rightly feared as incompatible with nationhood and political stability.

The Uses of Economic Conflict

This is a political view of economics—a view which explains economic events in terms of human behavior. This, more than anything else, distinguishes the traditional American approach to economics from the approach of the economist. The economist understands the behavior of commodities. And if he is naïve (as our present-day neo-Keynesians tend to be), he believes that human beings behave as commodities do. But even at his most profound and skeptical, he is likely to consider ordinary human behavior as economically nonrational behavior. In fact, the economist has, all along, been either suspicious or at the last contemptuous of the politician who, so it seems to him, subordinates the clear logic of economic rationality to the murky unreason of human emotions and vanities. From this starting point, the classical economists essentially arrived at a denial of politics. To them there was a pre-established harmony in the economic system, with economic self-regulation automatically producing the optimum for all groups and classes in society. Marx himself was no less contemptuous of the politician: no longer accepting the classical doctrine of harmony, but instead accepting the reality of economic conflict, he deduced therefrom the inevitability, beyond any politician's contriving, of class war and revolution.

The political economists of the American tradition never for one moment believed in pre-established economic harmony. Economic conflict was much too obvious for that. It had, after all, characterized relations between the infant colonies and the mother country, culminating finally in a violent upheaval of the political order in the Revolution. But at the same time, they saw in economic conflict their means to prevent the more dangerous ideological conflict. And they saw in economic conflict the means to establish order—*not* harmony, which they did not expect on this earth. Above all, they saw that economic conflict was the one clash within the body politic that could be managed. It could be managed because economic interests are divisible, whereas political or religious beliefs are not. One can always split an economic difference in two—and while half a loaf is better than no bread, half a child, as King Solomon long ago perceived, is no good at all. The same goes for half a religion, half a philosophy, or half a political principle.

Above all, their experience, unique at that time and quite at variance with what "common sense" would have taught elsewhere, had convinced the founding fathers that, unlike all other cleavages, economic conflicts tend to become less acute with time. They may not be self-healing, but they are capable of amelioration. If the fight is over "who gets how much," then one can satisfy both sides if the amount available for distribution increases. And their experience as colonists on the virgin continent had taught them that the economic pie is, indeed, capable of being increased by human action, rather than being fixed forever.

They may not have consciously thought this through. But Alexander Hamilton started out from the assumption that it is possible to increase the economic resources available. This assumption explains in large part why his countrymen, no matter how much they distrusted his politics, took to his economics at once, but also why he never attained full respectability as an economist. For the economist traditionally—until well past World War II—took it for granted that economic resources are given and limited, so that the problem is their most effective distribution in a system of equilibrium. In this respect Keynes, however much he otherwise might have differed from his predecessors, was as traditional as anyone. It was not until the most recent decades, until the advent of "economic development" as a goal of economic policy—with President Truman's "Point Four" declaration of the early 1950's being the crucial date—that proper economists accepted the purposeful creation of dynamic disequilibrium as possible and meaningful.

To the American—no matter how faithfully he repeated the teachings of the economist, no matter how faithfully he himself taught them in the classrooms of his colleges—it was obvious that in his country the economic resources had been proven to be capable of almost infinite expansion through human and, in large measure, political action. He may have agreed with the economist that this was purely the result of the rarest of accidents: the existence on this continent of vast areas of empty soil, ready for plow, ready to be appropriated and to be converted into an economic resource. But very early we find in the actual political behavior of the American strong evidence that, deep down, he knew differently. He knew what Hamilton had known in the last decade of the eighteenth century: that there is an economic dynamic and that economic resources are the creation of man rather than of Providence. This underlies quite clearly such bold measures as the Morrill Act of 1865 which, in creating the land-grant colleges, clearly assumed that the application of knowledge creates economic values and productive capacity well beyond that given in the existing resources. It underlay, from the beginning, all American trade-union movements. American "business unionism" assumes not only that the fight over the division of the pie is by itself likely to produce a larger pie—that, in other words, economic conflict by itself leads to economic growth and therefore, at the same time, to political and social unity.

There can be little doubt that the American concept of "economic interest" as an effective and unifying political force has served this country exceedingly well.

In fact, it is questionable whether there could have been an American nation without it. With the wisdom of hindsight, we have come to see in the frontier a source of strength. But, in reality, the experience of the frontier must have been an almost unbearable strain, as witness all other countries that have undergone a similar experience. It was not only the kind of strain on the physical resources and on the political energies which rapid, turbulent expansion produces. Above all, it was a strain on the unity of a country in which the new tidal wave of immigrants of different social background, national origin, and religious allegiance always arrived long before the preceding wave of immigrants had been absorbed. In such a country, growing at a frenetic speed, ideological, philosophical, or religious cleavages might have been fatal.

One should not forget that the immigrants, by and large, had themselves no tradition of self-government or even of political activity; it was not, after all, the respectable or well to do who arrived in the holds of the immigrant ships. And yet these vast heterogeneous masses had to become a nation under one government and with one set of basic values practically overnight, or else the American experiment would have floundered. If economic interests had not been available as the political organizer, it would either have been necessary to impose the most rigid authority on the population or else pluralism would have organized itself *against* the nation and its unity—with every imported tradition of religion and culture, every imported political value and belief, the focus of an ideology alien, if not hostile, to American nationhood (as is so clearly the case in Latin America).

The great phenomenon of the nineteenth century is not, after all, the rise of the American economy. It is the creation of the American nation. For a nation, as we are now finding out the world over, is not something one can easily create. It is, on the contrary, usually the fruit of long experience and of historical forces operating over many centuries. Neither the nations of Europe nor Japan were created overnight. That nationhood is difficult and takes a very long time to create is proven by the fact that, outside of these old nations—and of the United States—very few, if any, nations have yet come into being. In all of Latin America, for instance, despite centuries of political identity, only Mexico and, to a lesser extent, Brazil, can be said to be "nations"—and, in both, nationhood has come only in this century. But the United States achieved nationhood in a few short decades, or, at the most, within a century. This it owes to a very large extent to the tradition which used economic interests in their clashes and conflicts as the foundation for political issues, political alignments, and political conflicts. This has enabled the United States to tolerate, if not to encourage, pluralism in all other spheres, to survive the fiercest of civil wars, and to attain a unity of allegiance and of basic commitments which represent as strong a common bond and communion as centuries of common history, common language, and common experiences have given to any of the older nations.

The Bias Toward "Bipartisanship"

But the convention of economic interests has not only tended to prevent ideological issues from arising. It has forced the American political system into a nonpartisan approach to noneconomic problems.

A noneconomic issue threatens the existing political alignments. It is not easily encompassed within the American political system. The American politician shuns ideological stands, for the simple reason that they are certain to alienate a large proportion of a constituency brought together and held together by economic interests. Any noneconomic ideological stand would have at once, for instance, exploded the alliance between the lily-white, fundamentalist Protestant, and proudly Anglo-Saxon South and the cosmopolitan, largely Catholic or Jewish working class in the big cities on which the Democratic party was based for so long. The only thing that could hold them together— and could thereby get a Democratic politician into federal office—was their economic opposition to the manufacturing interest.

This built into the American political process a powerful incentive to handle noneconomic issues on a "bipartisan" basis, that is, to remove them essentially from party politics. Indeed, the greatest praise in our political system is reserved for the "patriot" who turns a potentially disruptive noneconomic issue into bipartisan consensus. It is for this act of patriotism at the expense, legend has it, of his presidential chances that American history primarily remembers Henry Clay. And a century later, Arthur Vandenberg earned himself a permanent place in the American political pantheon by similarly making American foreign policy after World War II "bipartisan," thereby again sacrific- ing whatever chance for the Presidential nomination of his party he might ever have had. A great many voters repudiated Barry Goldwater in 1964, not because they disagreed with his views, but because his decision to take a partisan stand on noneconomic issues seemed to them a greater threat to the Republic than a wrong, but bipartisan, consensus on the issues. And, in the context of the American political process, they were right.

Foreign affairs, religion, education, civil rights, and a host of other areas which, in any other country, are the bread and meat of party politics and political organization are, therefore, as much as possible, treated as "bipartisan" in the American political system. This does not mean that they are not controversial. It means that the system, as long as it works, uses them to bring together otherwise warring factions rather than to create new factions on each side of the controversy. In fact, we much prefer not even to tackle such issues unless there is available for them a broad coalition cutting across our conventional political alignments. Again and again, initiative in such an issue is left, as if by passive agreement, to the one body in the American political system that is outside the established party alignment, the Supreme Court. That in both the great constitutional issues of the recent years, civil rights and reapportionment of voting districts, a Supreme Court decision took the place of political action in these politically crucial areas was in full accord with the tradition that goes back to John

Marshall's Court. For a Supreme Court decision is the "law of the land" and thus establishes its own consensus.

The Limitations of "Consensus"

There are obvious and real limitations to the effectiveness of the convention of economic issues in American politics. Not every noneconomic—that is, ideological, political, or moral—issue can be either made to appear as economic or organized as "bipartisan." Indeed, the most important issues and decisions in the life of a nation cannot be handled in this fashion. The great example is slavery, of course, for in this country slavery did not primarily serve an economic function (as it did, for instance, in the plantation economy of Brazil). At least by the second quarter of the nineteenth century, the main function of slavery was to endow the "poor white" in the South with a feeling of human superiority, no matter how wretched his physical or moral condition. Even if slavery in its origin and spread was an economic institution, by the time it became an issue, that is, after 1820 or so, the main beneficiary was no longer the slave owner, but rather the nonowner who derived the psychic benefits of a slave society without having to carry the increasing economic burden of maintaining the slaves. In other words, the Abolitionists, as we now know, were right: slavery in this country was a sin rather than a crime. And for this reason the abolition of slavery by itself, without true civil rights for the blacks, settled so very little.

And for this reason, too, slavery could not be camouflaged as an economic issue, no matter how hard the politicians of the early nineteenth century tried. Moreover, as a truly "irreconcilable" issue, that is, as a spiritual and moral one, there could be no "consensus" on it, even though the Supreme Court of the time tried to establish it in the Dred Scott decision. Nor could the existing political organization handle such a noneconomic issue—and probably the political system of no country, no matter how organized, could have handled it. On the issue of slavery, then, the American political system floundered and sank into Civil War, almost destroying the country with it.

But even in less crucial and less sensitive areas, the American political system is not geared to handling the noneconomic issue. This is particularly true whenever foreign policy cannot be organized on the basis of "consensus" and "bipartisanship." Any such failure leaves deep and long-lasting scars. For any such failure endangers national cohesion. This was true of the War of 1812. It was true of the bitter foreign-policy conflict that preceded our entry into World War I. But for the Japanese attack on Pearl Harbor, the rift over foreign policy in the years before World War II might similarly have proven incapable of being organized within the American political system and might have torn apart American national unity. Today, Vietnam poses a similar threat.

But then there is also always a danger that our politicians may forget that the economic issue is a convention and fall into the error of believing that economics really controls politics. In domestic affairs, the danger is not very great. It is a very stupid politician who will not intuitively realize the limitations, as well as the uses, of the convention. Certainly no strong president—Jackson, Lincoln, the two Roosevelts, or Truman, for example—ever believed that economic interests and economic policy were by themselves sufficient, or that they necessarily prevailed. But in foreign affairs we have made this mistake again and again—and have paid dearly in every case. Again and again we have fallen for the illusion of "economic sanctions" as an effective tool in foreign affairs. And again and again we have found that they are ineffective. This holds true for the belief of the South that "King Cotton" would force the North to its knees and constrain Europe to ally itself with the Confederacy, but also for the balancing belief of the North that the Confederacy could be defeated by economic blockade. And we now also know, from the diaries of pro-Western Japanese leaders, that the economic sanctions which the United States and Great Britain imposed on Japan in 1940 and 1941 only strengthened the war party and deprived the moderates of all influence, just as the blockade of Germany in World War I emasculated the moderates in the German government and made the military extremists all-powerful. Ironically, we now also know that economic sanctions did not even work in the one case in which they seemed to have been successful—the case which probably explains the blind American belief in this policy: the "economic sanctions" of the American colonists against the first British attempt to tax them, a decade before the American Revolution. Recent historical research has made it reasonably certain that the

British Cabinet used the American boycott as an excuse for a retreat from a policy which had proven exceedingly unpopular among powerful backers at home, and not primarily for economic reasons.

The convention of economics as the ground of political action and organization is, in other words, just a tool. Like every tool, it has to be used with judgment. And like every tool, it has limitations. Whoever mistakes the convention for reality pays the heavy penalty one always has to pay for deceiving oneself.

But while not perfect, not infallible, and not a panacea by any means, the convention has served the American people remarkably well.

The question, however, is not really how well the convention has served in the past. It is: is it still useful, still serviceable? Can the common, ordinary, political business of the American people still be ordered by the traditional rule to formulate issues, as far as possible, as economic issues and to define alignments, to the greatest extent possible, as economic alignments?

It is just barely possible that this traditional convention of domestic American politics has a major future role to play in foreign and international affairs. But at the same time, it may be at the end of its usefulness at home.

Every one of the many new countries that have come into existence in the last two decades has yet to become a nation. Every one is less well equipped for this task, by history and tradition, than was the infant American Republic two hundred years ago. In every one, the cleavages between tribes, between religions, between races, run deep—and will have to be bridged fast if the country is to survive. No one of them could survive a conflict of ideologies. In this situation, a good many of them, if not all, will predictably take recourse in the elimination of politics—the vain promise of every dictatorship. Equally predictably, this will only worsen the conflicts and make them even less tractable. Predictably also, some of them at least will seek escape in braggart nationalism, if not in conquest abroad. This, too, history amply teaches, will not succeed. Only an approach to politics which allows conflicts to be productive and to create unity across the dividing line of tribe, religion, tradition, or race would seem to fit the needs of the new countries. In the traditional American approach, which makes politics turn on economic interests and economic issues, the new countries might well find what they need, ready-made and well tested.

Indeed, this approach might become increasingly more important for the international community altogether. The world today is threatened by a danger even greater than that of class war in the nineteenth century. It is threatened by the danger of a world-wide race war of the poor and largely colored majority against the largely white minority of the rich. At the same time, this is an infinitely smaller world than was that of the eighteenth century—a world in which everybody is everybody else's neighbor and in which, therefore, there is no alternative to living together. In such a world, a political concept which allows for productive conflict, but which also organizes unity beyond the ideologies and traditions that divide, might be of the greatest importance.

We are obviously very far from any such accomplishment; the "Alliance for Progress" in President Kennedy's original version was probably the closest to it. But, in retrospect, the development efforts of the 1950's and 1960's may well one day appear as the first uncertain and faltering steps toward a new, nonideological and yet unifying concept of international order, different alike in its flexibility and effectiveness from the world anarchy of sovereign states which has become a hopeless anachronism, and a world government which, if at all feasible, could today be only a world-wide tyranny.

The Need for Innovation

But at the same time, it seems likely that in domestic politics the traditional economic convention has come to the end of its usefulness. It is not that we are dissatisfied with it or that we hanker after ideological politics. It's just that the problems and challenges of American life no longer can be cast easily, if at all, into an economic mold. The civil-rights issue, in all likelihood, is typical of the issues that will be central to American politics from now on: the problems of the metropolis; the structure, values, and relationships of a society increasingly organized in large and powerful institutions; or the role, function, and limitations of science and technology. These questions cannot be converted into economic issues. Nor, despite President Johnson's attempts at "consensus," are they likely to admit of "bipartisanship." Bipartisanship is effective when the answers are known, at least in broad outline. But great political innovations, such as we need, are rarely the children of compromise.

And yet these are also issues which the traditional ideological alignments, the alignments of European politics, cannot tackle. To do as so many foreign and domestic critics of the American political system have urged us to do for well over a century—to organize our political life on the basis of "liberal" and "conservative," or "right" and "left"—would only add to the confusion. What is "liberal" in respect to the government of the metropolis? Or in respect to the relationship between the individual and the large organizations on which he depends for effectiveness, but on which he must not depend if he wants to be free? And what does it mean to be a "conservative" on these issues? There is obviously going to be violent disagreement in respect to these issues; in fact, there is need for such disagreement and for a diversity of approaches to their solution. But ideological alignments are bound to be as irrelevant to these issues as the traditional alignment by economic interests. The "New Left" is thus bound to be sterile and to be condemned to total frustration.

If, indeed, the world will permit us the luxury of domestic affairs in the next half-century or so, we will not only have to face up to new issues. We shall have to devise a new approach to domestic politics altogether. This signifies a greater upheaval in our political life, a greater strain on political sanity and stability, than the new issues themselves could possibly mean and a greater opportunity for creative political thought and effective political leadership than this country has known since the days of the founding fathers.

In seizing this opportunity, we may well have to abandon the traditional reliance on economic interests and economic issues as symbols of political intercourse and as means of political organization. I hope that we will not give up with them the principle underlying them: the mobilization of conflict to create unity, and the appeal to interest against the fanaticism of ideological faction. It is not only a civilized concept; it is a principle that makes politics productive for the common good. It has served the American people well—so well that doing without the Great Economist seems a very small price, indeed.

7

The Poverty of Economic Theory

What creates wealth? For the last 450 years, economists have either neglected the question, sought easy answers, or dismissed previous assessments. Nonetheless, we have something to learn from every interpretation.

The first generation of economists, the mercantilists, said, "Wealth is purchasing power." Their goal was to increase monetary wealth with an accumulation of bullion and a favorable trade balance. Another theory stated that wealth is not created by man, but by nature—that land creates wealth.

Yet another group of theorists related wealth to man. "Wealth," they said, "is created by human labor." This tenet signaled the beginning of economics as a discipline because it related wealth to something that man creates. And yet it was totally unsatisfactory. It could not be made to predict or analyze anything.

A little more than 100 years ago the field of economics split in two. The mainstream simply gave up the search for any answer to the question of the creation of wealth, became purely analytical, and stopped relating economics to the behavior of people. Economics was a discipline that governed the behavior of commodities. Ironically, analysis is a great strength of contemporary economics, but it also explains why the public at large is bored stiff by the field. It has nothing to say to them because it lacks a foundation in value.

First published in *New Management*, 1987.

Another Wrong Answer

Karl Marx understood this shortfall when he stuck with the labor theory of value. "Marxist economics" is a contradiction in terms—it has no analytical or predictive power—but it has a tremendous appeal precisely because it is grounded in a value. It defines the creators of wealth—human beings, labor. And yet, we know it's the wrong answer, too.

For the last 100 years, therefore, we have had a choice between an economics that has great analytical power but no foundation in value and an economics that wasn't economics at all, but a political manifest based on the human being. Today we've finally reached a point where that dilemma can be bridged, where we can begin to understand the right approach, if not the right answer. We now know that the source of wealth is something specifically human: knowledge. If we apply knowledge to tasks we already know how to do, we call it "productivity." If we apply knowledge to tasks that are new and different, we call it "innovation." Only knowledge allows us to achieve those two goals.

Tradition of Labor

This wasn't always true. Two hundred years ago when Adam Smith wrote about "the tradition of labor," his examples were people in what is now central Germany, who, because of the heavy winters with lots of snow, learned to be woodworkers and make clocks and violins. It takes 200 years to build such a tradition, Smith said, except for rare cases when refugees or immigrants bring their skills to a community.

Such was certainly the case when the U.S. won its independence. Every American consul had an unlimited slush fund (which probably meant $180) to bribe an English craftsman and supply him with false papers to come to this country and teach us how to build textile machinery and dye cotton. That's how New England became an industrial power around 1810.

During the nineteenth century, however, apprenticeship (a German invention) telescoped the 200 years into five, and during the twentieth century, training (an American invention) telescoped five years into six months or even ninety days. We invented training during World War I because we had no tradition of labor. After World War II, our invention spread worldwide, which is one reason why nations can no longer compete on the basis of a labor tradition.

Learning and Knowledge

Indeed, until recently, the quickest way for a person living in a developed country to made a decent living was to become a semi-skilled machine operator. After six weeks, he probably was better paid than associate professors, not to mention assistant deans. But that's over. Today he can make a middle-class living only through learning and knowledge.

(Consider that thirty years ago, there was hardly a single person in Korea who had any tradition of skill or craft, if only because Japan didn't allow its neighbors to acquire any for fifty years. Today, Korea can do almost anything any advanced industrial nation does, thanks to training.)

Of course, the realization that knowledge is the source of wealth has major implications for economics, which today is at a dead end. Economics used to be an enjoyable discipline because it was so humble. When someone asked an economist of 1925 a question, his answer was, "I don't know," which in many ways is a respectable answer. (At least it's a modest one.) And then he said, "We don't know and therefore the intelligent thing to do is as little as possible and pray. Keep taxes low, expenditures low, and pray."

New Philosophy

But my generation of economists became arrogant, largely as a result of an incredible performance during World War I. In military terms, World War I is the all-time low of performance, but the civilian accomplishment was incredible. Every country was bankrupt by December 1914, and according to the traditional rules of the game, should have had to stop fighting. But they kept on killing each other for four more years because the civilian administrators were so incredibly competent. and that competency corrupted and gave us delusions of grandeur.

When 1929 came along, suddenly there was a curious belief that government should be able to do something about the economy. That was totally unheard of in earlier days, but it became a popular demand, similar to the question, "If you can put a man on the moon, why can't you do something about AIDS?" And thus we saw the development of economics that knew the answers.

Keynes knew the answer: Whatever ails you, just create more purchasing power. Milton Friedman, who may be the last surviving member of the great generation, refined it and said, "You don't even have to do that. All you have to do is just make sure the money supply grows." For the supply-siders it was even easier: just cut taxes. What could appear nicer and more pleasant?

End of Euphoria

In the nineteenth century economics was known as the "dismal science," because it always forced us to make choices, and we always had to forgo something. Suddenly it became a euphoric science. For fifty years, it's been a euphoric science but believe me, that's over.

Economics hasn't worked. Whatever we tried, it failed. What's more, the basic assumptions of modern economic theories are unreasonable and invalid. All of them assume that the sovereign state is alone in this world and can control its destiny. If the five or six leading industrial nations would simply agree to hand over their economic policy to a czar, a commissioner, or a common organ, economic theory would work. But chances are it won't happen; by comparison, winning a million dollars at a Las Vegas slot machine is a cinch.

Most economists have assumed, too, that the velocity of the turnover of money is a social habit and a constant—against all evidence. When the U.S. tested the theory for the first time in 1935 and pumped a lot of purchasing power into American pockets, we did not spend, we hoarded. The economy collapsed the following year, and it was much worse than in 1930 or 1931 because the American public sabotaged the economic policy. The same thing happened under Mr. Carter and Mr. Reagan. The velocity of turnover of money is about as mercurial as teenage fashions and even less predictable.

Invalid Theory

In essence, macroeconomic theory is no longer a basis for economic policy because no one knows *what* is going to happen. Mr. Reagan came to power promising to cut the budget, but government expenses have never grown faster in the history of any country. He didn't betray his trust; politically he had no choice. Political leaders have no economic theory they can trust, a fact which escapes many businessmen.

The economics of tomorrow must do what economists have not been able to do: integrate the realms of the domestic and the world. (Note the use of the word *world* rather than *international*. *International* implies economies outside the domestic realm; *world* economies are "inside.")

Tomorrow's economics must also answer the question: How do we relate the way we run a business to results. What *are* results? The traditional answer—the bottom line—is treacherous. Under a bottom-line philosophy, we cannot relate the short term to the long term, and yet the balance between the two is a crucial test of management.

Two Guideposts

The beacons of productivity and innovation must be our guideposts. If we achieve profits at the cost of downgrading productivity or not innovating, they aren't profits. We're destroying capital. On the other hand, it we continue to improve productivity of all key resources and our innovative standing, we are going to be profitable. Not today, but tomorrow. In looking at knowledge applied to human work as the source of wealth, we also see the function of the economic organization.

For the first time we have an approach that makes economics a human discipline and relates it to human values, a theory that gives a businessman a yardstick to measure whether he's still moving in the right direction and whether his results a real or delusions. We are on the threshold of post-economic theory, grounded in what we now know and understand about the generation of wealth.

8

The Delusion of Profits

Businessmen habitually complain about the economic illiteracy of the public, and with good reason. The greatest threat to the free enterprise system in this country is not the hostility to business of a small, strident group, but the pervasive ignorance throughout our society in respect to both the structure of the system and its functioning.

But the same businessmen who so loudly complain about economic illiteracy are themselves the worst offenders. They don't seem to know the first thing about profit and profitability. And what they say to each other as well as to the public inhibits both business action and public understanding.

For the essential fact about profit is that there is no such thing. There are only costs.

What is called "profit" and reported as such in company accounts is genuine and largely quantifiable cost in three respects: as genuine cost of a major resource, namely capital; as a necessary insurance premium for the real—and again largely quantifiable—risks and uncertainties of all economic activity; and as cost of the jobs and pensions of tomorrow. The only exception, the only true surplus, is a genuine monopoly profit such as that now being extorted by the OPEC cartel in petroleum.

1. All economists have known for two hundred years that there are factors of production, that is, three necessary resources: labor, land (i.e., physical resources), and capital. And all of us should have

First published in the *Wall Street Journal* 1975.

learned in the last ten years that there are no "free" resources. They all have a cost. Indeed, the economists are way ahead of most business-men in their understanding and acceptance of a genuine "cost of capital." Some of them, such as Ezra Solomon, a former member of the Council of Economic Advisers now back at Stanford University, have worked out elegant methods both for determining the cost of capital and for measuring the performance of a business in earning it.

We know that in the post-World War II period, until the onset of global inflation in the mid-sixties, the cost of capital in all developed countries of the Free World ran somewhat above 10 percent a year (it is almost certainly much higher in communist economies). And we know that very few businesses actually earn enough to cover these genuine costs. But so far only a handful of businesses seem to know that there is such a cost. Fewer still seem to know whether they cover it or not. And even these few never talk about it and never in their published accounts subject their own performance to the test. Yet not to earn the cost of capital is as much a failure to cover costs as not to earn the costs of wages or of raw materials.

2. Economic activity is the commitment of existing resources to future expectations. It is a commitment, therefore, to risk and uncer-tainty—in respect to obsolescence of products, processes, and equip-ment; in respect to changes in markets, distributive channels, and consumer values; and in respect to changes in economy, technology, and society. The odds in any commitment to the future are always adverse; it is not given to human beings to know the future. The odds, therefore, are always in favor of loss rather than gain. And in a period of rapid change such as ours, the risks and uncertainties are surely not getting smaller.

These risks and uncertainties are not capable of precise determina-tion. But the *minimum* of risk in these commitments to the future is capable of being determined, and indeed quantified, with a fair degree of probability. Where this has been attempted in any business—and in both Xerox and IBM, for instance, it is known to have been done for years in respect to products and technologies—the risks have proven to be much higher than even conservative "business plans" assumed.

The risks of natural events—fire, for instance—have long been treated as normal business costs. A business that failed to set aside the appropriate insurance premiums for such risks would rightly be considered to be endangering the wealth producing assets in its keeping. Economic, technological, and social risks and uncertainties are no less real. They too require an adequate "insurance premium"—and to supply it is the function of profit and profitability.

Therefore, the proper question for any management is not "What is the *maximum* profit this business can yield?" It is "What is the *minimum* profitability needed to cover the future risks of this business?" And if the profitability falls short of this minimum—as it does in most companies I know—the business fails to cover genuine costs, endangers itself, and impoverishes the economy.

3. Profit is also tomorrow's jobs and tomorrow's pensions. Both are costs of a business and, equally, costs of the economy. Profit is not the only source of capital formation; there is also private savings, of course. But business earnings, whether retained in the business or paid out (returned to the capital market), are the largest single source of capital formation for tomorrow's jobs and, at least in the United States, the largest single source of capital formation for tomorrow's pensions.

The most satisfactory definition of "economic progress" is a steady rise in the ability of an economy to invest more capital for each new job and thereby to produce jobs that yield a better living as well as a better quality of work and life. by 1965, before inflation made meaningful figures increasingly difficult to obtain, investment per job in the American economy had risen from $35,000 to $50,000. The requirement will go up fairly sharply, for the greatest investment needs and opportunities are in industries: energy, the environment, transportation, health care, and, above all, increased food production, in which capital investment per job is far higher than the average in the consumer goods industries, which have dominated the economy these last twenty-five years.

At the same time, the number of jobs required is going up sharply— the aftermath of the "baby boom" between 1948 and 1960. We will have to increase the number of people at work by 1 percent, or almost a million people, each year until the early eighties to stay even with the demographics. At the same time, the number of people on pensions will also increase, if only because workers reaching retirement age will live longer, and so will the income expectations of the pensioners. Any company which does not produce enough capital, i. e., enough earnings, to provide for this expansion in jobs and pensions fails both to cover its own predictable and quantifiable costs and the costs of the economy.

These three kinds of costs—the costs of capital, the risk premium of economic activity, and the capital needs of the future—overlap to a considerable extent. But any company should be expected to cover adequately the largest of these three costs. Otherwise it operates at a genuine, certain, and provable loss.

There are three conclusions from these elementary premises:

1. Profit is not peculiar to capitalism. It is a prerequisite for any economic system. Indeed, the communists economies require a much higher rate of profit. Their costs of capital are higher. And central planning adds an additional and major economic uncertainty. In fact, the communist economies do operate at a substantially higher rate of profit than any market economy, no matter that for ideological reasons it is called "turnover tax" rather than "profit." And the only economies that can be considered as being based on profit planning are precisely communist economies in which the producer (state planner) imposes the needed profitability in advance rather than let market forces determine it.

2. The costs which are paid for out of the difference between current revenues and current expenses of production and distribution are fully as much economic reality as wages or payments for supplies. Since a company's accounts are supposed to reflect economic reality, these costs should be shown. They are, to be sure, not as precisely known or knowable as the accountants' "costs of doing business" supposedly are. But they are known and knowable within limits that are probably no wider or fuzzier than those of most cost accounting or depreciation figures—and they may be more important both for managing a business and for analyzing its performance. Indeed, it might not be a bad idea to tie executive bonuses and incentives to a company's performance in earning adequately these genuine costs rather than to profit figures that often reflect financial leverage as much as actual economic performance.

3. Finally, businessmen owe it to themselves and owe it to society to hammer home that there is no such thing as profit. There are only costs: costs of doing business and costs of staying in business; costs of labor and raw materials, and costs of capital; costs of today's jobs and costs of tomorrow's jobs and tomorrow's pensions.

There is no conflict between profit and social responsibility. To earn enough to cover the genuine costs, which only the so-called profit can cover, is economic and social responsibility—indeed it is the specific social and economic responsibility of business. It is not the business that earns a profit adequate to its genuine costs of capital, to the risks of tomorrow and to the needs of tomorrow's worker and pensioner, that "rips off" society. It is the business that fails to do so.

9

Schumpeter and Keynes

The two greatest economists of this century, Joseph A. Schumpeter and John Maynard Keynes, were born, only a few months apart, a hundred years ago: Schumpeter on February 8, 1883, in a provincial Austrian town; Keynes on June 5, 1883, in Cambridge, England. (And they died only four years apart—Schumpeter in Connecticut on January 8, 1950, Keynes in southern England on April 21, 1946.) The centenary of Keynes' birth is being celebrated with a host of books, articles, conferences, and speeches. If the centenary of Schumpeter's birth were noticed at all, it would be in a small doctoral seminar. And yet it is becoming increasingly clear that it is Schumpeter who will shape the thinking and inform the questions on economic theory and economic policy for the rest of this century, if not for the next thirty or fifty years.

The two men were not antagonists. Both challenged long-standing assumptions. The opponents of Keynes were the very "Austrians" Schumpeter himself had broken away from as a student, the neoclassical economists of the Austrian School. And although Schumpeter considered all of Keynes's answers wrong, or at least misleading, he was a sympathetic critic. Indeed, it was Schumpeter who established Keynes in America. When Keynes's masterpiece, *The General Theory of Employment, Interest and Money,* came out in 1936, Schumpeter, by then the senior member of the Harvard economics faculty, told his

First published in Forbes Magazine, 1983.

students to read the book and told them also that Keynes's work had totally superseded his own earlier writings on money.

Keynes, in turn, considered Schumpeter one of the few contemporary economists worthy of his respect. In his lectures he again and again referred to the works Schumpeter had published during World War I, and especially to Schumpeter's essay on the *Rechenpfennige* (that is, money of account) as the initial stimulus for his own thoughts on money. Keynes's most successful policy initiative, the proposal that Britain and the United States finance World War II by taxes rather than by borrowing, came directly out of Schumpeter's 1918 warning of the disastrous consequences of the debt financing of World War I.

Schumpeter and Keynes are often contrasted politically, with Schumpeter being portrayed as the "conservative" and Keynes the "radical." The opposite is more nearly right. Politically Keynes's views were quite similar to what we now call "neoconservative." His theory had its origins in his passionate attachment to the free market and in his desire to keep politicians and governments out of it. Schumpeter, by contrast, had serious doubts about the free market. He thought that an "intelligent monopoly"—the American Bell Telephone system, for instance—had a great deal to recommend itself. It could afford to take the long view instead of being driven from transaction to transaction by short-term expediency. His closest friend for many years was the most radical and most doctrinaire of Europe's left-wing socialists, the Austrian Otto Bauer, who, though staunchly anticommunist, was even more anticapitalist. And Schumpeter, although never even close to being a socialist himself, served during 1919 as minister of finance in Austria's only socialist government between the wars. Schumpeter always maintained that Marx had been dead wrong in every one of his answers. But he still considered himself a son of Marx and held him in greater esteem than any other economist. At least, so he argued, Marx asked the right questions, and to Schumpeter questions were always more important than answers.

The differences between Schumpeter and Keynes go much deeper than economic theorems or political views. The two saw a different economic reality, were concerned with different problems, and defined *economics* quite differently. These differences are highly important to an understanding of today's economic world.

Keynes, for all that he broke with classical economics, operated entirely within its framework. He was a heretic rather than an infidel. Economics, for Keynes, was the equilibrium economics of Ricardo's 1810 theories, which dominated the nineteenth century. This economics deals with a closed system and a static one. Keynes's key question was the same question the nineteenth-century economists had asked: "How can one maintain an economy in balance and stasis?"

For Keynes, the main problems of economics are the relationship between the "real economy" of goods and services and the "symbol economy" of money and credit; the relationship between individuals and businesses and the "macroeconomy" of the nation-state; and finally, whether production (that is, supply) or consumption (that is, demand) provides the driving force of the economy. In this sense Keynes was in a direct line with Ricardo, John Stuart Mill, the "Austrians," and Alfred Marshall. However much they differed otherwise, most of these nineteenth-century economists, and that includes Marx, had given the same answers to these questions: The "real economy" controls, and the money is only the "veil of things," the microeconomy of individuals and businesses determines, and government can, at best, correct minor discrepancies, and, at worst, create dislocations; and supply controls, with demand a function of it.

Keynes asked the same questions that Ricardo, Mill, Marx, the "Austrians," and Marshall had asked but, with unprecedented audacity, turned every one of the answers upside down. In the Keynesian system, the "symbol economy" of money and credit are "real," and goods and services dependent on it and are its shadows. The macroeconomy—the economy of the nation-state—is everything, with individuals and firms having neither power to influence, let alone to direct, the economy nor the ability to make effective decisions counter to the forces of the macroeconomy. And economic phenomena, capital formation, productivity, and employment are functions of demand.

By now we know, as Schumpeter knew fifty years ago, that every one of these Keynesian answers is the wrong answer. At least they are valid only for special cases and within fairly narrow ranges. Take, for instance, Keynes's key theorem: that monetary events—government deficits, interest rates, credit volume, and volume of money in circulation—determine demand and with it economic conditions. This assumes, as Keynes himself stressed, that the turnover velocity of money is constant and not capable of being changed over the short term by individuals or firms. Schumpeter pointed out fifty years ago that all evidence negates this assumption. And indeed, whenever tried, Keynesian economic policies, whether in the original Keynesian or in the modified Friedman version, have been defeated by the microeconomy of business and individuals, unpredictably and without warning, changing the turnover velocity of money almost overnight.

When the Keynesian prescriptions were initially tried—in the United States in the early New Deal days—they seemed at first to work. But then, around 1935 or so, consumers and businesses suddenly sharply reduced the turnover velocity of money within a few short months, which aborted a recovery based on government deficit spending and brought about a second collapse of the stock market in 1937. The best example, however, is what happened in this country in 1981 and 1982. The Federal Reserve's purposeful attempt to control the economy by controlling the money supply was largely defeated by consumers and businesses who suddenly and almost violently shifted deposits from thrifts into money-market funds and from long-term investments into liquid assets—that is, from low-velocity into high-velocity money—to the point where no one could really tell anymore what the *money supply* is or even what the term means. Individuals and businesses seeking to optimize their self-interest and guided by their perception of economic reality will always find a way to beat the "system"— whether, as in the Soviet bloc, through converting the entire economy into one gigantic black market or, as in the United States in 1981 and 1982, through transforming the financial system overnight despite laws, regulations, or economists.

This does not mean that economics is likely to return to pre-Keynesian neoclassicism. Keynes's critique of the neoclassic answers is as definitive as Schumpeter's critique of Keynes. But because we now know that individuals can and will defeat the system, we have lost the certainty which Keynes imposed on economics and which has made the Keynesian system the lodestar of economic theory and economic policy for fifty years. Both Friedman's monetarism and supply-side economics are desperate attempts to patch up the Keynesian system of equilibrium economics. But it is unlikely that either can restore the self-contained, self-confident equilibrium economics, let alone an economic theory or an economic policy in which one factor, whether government spending, interest rates, money supply, or tax cuts, controls the economy predictably and with near certainty.

That the Keynesian answers were not going to prove any more valid than the pre-Keynesian ones that they replaced was clear to Schumpeter from the beginning. But to him this was much less important than that the Keynesian questions—the questions of Keynes's predecessors as well—were not, Schumpeter thought, the important questions at all. To him the basic fallacy was the very assumption that the healthy, the "normal," economy is an economy in static equilibrium. Schumpeter, from his student days on, held that a modern economy is always in dynamic disequilibrium. Schumpeter's economy is not a closed system like Newton's universe—or Keynes's macroeconomy. It is forever growing and changing and is biological rather than mechanistic in nature. If Keynes was a "heretic," Schumpeter was an "infidel."

Schumpeter was himself a student of the great men of Austrian economics and at a time when Vienna was the world capital of economic theory. He held his teachers in lifelong affection. But his doctoral dissertation—it became the earliest of his great books, *The Theory of Economic Development* (which in its original German version came out in 1911, when Schumpeter was only twenty-eight years old)—starts out with the assertion that the central problem of economics is not equilibrium but structural change. This then led to Schumpeter's famous theorem of the innovator as the true subject of economics.

Classical economics considered innovation to be outside the system, as Keynes did, too. Innovation belonged in the category of "outside catastrophes" like earthquakes, climate, or war, which, everybody knew, have profound influence on the economy but are not part of economics. Schumpeter insisted that, on the contrary, *innovation*—that is, entrepreneurship that moves resources from old and obsolescent to new and more productive employments—is the very essence of economics and most certainly of a modern economy.

He derived this notion, as he was the first to admit, from Marx. But he used it to disprove Marx. Schumpeter's *Economic Development* does what neither the classical economists nor Marx nor Keynes was able to do: It makes profit fulfill an economic function. In the economy of change and innovation, profit, in contrast to Marx and his theory, is not a *Mehrwert,* a "surplus value" stolen from the workers. On the contrary, it is the only source of jobs for workers and of labor income. The theory of economic development shows that no one except the innovator makes a genuine "profit"; and the innovator's profit is always quite short-lived. But innovation in Schumpeter's famous phrase is also "creative destruction." It makes obsolete yesterday's capital equipment and capital investment. The more an economy progresses, the more capital formation will it therefore need. Thus what the classical economist—or the accountant or the stock exchange—considers "profit" is a genuine cost, the cost of staying in business, the cost of a future in which nothing is predictable except that today's profitable business will become tomorrow's white elephant. Thus, capital formation and productivity are needed to maintain the wealth-producing capacity of the economy and, above all, to maintain today's jobs and to create tomorrow's jobs.

Schumpeter's "innovator" with his "creative destruction" is the only theory so far to explain why there is something we call "profit." The classical economists very well knew that their theory did not give any rationale for profit. Indeed, in the equilibrium economics of a closed economic system there is no place for profit, no justification for it, no explanation of it. If profit is, however, a genuine cost, and especially if profit is the only way to maintain jobs and to create new ones, then capitalism becomes again a moral system.

Morality and profits: The classical economists had pointed out that profit is needed as the incentive for the risk taker. But is this not really a bribe and thus impossible to justify morally? This dilemma had driven the most brilliant of the nineteenth-century economists, John Stuart Mill, to embrace socialism in his later years. It had made it easy for Marx to fuse dispassionate analysis of the "system" with the moral revulsion of an Old Testament prophet against the exploiters. The weakness on moral grounds of the profit incentive enabled Marx at once to condemn the capitalist as wicked and immoral and assert "scientifically" that he serves no function and that his speedy demise is "inevitable." As soon, however, as one shifts from the axiom of an unchanging, self-contained, closed economy to Schumpeter's dynamic, growing, moving, changing economy, what is called profit is no longer immoral. It becomes a moral imperative. Indeed, the question then is no longer the question that agitated the classicists and still agitated Keynes: How can the economy be structured to minimize the bribe of the functionless surplus called profit that has to be handed over to the capitalist to keep the economy going? The question in Schumpeter's economics is always,. Is there sufficient profit? Is there adequate capital formation to provide for the costs of the future, the costs of staying in business, the costs of "creative destruction"?

This alone makes Schumpeter's economic model the only one that can serve as the starting point for the economic policies we need. Clearly the Keynesian—or classicist—treatment of innovation as being "outside," and in fact peripheral to, the economy and with minimum impact on it, can no longer be maintained (if it ever could have been). The basic question of economic theory and economic policy, especially in highly developed countries, is clearly: How can capital formation and productivity be maintained so that rapid technological change as well as employment can be sustained? What is the minimum profit needed to defray the costs of the future? What is the minimum profit needed, above all, to maintain jobs and to create new ones?

Schumpeter gave no answer; he did not much believe in answers. But seventy years ago, as a very young man, he asked what is clearly going to be the central question of economic theory and economic policy in the years to come.

And then, during World War I, Schumpeter realized, long before anyone else—and a good ten years before Keynes did—that economic reality was changing. He realized that World War I had brought about the monetarization of the economies of all belligerents. Country after country, including his own still fairly backward Austria-Hungary, had succeeded during the war in mobilizing the entire liquid wealth of the community, partly through taxation but mainly through borrowing. Money and credit, rather than goods and services, had become the "real economy."

In a brilliant essay published in a German economic journal in July 1918—when the world Schumpeter had grown up in and had known was crashing down around his ears—he argued that, from now on, money and credit would be the lever of control. What he argued was that neither supply of goods, as the classicists had argued, nor demand for goods, as some of the earlier dissenters had maintained, was going to be controlling anymore. Monetary factors—deficits, money, credit, taxes—were going to be the determinants of economic activity and of the allocation of resources.

This is, of course, the same insight on which Keynes later built his *General Theory*. But Schumpeter's conclusions were radically different from those Keynes reached. Keynes came to the conclusion that the emergence of the symbol economy of money and credit made possible the "economist-king," the scientific economist, who by playing on a few simple monetary keys—government spending, the interest rate, the volume of credit, or the amount of money in circulation—would maintain permanent equilibrium with full employment, prosperity, and stability. But Schumpeter's conclusion was that the emergence of the symbol economy as the dominant economy opened the door to tyranny and, in fact, invited tyranny. That the economist now proclaimed himself infallible, he considered pure *hubris*. But, above all, he saw that it was not going to be economists who would exercise the power, but politicians and generals.

And then, in the same year, just before World War I ended, Schumpeter published *The Tax State* ("The Fiscal State" would be a better translation). Again, the insight is the same Keynes reached fifteen years later (and, as he often acknowledged, thanks to Schumpeter): The modern state, through the mechanisms of taxation and borrowing, has acquired the power to shift income and, through "transfer payments," to control the distribution of the national product. To Keynes this power was a magic wand to achieve both social justice and economic progress, and both economic stability and fiscal responsibility. To Schumpeter—perhaps because he, unlike Keynes, was a student of both Marx and history—this power was an invitation to political irresponsibility, because it eliminated all economic safeguards against inflation. In the past the inability of the state to tax more than a very small proportion of the gross national product, or to borrow more than a very small part of the country's wealth, had made inflation self-limiting. Now the only safeguard against inflation would be political, that is, self-discipline. And Schumpeter was not very sanguine about the politician's capacity for self-discipline.

Schumpeter's work as an economist after World War I is of great importance to economic theory. He became one of the fathers of business cycle theory.

But Schumpeter's real contribution during the thirty-two years between the end of World War I and his death in 1950 was as a political economist. In 1942, when everyone was scared of a worldwide deflationary depression, Schumpeter published his best-known book, *Capitalism, Socialism and Democracy,* still, and deservedly, read widely. In this book he argued that capitalism would be destroyed by its own success. This would breed what we would now call the *new class:* bureaucrats, intellectuals, professors, lawyers, journalists, all of them beneficiaries of capitalism's economic fruits and, in fact, parasitical on them, and yet all of them opposed to the ethos of wealth production, of saving, and of allocating resources to economic productivity. The forty years since this book appeared have surely proved Schumpeter to be a major prophet.

And then he proceeded to argue that capitalism would be destroyed by the very democracy it had helped create and made possible. For in a democracy, to be popular, government would increasingly shift income from producer to nonproducer, would increasingly move income from where it would be saved and become capital for tomorrow to where it would be consumed. Government in a democracy would thus be under increasing inflationary pressure. Eventually, he prophesied, inflation would destroy both democracy and capitalism.

When he wrote this in 1942, almost everybody laughed. Nothing seemed less likely than an inflation based on economic success. Now, forty years later, this has emerged as the central problem of democracy and of a free-market economy alike, just as Schumpeter had prophesied.

The Keynesians in the 1940s ushered in their "promised land," in which the economist-king would guarantee the perfect equilibrium of an eternally stable economy through control of money, credit, spending, and taxes. Schumpeter, however, increasingly concerned himself with the question of how the public sector could be controlled and limited so as maintain political freedom and an economy capable of performance, growth, and change. When death overtook him at his desk, he was revising the presidential address he had given to the American Economic Association only a few days earlier. The last sentence he wrote was "The stagnationists are wrong in their diagnosis of the reason the capitalist process should stagnate; they may still turn out to be right in their prognosis that it will stagnate—with sufficient help from the public sector."

Keynes's best-known saying is surely "In the long run we are all dead." This is one of the most fatuous remarks ever made. Of course, in the long run we are all dead. But Keynes in a wiser moment remarked that the deeds of today's politicians are usually based on the theorems of long-dead economists. And it is a total fallacy that, as Keynes implies, optimizing the short term creates the right long-term future. Keynes is in large measure responsible for the extreme short-term focus of modern politics, of modern economics, and of modern business—the short-term focus that is now, with considerable justice, considered a major weakness of American policymakers, both in government and in business.

Schumpeter also knew that policies have to fit the short term. He learned this lesson the hard way—as minister of finance in the newly formed Austrian republic in which he, totally unsuccessfully, tried to stop inflation before it got out of hand. He knew that he had failed because his measures were not acceptable in the short term—the very measures that, two years later, a noneconomist, a politician and professor of moral theology did apply to stop the inflation, but only after it had all but destroyed Austria's economy and middle class.

But Schumpeter also knew that today's short-term measures have long-term impacts. They irrevocably make the future. Not to think through the futurity of short-term decisions and their impact long after "we are all dead" is irresponsible. It also leads to the wrong decisions. It is this constant emphasis in Schumpeter on thinking through the long-term consequences of the expedient, the popular, the clever, and the brilliant that makes him a great economist and the appropriate guide for today, when short-run, clever, brilliant economics—and short-run, clever, brilliant politics—have become bankrupt.

In some ways, Keynes and Schumpeter replayed the best-known confrontation of philosophers in the Western tradition—the Platonic dialogue between Protagoras, the brilliant, clever, irresistible sophist, and the slow-moving and ugly, but wise Socrates. No one in the interwar years was more brilliant, more clever than Keynes. Schumpeter, by contrast, appeared pedestrian—but he had wisdom. Cleverness carries the day. But wisdom endureth.

10

Keynes: Economics as a Magical System

The influence and reputation of John Maynard Keynes are not explained by his having been a great economist, nor did his importance lie primarily in his economic theories. He was, indeed, a very great economist, in all likelihood the last of the "pure" economists of the classical school, at once the legitimate heir and the liquidator of Adam Smith. But he was, above all, the representative political thinker of the interwar period; he expressed perfectly its attempt to master what it knew to be a new world by pretending that it was the old one. Keynes's work was built on the realization that the fundamental assumptions of nineteenth-century laissez-faire economics no longer hold true in an industrial society and a credit economy. But it aimed at the restoration and preservation of the basic beliefs, the basic institutions of nineteenth-century laissez-faire politics; above all, it aimed at the preservation of the autonomy and automatism of the market. The two could no longer be brought together in a rational system; Keynes's policies are magic—spells, formulae, and incantations, to make the admittedly irrational behave rationally.

Keynes's theoretical analysis of the new social and economic reality is a masterpiece that will endure. His conclusions from this analysis proved wrong, however; the economic policies which gave him his reputation and influence have failed. When he died in the spring of 1946 he was apparently at the peak of success and power: the chief financial adviser of his government, a peer of the realm, the almost

First published in *Virginia Quarterly Review,* shortly after Keynes's death, 1946.

undisputed master of the schools, especially in this country. But his very disciples, while using Keynesian terms, methods, and tools, were actually abandoning fast both his economic policies and his aim.

1

Keynes, who could write prose of a rare lucidity if he wanted to, chose to present his theories in the most technical and most jargon-ridden language, but his central ideas are quite simple.

Classical economics knew neither money nor time as factors in the economics process. Money was the "universal commodity," the symbol of all other commodities, but without any life or effect of its own. It was convenient and necessary, but only an accounting unit to keep track of what went on in the economy of "real" goods and "real" labor; price was simply the rate at which one commodity could be exchanged against all others. The money of classical economics is very much like the ether of classical physics: it pervades all and carries all, but it has neither properties nor effects of its own. And the classical economist was also very much like the physicist of his age in his concept of time: while everything happens in time, time itself is not a factor in the events themselves. This is no accident; classical economics were patterned consciously on the model of Newtonian physics, in structure as well as in its basic assumption of a mechanical and static economic universe.

Keynesian theory is based on the assertion, axiomatic in an industrial age, that the economic process is not only in time but largely determined by time and that the economic expression of the time factor is money. To the classical economist, money was the shadow of existing goods. Actually money, especially the bank deposits which are the money of a credit economy, is created and comes into being in anticipation of goods to be produced, of work to be performed. This means that money is not determined mechanically and according to economic rationality, but psychologically and socially on the basis of confidence in the future. Time thus enters into every economic transaction in the form of fixed money obligations for the investments of the past on which the present is based. These money obligations for the past actually are the largest factor in the economic transactions of every member of an industrial society; for the cost of everything we use, whether a house, a loaf of bread, or a hired man's labor, is made up very largely of the money obligations for the past. Money, instead of being an inert and propertyless expression of economic transactions, influences, molds, and directs economic life; changes in the money sphere cause changes in the "real" economy. We live at the same time in two closely interwoven but distinct economic systems: the "real" economy of the classics—an economy of goods, services, and labor, existing in the present and determined mechanically; and the "symbol" economy of money, heavy with the obligations of the past and determined psychologically by our confidence in the future.

It is no belittling of Keynes to say that these insights did not originate with him, but were the work of a whole generation of economic thinkers before him, especially of his two countrymen Hawtrey and Withers, of the Swedes Cassel and Wicksell, and of the German Knapp. But Keynes synthesized their isolated observations and thoughts into one system and developed a theory of the dynamics of the economic process from them; and it is this theory we usually mean when we talk of "Keynesian economics." The assumptions of the classics had made it virtually impossible for them to understand how a depression could ever happen except as a result of physical catastrophes such as an earthquake or the destruction of war. Also, they were entirely unable to explain how a depression could last; if only left alone it had to correct itself. With the new understanding of the autonomy of the monetary sphere as his starting point, Keynes could give the first adequate theoretical explanation of the vital phenomena of depression and unemployment.

The first answer was Keynes's most famous theory—the theory of oversaving. Any saving is by definition a surplus of productive resources—goods, labor, equipment—over current consumption. For the classics that meant that, unless physically destroyed, any saving must automatically be "invested," that is, used for future production. This, however, ceases to be true as soon as we bring in money as autonomous, as having an economic reality of its own. Then it becomes possible for savings not to be invested but to become mere money savings, with the productive resources they represent left unused and unemployed. Keynes asserted that the modern economy has an inherent tendency toward oversaving.

Of even greater importance was his explanation of the unemployment of a chronic depression. In the universe of classical economics a long-term depression simply could not happen; before a maladjustment could reach depression proportions it would have been corrected by the infallible and automatic mechanism of falling prices and falling costs. Yet long before 1929 long-term depressions had become far too familiar for their existence to be denied except by the most bigoted academician. Hence, orthodox economics had to engage in a search for the criminal conspiracy that prevented "natural" adjustment and correction. Price monopolies, unions, government intervention through relief payments, subsidies, and tariffs—these and all the other measures by which society seeks to protect itself against the social destruction of a depression—became diabolical forces; and the resulting persecution mania of the economists who saw the cloven hoof in the mildest attempt at controlling economic forces soon made it impossible to base economic policy on the classical theories, even though these theories themselves were still generally accepted. From 1870 to 1930 economic policy was without proper theoretical basis. The ruling theory could not justify any of the measures actually taken; and as any economic policy that was possible politically was open to condemnation on theoretical grounds, theory furnished no guidance to distinguish between beneficial and destructive policies. Out of social necessity every economist in office had to do things he opposed in his writings; the resulting blend of cynicism and bad conscience finally gave us that evil genius of old-school financial economists, Dr. Schacht, Hitler's financial manipulator.

But with money a factor, the automatic and infallible adjustment becomes the exception rather than the rule. In a credit economy prices and wages cannot adjust themselves very readily; they must be comparatively inflexible. For a very large—indeed, the largest—part of all costs is the money obligation for the past. This obligation is unaffected by changes in the present value of money as the goods and services it represents were produced in the past at past prices and wages. We may add—though this may be going beyond Keynes—that prices and wages are also hard to adjust downward because money has a social meaning, independent largely of its purchasing power; it buys not only goods but prestige. This is especially true of the lower income levels where the weekly money wage represents a definite social position.

For these reasons, the adjustment in a depression will not, in the modern economy, take the form of lower prices and lower wages. Prices and wages will tend to stay up. Hence the adjustment will take the only form possible: lower employment both of men and of capital equipment. And, unlike the adjustment through lower prices and wages, unemployment not only does not tend to correct the depression; it tends to make the disequilibrium permanent.

Actually, under modern conditions prices will fall, though not as evenly as they should—and with significant exceptions in the capital-goods fields. But wage rates will not go down. In the first place, the wage earner has usually much less margin between his income and his fixed obligations than the industrialist; hence the economic factors militate against wage cuts. Second, the political pressure of organized labor is much more effective in modern society than the economic power even of the strongest monopolist. Hence the maladjustment will not only not be corrected in the "normal course of events"; it will become worse. The point at which new investments again become profitable will recede into the distance. From this follows one of Keynes's most important and, at first sight, most paradoxical conclusions: that we have to raise prices in a depression in order to obtain the very effect orthodox economics expected to get from falling prices.

2

These general theories have been justly criticized for their narrow emphasis on the monetary phenomenon to the exclusion of everything else. The monetary factor is probably only one of the causes of a depression, though perhaps a central one, rather than, as Keynes asserts, always and everywhere the only cause. But aside from this not unimportant question of emphasis, the Keynesian theories have been almost universally accepted. And with these basic theories a great many economists also accepted at first his economic policies. But, as most of the disciples have begun to find out, Keynes's economic policies do not follow from his basic theories; indeed, they are hardly compatible with them. His policies were really dictated by his political aim, not by his economic observations. His attempt to bring the two together into one whole, to make the policies emerge as the inevitable conclusion from the theories, may very well explain the tortuous and tortured style of his later writing, his increasing reliance on purely formal arguments, and his uncritical use of mathematical techniques.

According to Keynes the economic theorist, the level of business activity is determined by the amount of investment in capital goods, which in turn is determined by the confidence which leads businessmen to borrow for expansion. Business activity depends in the last analysis on psychological, that is on economically irrational, factors. According to Keynes the economic politician, the very confidence which creates credit is itself strictly determined by credit. Keynes offers two answers to the question of what causes confidence. He asserts that confidence is a function of the interest rate: the lower the rate, the greater the confidence. He also asserts that confidence is a function of consumer spending: the higher consumer purchases, the higher the investment in capital goods. In his theories Keynes seems to have wavered between these two answers; politically, it does not make too much difference, however, which explanation is preferred. Both lead to pretty much the same conclusion: the quantity of money or credit available determines the degree of confidence, with it the rate of investment, and thus the level of business activity and of employment. Hence Keynes's monetary panacea for booms and depression: in a boom, prevent maladjustment through "draining off" purchasing power into a budget surplus; in a depression, cure maladjustment by creating purchasing power through budget deficits. In either case the quantity of money automatically and infallibly regulates confidence.

Keynes starts out with the statement that human behavior in economic life is not, as the classical economists assumed, determined by objective economic forces, but that, on the contrary, economic forces are directed, if not determined, by human behavior. He ends by asserting as rigid an economic determinism of human behavior and actions as any Ricardo or Malthus ever proclaimed. By this assertion Keynes's entire economic policy stands and falls. And it is this assertion that was conclusively refuted by the experience of the New Deal. The New Deal—at least from 1935 to 1939—was based on deficit spending which created consumer purchasing power and forced down the level of interest rates. Neither brought about a resumption of investment or a significant cut in unemployment. With the credit pumped into the banks, business promptly repaid its old debts instead of borrowing for new investments; and the money paid out by the government to the consumers flowed back to the banks almost at once to become "oversavings."

The faithful Keynesians have been hard pressed to explain away what happened. Their favorite argument is that political opposition to the New Deal offset the economically created confidence. But this defense is not permissible, let alone convincing. Either confidence can be created by creating credit and purchasing power regardless of the way business or any other group feels about governmental policy, or Keynes's economic policies are wrong. And confidence has been conclusively proved not to be producible by a check-writing machine.

Most of the disciples of the earlier years have drawn the conclusions from this experience. They are still Keynesians in their theoretical analysis, but no longer so in their policies. They continue to express their thoughts in monetary terms, but they no longer talk about the interest rate or even about budget deficits or surpluses. Consumer purchasing power and "confidence" have all but disappeared from their vocabulary. The program of the most influential group of Keynesian economists in this country—as written by Alvin Hansen of Harvard into the original draft of the Full Employment bill—provided that, in times of depression and unemployment, the government shall *produce* capital goods through public works and government orders in a quantity sufficient to bring the total capital goods production to a level which gives full employment. Whether this is done with or without a deficit, at a high or at a low interest rate, is of very minor importance; what matters to the neo-Keynesian of today is not monetary policy, but capital-goods production. This shift denies both Keynes's economic concepts and his overall political goal.

In fact, this shift even caught up with Keynes himself. Ironically enough, the very event which brought him official recognition and honors showed up the shortcomings of his theories and policies. In the course of the war, Keynes became the official financial adviser to his government, a director of the Bank of England, a member of the peerage. For the first time his native country officially adopted Keynesian ideas as the basis of its financial policy; the measures outlined in his little pamphlet, *How to Pay for the War*, were adopted almost unchanged by the British government. But the war also showed, especially in Britain, that monetary policy is quite subordinate and that, by itself, it achieves very little. England's war production was obtained not by directing the flow of money, credit, and purchasing power, but through physical controls of men, raw materials, plant equipment, and output which could have worked almost as well with a different monetary policy; in Nazi Germany and in the United States they worked without any monetary policy at all.

3

In the popular mind Keynes stands for government intervention in business. This may well be a correct evaluation of the ultimate effect of Keynesian economics; but if so, Keynes achieved precisely the opposite of what he intended to achieve. For the one passionate aim of his policies was to make possible an economic system free from government interference, a system determined exclusively by objective and impersonal economic forces. "The free market is dead, long live the free market," would be a fitting motto for his entire work.

Keynes's basic insight was the realization that the free market of the classical economists fails to adjust itself automatically as predicted because the economic forces of demand and supply, cost and price, are overridden by the psychological forces of money and credit. From this basis several conclusions as to economic policy would have been logically possible.

Keynes could have argued that conscious political action had to achieve by breaking through the money wall what the market forces should have achieved by themselves. That would have been an economic policy of direct government intervention into *production* through public works and public orders rather than a policy of credit creation; and it is precisely what most of the neo-Keynesians have advocated. Such a policy would restore the supremacy of the "real" system, but at the price of its political independence.

Keynes might also have arrived in logical development from his premises at a policy in which governmental action is used only to induce private business to build up reservoirs of capital-goods production for use in a depression, for instance, through a system of tax rewards for building up reserves in good years to be coupled with stiff tax penalties incurred if these reserves are not used for employment—creating new investments in a depression. He might even have come to the conclusion that the proper policy is psychological rather than economic, i.e., propaganda to create confidence; the German economist Knapp, whose ideas had great influence on Keynes, actually gave this answer.

The one conclusion which logically and theoretically it seems impossible to derive from Keynes's premise is the one he actually did derive. But it was the one and only conclusion which gave Keynes the desired political result: the maintenance of a laissez-faire political system in which only objective economic factors determine the economy and in which man's economic activities are entirely under the control of the individual, not under that of the government.

If the liberal state of nineteenth-century laissez faire was a night watchman protecting the peaceful and law-abiding burgher against thieves and disturbers of the peace, Keynes's state was a thermostat protecting the individual citizen against sharp changes in the temperature. And it was to be a fully automatic thermostat. A fall in economic activity would switch on credit; a rise would cut it off again; and in a boom the mechanism would work in reverse. In contrast to the nineteenth-century state, Keynes's state was, indeed, to act positively; but the actions as well as their timing were to be determined strictly by economic statistics, not subject to political manipulation. The only purpose of these actions was to restore the individual's freedom in the economic sphere, that is, the freedom from all but economic factors, from all but economic considerations—with "economic" referring to the "real" economy of the classics.

The economic system of orthodox economics had been a machine built by the "divine watchmaker," hence without friction and in perpetual motion and perpetual equilibrium. Keynes's system was a clock, a very good and artful clock, but still one built by a human watchmaker, and thus subject to friction. But the only actions required of the watchmaker were to wind, to oil, and, where necessary, to regulate the clock. He was not to run it; he was only to make it fit to run itself; and it was to run according to mechanical laws, not according to political decisions.

Keynes's basic aim of restoring by unorthodox methods the orthodox automatic market system, his basic belief that his methods were objective, nonpolitical, and capable of determination by the impersonal yardstick of statistics, show best in his last major work: the

"Keynes Plan" of an international currency and credit system proposed in 1943. This plan projected his policies from the national into the international sphere. It proposed to overcome international depressions and maladjustments by the adjustment of prices and purchasing power through international credit creation. The agency which was to be in control of this international currency and credit was not to be a world government, but an international body of economic statisticians governed by index numbers and almost entirely without discretionary power. The result of this international system and its main justification was to be the restoration of the full freedom and equilibrium in international trade and currency movements.

Critics have rightly pointed out that Keynes was naïve, to a degree amazing in such an accomplished and experienced political practitioner, in believing that his system could really be immune to political manipulation. It may be possible to obtain objective statistics. But to be meaningful, statistics have to be interpreted by human beings; and interpretations will differ radically with the political beliefs and desires of the interpreter—as witness the widely accepted 1945 forecast of ten million unemployed in the United States by the spring of 1946, made in food faith by government experts interested in setting up a planned economy. Also, even impersonal and objective control is still control; and it is an old political axiom that a government that controls the national income, i.e., the livelihood of the people, inevitably controls the souls of the people. Keynes's political system, in which the state has the power to interfere in the individual's economic activities but refrains from using it, is thus a very different thing from his ideal, the state of nineteenth-century liberalism which was without power of interference.

But the decisive criticism of Keynes's argument is not that there are flaws in it, but that it is an irrational argument. It says in brief: we have proved that the factors that control economic activity are economically irrational, i.e., psychological factors; *therefore* they themselves must be controllable by an economic mechanism. But this "therefore" is not of the vocabulary of reason, not even of that of faith; it is the "therefore" of magic. It is on this very belief—that the admittedly irrational can be controlled and directed by mechanical means—that every system of magic is based. The Keynesian "policies," in spite— or perhaps because—of their elaborate apparatus of mathematical formulae and statistical tables, are spells. Because of this, the fact that they failed once, in the New Deal, means that they have failed forever. For it is of the nature of a spell that it ceases altogether to be effective as soon as it is broken once.

But it was precisely its irrationality that made Keynes's policy so convincing to the generation of the long armistice. After World War I the Western world suddenly awakened to the realization that the basic nineteenth-century assumptions no longer applied. But it refused to face the necessity of new thought and decision. The timid pulled the featherbed of normalcy over their eyes and ears to sleep on a little longer. The courageous accepted the new situation, but attempted to avoid facing it by finding a formula, a mechanical gadget, a spell, in other words, which would make the new behave as if it were the old. One example would be the labor policy of the New Deal. It started with the realization that social and political relationships, rather than the purely economic nexus of the pay check, are the essence of modern industry. But it concluded that the mechanical device of "equal bargaining power" on economic issues would do the trick. Another example is in the field of international relations. Here World War I had clearly shown that peace cannot be based on the concept of equal sovereign states whose internal and external policies are nobody's business but their own. The answer was a strictly mechanical formula, the League of Nations, which represented nothing but the equal sovereign states in their fullest equality and sovereignty, which was neither a supergovernment nor a supercourt, not even an alliance of the Great Powers, but which was expected, in some magical way, to overcome sovereignty.

We can trace this desire for a mechanical formula to make the new function like the old, to make what was irrational in the old assumptions again behave rationally, into fields far removed from politics. It explains, for instance, the tremendous appeal of Freudian psychoanalysis as a cure-all. Freud had had the insight to see the fallacy of traditional, mechanist psychology—the same psychology on which the classical economists had based themselves, incidentally. He realized that man is not a bundle of mechanical reflexes and reactions, but a personality. But he avoided facing the problem—a philosophical or religious one—by asserting that this personality is determined biologically, that it operates through the grossly mechanical forces of repression and sublimation, and that it can be controlled by the mechanical technique of analysis.

But the area in which the desire for a magical system was greatest was that of politics. And in the political field Keynes's economic policy was the most accomplished, the most brilliant, the most elegant attempt to make the impossible again possible, the irrational again rational.

4

In the field of economic thought, Keynes was both a beginning and an end. He showed that classical economics no longer apply and why. He showed that economic theory has to give an answer to a new problem: the impact of man, acting as a human being and not as an economic machine, on the economy. But he contributed little or nothing to the solution of these new problems; he himself never went beyond the classical methods and the classical analysis. Indeed, it may be said that he held back economic thought. Before he came to dominate the scene we had made promising beginnings toward an understanding of the human factor in economic life in such books as Knight's *Risk, Uncertainty, and Profit* and Schumpeter's *Theory of Economic Dynamics,* both written around World War I; and at the Harvard Business School, Elton Mayo had begun his pioneering studies of the relationship between worker and production. Keynes's influence, his magnetic attraction on young men in the field, made theoretical economics again focus on the mechanical equilibrium and on a mechanical concept of economic man determined by impersonal and purely quantitative forces.

Keynes's main legacy is in the field of economic policy. He has formulated our tasks here; even the term "full employment" is his. But his only contribution to a solution—by no means an unimportant one—was to show us which way we cannot go; we cannot, as he did, assert that economic policy is possible without a political decision. We may decide perhaps that the state has to assume direct economic control of production—the decision of most of the neo-Keynesians. This decision raises the question of how political freedom can be maintained in such a state. It also brings up the equally difficult question of what the state is to produce and who is to decide on it; so far no state, whether capitalist, fascist, or socialist, has been able to overcome unemployment by direct government intervention except through producing armaments and armaments plants, that is, through a war economy.

Or we can decide that the state has to create by political means the conditions in which a free-enterprise economy will itself prevent and overcome depressions. Such a policy is not impossible to devise on paper. We would need a fiscal policy that recognizes that industrial production extends over the business cycle, rather than one based on the fiction of the annual profit. We would need a policy of definite encouragement of new ventures and of capital investment in bad years. And we would need a labor policy which, while restoring the flexibility of wages through tying them in with productive efficiency and with business profits, gives security through such employment guarantees as an annual wage. But all this raises the question of how such policies, which demand of all groups that they subordinate their short-term interests to the long-term good of the whole, can be realized in a popular government based on sectional groups and subject to their constant pressure.

But Keynes himself cannot help us to make these decisions, or to answer these questions.

Part Three

The Social Function
of Management

Introduction to Part Three

Most people still hear "*business* management" when they hear the word "management." So did I when I first began to study management, in the early 1940s. But even then I studied management not because I was interested in business but because I was interested in society, community, and organization. And management, I knew from the start, was the functioning, governing organ of that new phenomenon, "organization"—a phenomenon so new that the word did not even exist in its present meaning when I started to work in the late 1920s and did not become common usage until after World War II. Within a few years—surely no later than the early 1950s—I then had come to realize that management is the distinctive function of *all* organizations, whether business or not, and that its function is not an economic, but a social one. This was by no means a popular idea at the time. In fact the idea of management itself was then by no means a popular one. The Harvard Business School, for instance, wanted me to join its faculty in the late 1940s and I needed a job. But the then dean did not want me to teach management—he couldn't figure out what that could be. Still, it was a business school—the Graduate Business School at New York University—that, in the end, did invite me to teach management; neither government departments nor sociology departments nor economics departments—the faculties which, I then thought should be interested in management—had the slightest use for the bastard child.

It then took twenty years for management to gain acceptance as a social function and as for management as governance, it was twenty years before even the management people would listen to such an outlandish idea. The first essay in this part, "Management's New Role," was written in 1969 in response to a specific invitation from the

135

International Management Movement (first established in 1920 by such luminaries as Herbert Hoover and Thomas Masaryk, the first president of newly independent Czechoslovakia) to present these new ideas to a worldwide forum. It forced me to develop both the new— that is, *my*—view of management, and to contrast it with the traditional theory of management as being "business" and "economics." This is therefore a highly structured essay, and may even be called a "theory" of management.

The remaining two essays in this part were both written almost twenty years later, and even then most people, including most of the by then plentiful management professors, viewed management still as "business management." The first of these essays, "Management: The Problems of Success," was delivered as the keynote speech at a 1986 meeting that celebrated the fiftieth anniversary of the founding of the American Academy of Management. It focuses on the unfinished business—and there is plenty of it. The third and last essay, "Social Innovation: Management's New Dimension," takes issue with the still popular notion that innovation is technical and means new processes, new tools, new products. It points out the importance of *social* innovation, but it also asserts that social innovation, in the nineteenth century the province of government, has, in the twentieth century, become the province of—and the opportunity for—the managements of the autonomous institutions of society, businesses and nonbusiness, nonprofit, institutions alike.

11

Management's Role

The major assumptions on which both the theory and the practice of management have been based these past fifty years are rapidly becoming inappropriate. A few of these assumptions are actually no longer valid and, in fact, are obsolete. Others, while still applicable, are fast becoming inadequate; they deal with what is increasingly the secondary, the subordinate, the exceptional, rather than with the primary, the dominant, the ruling function and reality of management. Yet most men of management, practitioners and theoreticians alike, still take these traditional assumptions for granted.

To a considerable extent the obsolescence and inadequacy of these assumed verities of management reflect management's own success. For management has been the success story par excellence of these last fifty years—more so even than science. But to an even greater extent, the traditional assumptions of management scholar and management practitioner are being outmoded by independent—or at least only partially dependent—developments in society, in economy, and in the world view of our age, especially in the developed countries. To a large extent objective reality is changing around the manager—and fast.

Managers everywhere are very conscious of new concepts and new tools of management, of new concepts of organization, for instance, or of the "information revolution." These changes within management

Keynote address given at the 15th CIOS International Management Congress, Tokyo, 1969.

are indeed of great importance. But more important yet may be the changes in the basic realities and their impact on the fundamental assumptions underlying management as a discipline and as a practice. The changes in managerial concepts and tools will force managers to change their behavior. The changes in reality demand, however, a change in the manager's role. The changes in concepts and tools means changes in what a manager *does* and *how* he does it. The change in basic role means a change in what a manager *is*.

The Old Assumptions

Six assumptions may have formed the foundation of the theory and practice of management this last half-century. Few practitioners of management have, of course, ever been conscious of them. Even the management scholars have, as a rule, rarely stated them explicitly. But both practitioners and theorists alike have accepted these assumptions, have, indeed, treated them as self-evident axioms and have based their actions, as well as the theories, on them.

These assumptions deal with

• the scope,
• the task,
• the position, and
• the nature of management.

Assumption one. Management is management of business, and business is unique and the exception in society.

This assumption is held subconsciously rather than in full awareness. It is, however, inescapably implied in the view of society which most of us still take for granted whether we are "Right" or "Left," "conservative," "liberal," or "radical," "capitalists" or "communists:" the view of European (French and English) seventeenth-century social theory which postulates a society in which there is only one organized power center, the national government, assumed to be sovereign though self-limited, with the rest of society essentially composed of the social molecules of individual families. Business, if seen at all in this view, is seen as the one exception, the one organized institution. Management, therefore, is seen as confined to the special, the atypical, the isolated institution of the economic sphere, that is, to business enterprise. The nature as well as the characteristics of management are, in the traditional view, thus very largely grounded in the nature and the characteristics of business activity. One is "for management" if one is "for business," and vice versa. And somehow, in this view, economic activity is quite different from all other human concerns—to the point where it has become fashionable to speak of "the economic concern" as opposed to "human concerns."

Assumption two. "Social responsibilities" of management, that is, concerns that cannot be encompassed within an economic calculus, are restraints and limitations imposed on management rather than management objectives and tasks. They are to be discharged largely without the enterprise and outside of management's normal working day. At the same time and because business is assumed to be the one exception, only business has social responsibilities; indeed, the common phrase is "the social responsibilities of business." University, hospital and government agencies are clearly not assumed, in the traditional view, to have any social responsibilities.

This view derives directly from the belief that business is the one, the exceptional institution. University and hospital are not assumed to have any social responsibility primarily because they are not within the purview of the traditional vision—they are simply not seen at all as "organizations." Moreover, the traditional view of a social responsibility peculiar, and confined, to business derives from the assumptions that economic activity differs drastically from other human activities (if, indeed, it is even seen as a 'normal' human activity), and that "profit" is something extraneous to the economic process and imposed on it by the "capitalist" rather than an intrinsic necessity of any economic activity whatever.

Assumption three. The primary, perhaps the only, task of management is to mobilize the energies of the business organization for the accomplishment of known and defined tasks. The tests are efficiency in doing what is already being done, and adaptation to changes outside.

Entrepreneurship and innovation—other than systematic research—lie outside the management scope.

To a large extent this assumption was a necessity during the last half-century. The new fact then was, after all, not entrepreneurship and innovation with which the developed countries had been living for several hundred years. The new fact of the world of 1900, when concern with management first arose, was the large and complex organization for production and distribution with which the traditional managerial systems, whether of workshop or of local store, could not cope. The invention of the steam locomotive was not what triggered concern with management. Rather it was the emergence, some fifty years later, of the large railroad company which could handle steam locomotives without much trouble but was baffled by the problem of coordination between people, of communication between them, and of their authorities and responsibilities.

But the focus on the *managerial* side of management—to the almost total neglect of entrepreneurship as a function of management—also reflects the reality of the economy in the half-century since World War I. It was a period of high technological and entrepreneurial continuity, a period that required adaptation rather than innovation, and ability to do better rather than courage to do differently.[1]

The long and hard resistance against management on the part of the German *Unternehmer* or French *patron* reflects in large measure a linguistic misunderstanding. There is no German or French word that adequately renders *management,* just as there is no English word for *entrepreneur* (which has remained a foreigner after almost two hundred years of sojourn in the English-speaking world). In part this resistance arises out of peculiarities of economic structure, for example, the role of the commercial banks in Germany which makes the industrialist concerned for his autonomy stress the "charisma" of the *Unternehmer* against the impersonal professionalism of the "manager." In part also "management" is classless and derives its authority from its objective function rather than, as does German *Unternehmer* or French *patron,* from ownership or social class. But surely one of the main reasons for the resistance against "management"—both as a term and as a concept—on the continent of Europe has been the—

largely subconscious—emphasis on the managerial internal task as against the external, entrepreneurial, innovation function.

Assumption four. It is the manual worker—skilled or unskilled—who is management's concern as resource, as a cost center, and as a social and individual problem.

To have made the manual worker productive is, indeed, the greatest achievement of management to date. Frederick Winslow Taylor's "Scientific Management" is often attacked these days (though mostly by people who have not read Taylor). But it was his insistence on studying work that underlies the affluence of today's developed countries; it raised the productivity of manual work to the point where yesterday's "laborer"—a proletarian condemned to an income at the margin of subsistence by "the iron law of wages" and to complete uncertainty of employment from day to day—has become the "semi-skilled worker" of today's mass-production industries with a middle-class standard of living and guaranteed job or income security. And Taylor thereby found the way out of the apparently hopeless impasse of nineteenth-century "class war" between the "capitalist exploitation" of the laboring man and the "proletarian dictatorship."

As late as World War II, the central concern was still the productivity and management of manual work; the central achievement of both the British and the American war economies was the mobilization, training, and managing of production workers in large numbers. Even in the postwar period one major task—in all developed countries other than Great Britain—was the rapid conversion of immigrants from the farm into productive manual workers in industry. On this accomplishment—made possible only because of the "Scientific Management" which Taylor pioneered seventy years ago—the economic growth and performance of Japan, of Western Europe, and even of the United States largely rest.

Assumption five. Management is a "science" or at least a "discipline," that is, it is as independent of cultural values and individual beliefs as are the elementary operations of arithmetic, the laws of physics, or the stress tables of the engineer. But all management is being practiced within one distinct national environment and embedded in one national culture, circumscribed by one legal code and part of one national economy.

These two propositions were as obvious to Taylor in the United States as they were to Fayol in France. Of all the early management authorities only Rathenau in Germany seemed to have doubted that management was an "objective," that is, culture-free, discipline—and no one listened to him. The Human Relations school attacked Taylor as "unscientific;" they did not attack Taylor's premise that there was an objective "science" of management. On the contrary, the Human Relations school proclaimed its findings to be "true" scientific psychology and grounded in the "Nature of Man." It refused even to take into account the findings of its own colleagues in the social sciences, the cultural anthropologists. In so far as cultural factors were considered at all in the traditional assumptions of management, they were "obstacles." It is still almost axiomatic in management that social and economic development requires the abandonment of "non-scientific," that is, traditional cultural beliefs, values, and habits. And the Russians, for example, in their approach to Chinese development under Stalin, differ no whit from the Americans or the Germans in respect to this assumption. That, however, the assumption is little but Western narrowness and cultural egocentricity, one look at the development of Japan would have shown.

At the same time, management theory and practice saw in the national state and its economy the "natural" habitat of business enterprise—as did (and still does), of course, all our political, legal, and economic theory.

Assumption six. Management is the result of economic development.

This had, of course, been the historical experience in the West (though not in Japan where the great organizers such as Mitsui, Iwasaki, Shibusawa, came first, and where economic development, clearly, was the result of management). But even in the West the traditional explanation of the emergence of management was largely myth. As the textbooks had it (and still largely have it), management came into being when the small business outgrew the owner who had done everything himself. In reality, management evolved in enterprises that started big and could never have been anything but big—the railways in particular, but also the postal service, the steamship companies, the steel mills, and the department stores. To industries that could start small, management came very late; some of those, eg, the textile mill or the bank, are still often run on the pattern of the "one boss" who does everything and who, at best, has 'helpers.' But even where this was seen—and Fayol as well as Rathenau apparently realized that management was a function rather than a stage—management was seen as a result rather than as a cause, and as a response to needs rather than as a creator of opportunity.

I fully realize that I have oversimplified—grossly so. But I do not believe that I have misrepresented our traditional assumptions. Nor do I believe that I am mistaken that these assumptions, in one form or another, still underlie both the theory and the practice of management, especially in the industrially developed nations.

And The New Realities

Today, however, we need quite different assumptions. They, too, of course, oversimplify—and grossly, too. But they are far closer to today's realities than the assumptions on which theory and practice of management have been basing themselves these past fifty years.

Here is a first attempt to formulate assumptions that correspond to the management realities of our time.

Assumption one. Every major task of developed society is being carried out in and through an organized and managed institution. Business enterprise was only the first of those and, therefore, became the prototype by historical accident. But while it has a specific job— the production and distribution of economic goods and services—it is neither the exception nor unique. Large-scale organization is the rule rather than the exception. Our society is one of pluralist organizations rather than a diffusion of family units. And management, rather than the isolated peculiarity of one unique exception, the business enterprise, is generic and the central social function in our society.[2]

A recent amusing book[3] points out that management is a form of government and applies to it Machiavelli's classic insights. But this is really not a very new idea, far from it. It underlay a widely read book of 1941—James Burnham's *Managerial Revolution* (though Mr. Burnham applied Marx rather than Machiavelli to management). It was treated in considerable detail in three of my books—*The Future of Industrial Man* (1942), *Concept of the Corporation* (1946)[4] and *The New Society* (1950). Mr. Justice Brandeis knew this well before World War I when he coined the term "Scientific Management" for Frederick Taylor's investigations into manual work. That a business organization is a form of government was also perfectly obvious to the entire tradition of American institutional economics from John R. Commons on, that is, from around the turn of the century. And across the Atlantic, Walter Rathenau saw the same thing clearly well before 1920.

But what is new is that nonbusiness institutions flock in increasing numbers to business management to learn from it how to manage themselves. The hospital, the armed services, the Catholic diocese, the civil service—all want to go to school for business management. And where Britain's first postwar Labour government nationalized the Bank of England to prevent its being run like a business, the next Labour government hired in 1968 a leading American firm of management consultants (McKinsey & Company) to reorganize the Bank of England to make sure it would be managed as a business.

This does not mean that business management can be transferred to other, nonbusiness institutions. On the contrary, the first thing these institutions have to learn from business management is that management begins with the setting of objectives and that, therefore, noneconomic institutions, such as a university or a hospital, will also need very different management from that of a business. But these institutions are right in seeing in business management the prototype. What we have done in respect to the management of a business we increasingly will have to do for the other institutions, including the government agencies. Business, far from being exceptional, is, in other words, simply the first of the species and the one we have studied the most intensively. And management is generic rather than the exception.

Indeed, what has always appeared as the most exceptional characteristic of business management, namely, the measurement of results in economic terms, that is, in terms of profitability, now emerges as the exemplar of what all institutions need: an objective outside measurement of the allocation of resources to results and of the rationality of managerial decisions. Noneconomic institutions need a yardstick that does for them what profitability does for the business—this underlies the attempt of Robert McNamara, while Secretary of Defense of the United States, to introduce "cost-effectiveness" into the government and to make planned, purposeful, and continuous measurement of programmes by their results, as compared to their promises and expectations, the foundation for budget and policy decisions. "Profitability," in other words, rather than being the "exception" and distinct from "human" or "social" needs, emerges, in the pluralist society of organizations, as the prototype of the measurement needed by every institution to be managed and manageable.[5]

Assumption two. Because our society is rapidly becoming a society of organizations, all institutions, including business, will have to hold themselves accountable for the "quality of life" and will have to make fulfillment of basic social values, beliefs, and purposes a major objective of their continuing normal activities rather than a "social responsibility" that restrains or that lies outside of their normal main functions. They will have to learn to make the "quality of life" into an opportunity for their own main tasks. In the business enterprise, this means that the attainment of the "quality of life" increasingly will have to be considered a business opportunity and will have to be converted by management into profitable business.

This will apply increasingly to fulfillment of the individual. It is the organization which is today our most visible social environment. The family is "private" rather than "community"—not that this makes it any less important. The "community" is increasingly in the organization, and especially in the one in which the individual finds his livelihood and through which he gains access to function, achievement, and social status. (On this see my 1949 *The New Society* (Reissued as a *Transaction* book in 199?) It will increasingly be the job of management to make the individual's values and aspirations redound to organizational energy and performance. It will simply not be good enough to be satisfied—as Industrial Relations and even Human Relations traditionally have been—with "satisfaction," that is, with the absence of discontent. Perhaps one way to dramatize this is to say that we will, within another ten years, become far less concerned with "management development" as a means of adapting the individual to the demands of the organization and far more with "organizational development" to adapt the organization to the needs, aspirations, and potential of the individual.

Assumption three. Entrepreneurial innovation will be as important to management as the managerial function, both in the developed and in the developing countries. Indeed, entrepreneurial innovation may be more important in the years to come. Unlike the nineteenth century, however, entrepreneurial innovation will increasingly have to be carried out in and by existing institutions such as existing businesses. It will, therefore, no longer be possible to consider it as lying outside of management or even as peripheral to management. Entrepreneurial innovation will have to become the very heart and core of management.

There is every reason to believe that the closing decades of the twentieth century will see changes as rapid as those that characterized the fifty years between 1860 and 1914, when a new major invention ushering in almost immediately a new major industry with new big businesses appeared on the scene every two or three years. But unlike the last century, these innovations of our century will be as much social innovations as they will be technical; a metropolis, for instance, is clearly as much of a challenge to the innovator today as the new science of electricity was to the inventor of 1870. And unlike the last century, innovation in this century will be based increasingly on knowledge of any kind rather than on science alone.

At the same time, innovation will increasingly have to be channeled in and through existing businesses, if only because the tax laws in every developed country make the existing business the center of capital accumulation. And innovation is capital-intensive, especially in the two crucial phases, the development phase and the market introduction of new products, processes, or services. We will, therefore, increasingly have to learn to make existing organizations capable of rapid and continuing innovation. How far we are still from this is shown by the fact that management still worries about "resistance to change." What existing organizations will have to learn is to reach out for change as an opportunity and to resist continuity.

Assumption four. A primary task of management in the developed countries in the decades ahead will increasingly be to make knowledge productive. The manual worker is "yesterday"—and all we can fight on that front is a rearguard action. The basic capital resource, the fundamental investment, but also the cost center of a developed economy, is the knowledge worker who puts to work what he has learned in systematic education, that is, concepts, ideas, and theories, rather than the man who puts to work manual skill or muscle.

Taylor put knowledge to work to make the manual worker productive. But Taylor himself never asked the question, 'What constitutes "productivity" in respect to the industrial engineer who applies "Scientific Management"'? As a result of Taylor's work, we can answer what productivity is in respect to the manual worker. But we still cannot answer what productivity is in respect to the industrial engineer, or to any other knowledge worker. Surely the measurements which give us productivity for the manual worker, such as the number of pieces turned out per hour or per dollar of wage, are quite irrelevant if applied to the knowledge worker. There are few things as useless and unproductive as the engineering department which with great dispatch, industry, and elegance turns out the drawings for an unsalable product. Productivity in respect to the knowledge worker is, in other words, primarily quality. We cannot even define it yet.

One thing is clear: to make knowledge productive will bring about changes in job structure, careers, and organizations as drastic as those which resulted in the factory from the application of Scientific Management to manual work. The entrance job, above all, will have to be changed drastically to enable the knowledge worker to become productive. For it is abundantly clear that knowledge cannot be productive unless the knowledge worker finds out who he himself is, what kind of work he is fitted for, and how he best works. In other words, there can be no divorce of "planning" from "doing" in knowledge work. On the contrary, the knowledge worker must be able to "plan" himself.

And this the present entrance jobs, by and large, do not make possible. They are based on the assumption—valid for manual work but quite inappropriate to knowledge work—that an outsider can objectively determine the "one best way" for any kind of work. For knowledge work, this is simply not true. There may be "one best way," but it is heavily conditioned by the individual and not entirely determined by physical, or even by mental, characteristics of the job. It is temperamental as well.

Assumption five. There are management tools and techniques. There are management concepts and principles. There is a common language of management. And there may even be a universal "discipline" of management. Certainly there is a worldwide generic function which we call management and which serves the same purpose in any and all developed societies. But management is also a culture and a system of values and beliefs. It is also the means through which a given society makes productive its own values and beliefs. Indeed, management may well be considered the bridge between a "civilization" that is rapidly becoming worldwide, and a "culture" which express divergent traditions, values, beliefs, and heritages. Management must, indeed, become the instrument through which cultural diversity can be made to serve the common purposes of mankind. At the same time, management increasingly is not being practiced within the confines of one national culture, law, or sovereignty but "multinationally." Indeed, management increasingly is becoming an institution—so far, the only one—of a genuine world economy.

Management we now know has to make productive values, aspirations, and traditions of individuals, community, and society for a common productive purpose. If management, in other words, does not succeed in putting to work the specific cultural heritage of a country and of a people, social and economic development cannot take place. This is, of course, the great lesson of Japan—and the fact that Japan succeeded, a century ago, in putting to work her own traditions of community and human values for the new ends of a modern industrialized state explains why Japan succeeded while every other non-Western country has so far failed. Management, in other words, will have to be considered both a science and a humanity, both a statement of findings that can be objectively tested and validated and a system of belief and experience.

At the same time, management—and here we speak of business management alone, so far—is rapidly emerging as the one and only institution that is common and transcends the boundaries of the national state. The multi-national corporation does not really exist so far. What we have, by and large, are still businesses that are based on one country with one culture and, heavily, one nationality, especially in top management. But it is also becoming clear that this is a transition phenomenon and that continuing development of the world economy both requires and leads to genuinely multinational companies in which not only production and sales are multinational, but ownership and management as well—all the way from the top down.

Within the individual country, especially the developed country, business is rapidly losing its exceptional status as we recognize that it is the prototype of the typical, indeed, the universal, social form, the organized institution requiring management. Beyond the national boundary, however, business is rapidly acquiring the same exceptional status it no longer has within the individual developed country. Beyond the national boundary, business is rapidly becoming the unique, the exceptional, the one institution which expresses the reality of a world economy and of a worldwide knowledge society.

Assumption six. Management creates economic and social development. Economic and social development is the *result* of management.

It can be said without too much oversimplification that there are no "underdeveloped countries." There are only "undermanaged" ones. Japan a hundred years ago was an underdeveloped country by every material measurement. But it very quickly produced management of great competence, indeed, of excellence. Within twenty-five years Meiji Japan had become a developed country, and, indeed, in some aspects, such as literacy, the most highly developed of all countries. We realize today that it is Meiji Japan, rather than eighteenth-century England—or even nineteenth-century Germany—which has to be the model of development for the underdeveloped world. This means that management is the prime mover and that development is a consequence.

All our experience in economic development proves this. Wherever we have only contributed the economic 'factor of production,' especially capital, we have not achieved development. In the few cases where we have been able to generate management energies (e.g., in the Cauca Valley in Colombia[6]) we have generated rapid development. Development, in other words, is a matter of human energies rather than of economic wealth. And the generation and direction of human energies is the task of management.

Admittedly, these new assumptions oversimplify; they are meant to. But I submit that they are better guides to effective management in the developed countries today, let alone tomorrow, than the assumptions on which we have based our theories as well as our practice these last fifty years. We are not going to abandon the old tasks. We still, obviously, have to manage the going enterprise and have to create internal order and organization. We still have to manage the manual worker and make him productive. And no one who knows the reality of management is likely to assert that we know everything in these and similar areas that we need to know; far from it. But the big jobs waiting for management today, the big tasks requiring both new theory and new practice, arise out of the new realities and demand different assumptions and different approaches.

More important even than the new tasks, however, may be management's new role. Management is fast becoming the central resource of the developed countries and the basic need of the developing ones. From being the specific concern of one, the economic institutions of society, management and managers are becoming the generic, the distinctive, the constitutive organs of developed society. What management is and what managers do will, therefore—and properly— become increasingly a matter of public concern rather than a matter for the "experts." Management will increasingly be concerned as much with the expression of basic beliefs and values as with the accomplishment of measurable results. It will increasingly stand for the quality of life of a society as much as for its standard of living.

There are many new tools of management the use of which we will have to learn, and many new techniques. There are, as this paper points out, a great many new and difficult tasks. But the most important change ahead for management is that increasingly the aspirations, the values, indeed, the very survival of society in the developed countries will come to depend on the performance, the competence, the earnestness and the values of their managers. The task of the next generation is to make productive for individual, community, and society the new organized institutions of our New Pluralism. And that is, above all, the task of management.

Notes

1. As documented in some detail in my 1969 book, *The Age of Discontinuity* (Reissued in a Transaction edition in 1992).
2. On this New Pluralism, see *The Age of Discontinuity,* especially Part Three: 'A Society of Organizations.'
3. Anthony Jay: *Management and Machiavelli* (London: Hodder and Stoughton, 1967).
4. Published in the UK as *Big Business.*
5. This, of course, also underlies the rapid return to profit and profitability as yardsticks and determinants of allocation decisions in the developed communist countries, that is, Russia and the European satellites.
6. For a description, see my *Age of Discontinuity.*

12

Management: The Problems of Success

The best-kept secret in management is that the first systematic applications of management theory and management principles did not take place in business enterprise. They occurred in the public sector. The first systematic and deliberate application of management principles in the United States—undertaken with full consciousness of its being an application of management—was the reorganization of the U.S. Army by Elihu Root, Teddy Roosevelt's secretary of war. Only a few years later, in 1908, came the first "city manager" (in Staunton, Virginia), the result of a conscious application of such then-brand-new management principles as the separation of "policy" (lodged in an elected and politically accountable city council) from "management" (lodged in a nonpolitical professional, accountable managerially). The city manager, by the way, was the first senior executive anyplace called a *manager;* in business, this title was still quite unknown. Frederick W. Taylor, for instance, in his famous 1911 testimony before the U.S. Congress never used the term but spoke of "the owners and their helpers." And when Taylor was asked to name an organization that truly practiced "Scientific Management," he did not name a business but the Mayo Clinic.

Thirty years after, the city manager, Luther Gulick applied management and management principles to the organization of a federal government that had grown out of control in the New Deal years. It

First delivered as a keynote address at the Academy of Management Congress in Chicago in 1986.

was not until 1950 and 1951, that is, more than ten years later, that similar management concepts and principles were systematically applied in a business enterprise to a similar task: the reorganization of the General Electric Company after it had outgrown its earlier, purely functional organization structure.

Today, surely, there is as much management outside of business as there is in business—maybe more. The most management-conscious of our present institutions are probably the military, followed closely by hospitals. Forty years ago the then-new management consultants considered only business enterprises as potential clients. Today half of the clients of a typical management consulting firm are nonbusiness: government agencies, the military, schools and universities, hospitals, museums, professional associations, and community agencies like the Boy Scouts and the Red Cross.

And increasingly, holders of the advanced degree in business administration, the MBA, are the preferred recruits for careers in city management, in art museums, and in the federal government's Office of Management and Budget.

Yet most people still hear the words *business management* when they hear or read *management*. Management books often outsell all other nonfiction books on the bestseller lists; yet they are normally reviewed on the business page. One "graduate business school" after another renames itself "School of Management." But the degree it awards has remained the MBA, the Master of *Business* Administration. Management books, whether textbooks for use in college classes or books for the general reader, deal mainly with business and use business examples or business cases.

That we hear and read *business management* when the word *management* is spoken or printed has a simple explanation. The business enterprise was not the first of the managed institutions. The modern university and the modern army each antedate the modern business enterprise by a half century. They emerged during and shortly after the Napoleonic Wars. Indeed, the first "CEO" of a modern institution was the chief of staff of the post-Napoleonic Prussian army, an office developed between 1820 and 1840. In spirit as well as in structure, both the new university and the new army represented a sharp break with their predecessors. But both concealed this—deliberately—by using the old titles, many of the old rites and ceremonies and, especially, by maintaining the social position of the institution and of its leaders.

No one could, however, have mistaken the new business enterprise, as it arose in the third quarter of the nineteenth century, for a direct continuation of the old and traditional "business firm"—the "counting house" consisting of two elderly brothers and one clerk that figures so prominently in Charles Dickens's popular books published in the 1850s and 1860s, and in so many other nineteenth-century novels, down to Thomas Mann's *Buddenbrooks* published in 1906.

For one, the new business enterprise—the *long-distance railroad* as it developed in the United States after the Civil War, the *universal bank* as it developed on the European Continent, or the *trust* such as United States Steel, which J. P. Morgan forged in the United States at the turn of the twentieth century—were not run by the "owners." Indeed, they had no owners, they had "shareholders." Legally, the new university or the new army was the same institution it had been since time immemorial, however much its character and function had changed. But to accommodate the new business enterprise, a new and different legal *persona* had to be invented, the "corporation." A much more accurate term is the French *Société Anonyme,* the anonymous collective owned by no one and open to investment by everyone. In the corporation, shares become a claim to profits rather than to property. Share ownership is, of necessity, separate from control and management, and easily divorced from both. And in the new corporation capital is provided by large, often by very large, numbers of outsiders, with each of them holding only a minute fraction and with none of them necessarily having an interest in, or—a total novelty—any liability for, the conduct of the business.

This new "corporation," this new *Société Anonyme,"* this new *"Aktiengesellschaft,"* could not be explained away as a *reform,* which is how the new army, the new university, and the new hospital presented themselves. It clearly was a genuine innovation. And this innovation soon came to provide the new jobs—at first, for the rapidly growing urban proletariat, but increasingly also for educated people. It soon came to dominate the economy. What in the older institutions could be explained as different procedures, different rules, or different regulations became in the new institution very soon a new function, management, and a new kind of work. And this then invited study; it invited attention and controversy.

But even more extraordinary and unprecedented was the position of this newcomer in society. It was the first new autonomous institution in hundreds of years, the first to create a power center that was within society yet independent of the central government of the national state. This was an offense, a violation of everything the nineteenth century (and the twentieth-century political scientists still) considered "law of history," and frankly a scandal.

Around 1860 one of the leading social scientists of the time, the Englishman Sir Henry Maine, coined the phrase in his book *Ancient Law* that the progress of history is "from status to contract." Few phrases ever have become as popular and as widely accepted as this one.

And yet, at the very time at which Maine proclaimed that the law of history demands the elimination of all autonomous power centers within society, the business enterprise arose. And from the beginning it was clearly a power center within society and clearly autonomous.

To many contemporaries it was, and understandably so, a totally unnatural development and one that bespoke a monstrous conspiracy. The first great social historian America produced, Henry Adams, clearly saw it this way. His important novel, *Democracy,* which he wrote during the Grant administration, portrays the new economic power as itself corrupt and, in turn, as corrupting the political process, government, and society. Henry's brother, Brooks Adams, a few decades later, further elaborated on this theme in one of the most popular political books ever published in the United States, *The Degeneration of the Democratic Dogma.*

Similarly, the Wisconsin economist, John R. Commons—the brain behind the *progressive movement* in Wisconsin, the father of most of the "reforms" that later became the social and political innovations of the New Deal, and, last but not least, commonly considered the father of America's "business unionism"—took very much the same tack. He blamed business enterprise on a lawyers' conspiracy leading to a misinterpretation of the Fourteenth Amendment to the Constitution by which the corporation was endowed with the same "legal personality" as the individual.

Across the Atlantic in Germany, Walter Rathenau—himself the successful chief executive of one of the very large new "corporations" (and later on to become one of the earliest victims of Nazi terror when he was assassinated in 1922 while serving as foreign minister of the new Weimar Republic)—similarly felt that the business enterprise was something radically new, something quite incompatible with prevailing political and social theories, and indeed a severe social problem.

In Japan, Shibusawa Eiichi, who had left a promising government career in the 1870s to construct a modern Japan through building businesses, also saw in the business enterprise something quite new and distinctly challenging. He tried to tame it by infusing it with the Confucian ethic; and Japanese big business as it developed after World War II is very largely made in Shibusawa's image.

Everyplace else, the new business enterprise was equally seem as a radical and dangerous innovation. In Austria, for instance, Karl Lueger, the founding father of the "Christian" parties that still dominate politics in Continental Europe, was elected lord mayor of Vienna in 1897 on a platform that defended the honest and honorable small businessman—against the evil and illegitimate corporation. A few years later, an obscure Italian journalist, Benito Mussolini, rose to national prominence by denouncing "the soulless corporation."

And thus quite naturally, perhaps even inevitably, concern with management, whether hostile to it or friendly, concentrated on the business enterprise. No matter how much management was being applied to other institutions, it was the business enterprise that was visible, prominent, controversial, and above all, new, and therefore significant.

By now, however, almost a hundred years after management arose in the early large business enterprises of the 1870s, it is clear that management pertains to every single social institution. In the last hundred years every major social function has become lodged in a large and managed organization. The hospital of 1870 was still the place where the poor went to die. By 1950 the hospital had become one of the most complex organizations, requiring management of extraordinary competence. The labor union in developed countries is run today by a paid managerial staff, rather than by the politicians who are nominally at the head. Even the very large university of 1900 (and the largest then had only five thousand students) was still simple, with a faculty of, at most, a few hundred, each professor teaching his own specialty. It has by now become increasingly complex—including undergraduate, graduate, and postgraduate students—with research institutes and research grants from government and industry and, increasingly, with a large administrative superstructure. And in the modern military, the basic question is the extent to which management is needed and the extent to which it interferes with leadership—with management apparently winning out.

The identification of management with business can thus no longer be maintained. Even though our textbooks and our studies still focus heavily on what goes on in a business—and typically, magazines having the word *management* in their title (for example, Britain's *Management Today* or Germany's *Manager Magazin*) concern themselves primarily if not exclusively with what goes on in business enterprises—management has become the pervasive, the universal organ of a modern society.

For modern society has become a "society of organizations." The individual who conforms to what political and social theorists still consider the norm has become a small minority: the individual who stands in society directly and on his own, with no intermediary institution of which he is a member and an employee between himself and the sovereign government. The overwhelming majority of all people in developed societies are employees of an organization; they derive their livelihood from the collective income of an organization; see their opportunity for career and success primarily as opportunity within an organization; and define their social status largely through their position within the ranks of an organization. Increasingly, especially in the United States, the only way in which the individual can amass a little property is through the pension fund, that is, through membership in an organization.

And each of these organization, in turn, depends for its functioning on management. Management makes an organization out of what otherwise would be a mob. It is the effective, integrating, life-giving organ.

In a society of organizations, managing becomes a key social function and management the constitutive, the determining, the differential organ of society.

The New Pluralism

The dogma of the "liberal state" is still taught in our university departments of government and in our law schools. According to it, all organized power is vested in one central government. But the society of organizations is a *pluralist* society. In open defiance of the prevailing dogma, it contains a diversity of organizations and power centers. And each has to have a management and has to be managed. The business enterprise is only one; there are the labor unions and the farm organizations, the health-care institutions and the schools and universities, not to mention the media. Indeed, even government is increasingly becoming a pluralist congeries of near-autonomous power centers, very different indeed from the branches of government of the American Constitution. There is the civil service, for instance. The last president of the United States who had effective control of the civil service was Franklin D. Roosevelt fifty years age; in England it was Winston Churchill; in Russia, Stalin. Since their time the civil service in all major countries has become an establishment in its own right. And so, increasingly, has the military.

In the nineteenth century the "liberal state" had to admit the parties, though it did so grudgingly and with dire misgivings. But the purpose of the parties was the conquest of government. They were, so to speak, gears in the governmental machine and had neither existence nor justification outside of it.

No such purpose animates the institutions of the new pluralism.

The institutions of the old pluralism, that is, of medieval Europe or of medieval Japan (the princes and the feudal barons, the free cities, the artisans, the bishoprics and abbeys) were themselves governments. Each indeed tried to annex as much of the plenitude of governmental power as it could get away with. Each levied taxes and collected customs duties. Each strove to be granted the right to make laws, and to establish and run its own law courts. Each tried to confer knighthoods, patents of nobility, or titles of citizenship. And each tried to obtain the most coveted right of them all, the right to mint its own coins.

But the purpose of today's pluralist institution is nongovernmental: to make and to sell goods and services, to protect jobs and wages, to heal the sick, to teach the young, and so on. Each only exists to do something that is different form what government does or, indeed, to do something so that government need not do it.

The institutions of the old pluralism also saw themselves as total communities. Even the craft guild, the powerful woolen weavers of Florence, for instance, organized itself primarily to control its members. Of course, weavers got paid for selling woolen goods to other people. But their guild tried as hard as possible to insulate the members against economic impacts from the outside by severely restricting what could be made, how much of it, and how and at what price it could be sold, and by whom. Every guild gathered its members into its own quarter in the city, over which it exerted governmental control. Every one immediately built its own church with its own patron saint. Every one immediately built its own school; there is still "Merchant Taylor's" in London. Every one controlled access to membership in the guild. If the institutions of the old pluralism had to deal with the outside at all, they did so as "foreign relations" through formal pacts alliances, feuds, and, often enough, open war. The outsider was a foreigner.

The institutions of the new pluralism have no purpose except outside of themselves. They exist in contemplation of "customer" or a "market." Achievement in the hospital is not a satisfied nurse, but a cured *former* patient. Achievement in business is not a happy work force, however desirable it may be; it is a satisfied customer who reorders the product.

All institutions of the new pluralism, unlike those of the old, are single-purpose institutions. They are tools of society to supply one specific social need, whether making or selling cars, giving telephone service, curing the sick, teaching children to read, or providing benefit checks to unemployed workers. To make this single, specific contribution, they themselves need a considerable measure of autonomy, however. They need to be organized in perpetuity, or at least for long periods of time. They need to dispose of a considerable amount of society's resources, of land, raw materials, and money, but above all of people, and especially of the scarcest resource of them all, highly trained and highly education people And they need a considerable amount of power over people, and coercive power at that. It is only too easy to forget that in the not-so-distant past, only slaves, servants, and convicts had to be at the job at a time set for them by someone else.

This institution has—and has to have—power to bestow or to withhold social recognition and economic rewards. Whichever method we use to select people for assignments and promotions—appointment from above, selection by one's peers, even rotation among jobs—it is always a power decision made for the individual rather than by him, and on the basis of impersonal criteria that are related to the organization's purpose rather than to the individual's purpose. The individual is thus, of necessity, subjected to a power grounded in the value system of whatever specific social purpose the institution has been created to satisfy.

And the organ through which this power is exercised in the institution is the organ we call *management.*

This is new and quite unprecedented. We have neither political nor social theory for it as yet.

This new pluralism immediately raises the question, Who takes care of the commonweal when society is organized in individual power centers, each concerned with a specific goal rather than with the common good?

Each institution in a pluralist society sees its own purpose as the central and the most important one. Indeed, it cannot do otherwise. The school, for instance, or the university could not function unless they saw teaching and research as what makes a good society and what makes a good citizen. Surely nobody chooses to go into hospital administration or into nursing unless he or she believes in health as an absolute value. And as countless failed mergers and acquisitions attest, no management will do a good job running a company unless it believes in the product or service the company supplies, and unless it respects the company's customers and their values.

Charles E. Wilson, GM's chairman (later President Eisenhower's secretary of defense), never said, "What is good for General Motors is good for the country." What he actually said is "What is good for the country is good for General Motors, and vice versa." But that Wilson was misquoted is quite irrelevant. What matters is that everybody believed that he not only said what he was misquoted to have said, but that he actually believed it. And indeed no one could run General

Motors—or Harvard University, or Misericordia Hospital, or the Bricklayers Union, or the Marine Corps—unless he believed that what is good for GM, or Harvard, or Misericordia, or the Bricklayers, or the Marines is indeed good for the country and is indeed a "mission," that if not divinely ordained, is still essential to society.

Yet each of these missions is one and only one dimension of the common good—important yes, indispensable perhaps, and yet a relative rather than an absolute good. As such, it must be limited, weighed in the balance with, and often subordinated to, other considerations. Somehow the common good must be made to emerge out of the clash and clamor of special interests.

The old pluralism never solved this problem. This explains why suppressing it became the "progressive cause" and the one with which the moral philosophers of the modern age (that is, of the sixteenth through the nineteenth centuries) aligned themselves.

Can the new pluralism do any better? One solution is, of course, to suppress the pluralist institutions. This is the answer given by totalitarianism and is indeed its true essence. The totalitarian state, whether it calls itself Fascist, Nazi, Stalinist, or Maoist, makes all institutions subservient to and extensions of the state (or of the omnipotent party). This saves the "state" of modern political theory, but at the sacrifice of individual freedom, of free thought and free expression, and of any limitation on power altogether. The state (or the party) is then indeed the *only* power center, as traditional theory preaches. But it can maintain its monopoly on power only by being based on naked terror, as Lenin was the first to realize. And even at that horrible price, it does not really work. As we now know—and the experience of all totalitarian regimes is exactly the same, whether they call themselves Right or Left—the pluralist institutions persist behind the monolithic facade. They can be deprived of their autonomy only if they and society together are rendered unable to perform, for instance, through Stalin's purges or Mao's Cultural Revolution. What the totalitarian regimes have proved is that modern society *has* to be a "society of organizations," and that means a pluralist society. The only choice is whether individual freedom is being maintained or is being suppressed and destroyed, albeit to no purpose other than naked power.

The opposite approach to that of the totalitarian is the American one.

The United States, alone among modern nations, never fully accepted the dogma of the liberal state. It opposed to it, quite early in its history, a pluralist political theory, that of John C. Calhoun's "concurrent majority." In the way in which Calhoun presented his theory in the 1830s and 1840s, that is, as a pluralism exercised through the individual states and intended to prevent the breakup of the Union over slavery, the "concurrent majority" did not survive the Civil War. But thirty years later, Mark Hanna, the founder of the modern Republican party and of modern American politics altogether, reformulated Calhoun's pluralism as a concurrent majority of the major "interests": farmers, workers, business. Each of these three "estates of the realm" can effectively veto the majority. It must not impose its will on the others. But it must be able to prevent the others from imposing their will on it. Another thirty years later, Franklin D. Roosevelt made this the basic political creed of the New Deal. In Roosevelt's system government became the arbiter whose job it is to make sure that no one interest gets too powerful. When Roosevelt came in, "*capital*"—*business* as a term came later, and *management* later still—appeared to be far too powerful. Farmers and workers were thus organized to offset the business power. And then, not so many years later, when the labor power seemed to become too great, farmers and business were organized to offset and balance labor power, and so on.

Each of the "interests" is free to pursue its own goals regardless of the common good; it is indeed expected to do so. In the darkest days of World War II, in 1943 when American troops still lacked arms and ammunition, John L. Lewis, the founder of the Congress of Industrial Organizations (that is, of modern American unionism) and the powerful head of the coal miners' union, called a coal strike to get higher wages for his men, defying national wage controls. President Roosevelt attacked him publicly for endangering the nation's survival. Lewis retorted: "The President of the United States is paid to look after the nation's survival. I am paid to look after the interests of the coal miners." And while the newspapers attacked Lewis harshly, public opinion apparently felt that Lewis had only said out aloud what the Roosevelt administration had practiced all along. It gave Lewis enough support to win the strike.

This example, however, shows that the American pluralist doctrine is hardly adequate. Indeed, just as the old pluralism did, it has given birth to so many vested interests and pressure groups that it is almost impossible to conduct the business of government, let alone to conduct it for the common good.

In 1984–85 practically everyone in the United States agreed that the country needed a drastic tax reform to replace an increasingly complicated and irrational tax code, with a few tax rates and with exemptions eliminated. But no such code could be enacted. Every single exemption became the sacred cause of a vested interest. And even though some of them represented only a few hundred or a few thousand voters, each of them could and did block tax reform.

Is there a way out? The Japanese seem to be the only ones so far able to reconcile a society of organizations with the pursuit of the common good. It is expected of the major Japanese interests that they take their cue from "what is good for the country": Then they are expected to fit what is good for themselves into the framework of a public policy designed to serve the national interest.

It is doubtful, however, whether even Japan can long maintain this approach. It reflects a past in which Japan saw herself as isolated in a hostile and alien world—so that all of Japan, regardless of immediate interests, had to hang together lest it hang separately. Will this attitude survive Japan's success? And could such an approach have a chance in the West, where interests are expected to behave as interests?

Is this a problem of management, it will be asked? Is it not a problem of politics, of government, or political philosophy? But if management does not tackle it, then almost inevitably there will be imposed political solutions. When, for instance, the health-care institutions in America, the hospitals and the medical profession, did not take responsibility for spiraling health-care costs, government imposed restrictions on them, for example, the Medicare restrictions on the care of the aged in hospitals. These rules clearly are not concerned with health care at all and may even be detrimental to it. They are designed to serve short-run fiscal concerns of government and employers, that is, designed to substitute a different but equally one-sided approach for the one-sided, self-centered approach of the health-care "interests."

This must be the outcome unless the managements of the institutions of the new pluralism see it as their job to reconcile concern for the common good with the pursuit of the special mission for the sake of which their institution exists.

The Legitimacy of Management

Power has to be legitimate. Otherwise it has only force and no authority, is only might and never right. To be legitimate, power has to be grounded outside of it in something transcending it that is accepted as a genuine value, if not as a true absolute by those subject to the power—whether descent from the gods or apostolic succession; divine institution or its modern, totalitarian counterpart the scientific laws of history; the consent of the governed, popular election or, as in so much of modern society, the magic of the advanced degree. If power is an end in itself, it becomes despotism and both illegitimate and tyrannical.

Management has to have power to do its job, whatever the organization. In that respect there is little difference between the Catholic diocese, the university, the hospital, the labor union, and the business enterprise. And because the governing organ of each of these institutions has to have power, it has to have legitimacy.

And here we encounter a puzzle. The management of the key institutions of our society of organizations is by and large accepted as legitimate. The single exception is the management of the business enterprise. Business enterprise is seen as necessary and accepted as such. Indeed, society is often more concerned with the survival of a large business or an industry than it is with that of any other single institution. If a major business is in trouble, there is a crisis and desperate attempts to salvage the company. But at the same time, business management is suspect. And any exercise of management power is denounced as usurpation, with cries from all sides for legislation or for judicial action to curb if not to suppress managerial power altogether.

One common explanation is that the large business enterprise wields more power than any other institution. But this simply does not hold water. Not only is business enterprise hemmed in in its power on all sides—by government and government regulations, by labor unions, and so on. The power of even the largest and wealthiest business enterprise is insignificant next to that of the university now that a college degree has become a prerequisite for access to any but the most menial jobs. The university and its management are often criticized, but their legitimacy is rarely questioned.

The large labor union in Western Europe and in American mass-production industries surely has more power than any single business enterprise in its country or industry. Indeed in Western Europe, both in Britain and on the Continent, the large labor union became society's most powerful institution in the period after World War II, more powerful sometimes than the nation's government. The unions' exercise of their power during this period was only too often self-serving, if not irresponsible. But even their bitterest critics in Western Europe and in the United States rarely questioned the unions' legitimacy.

Another explanation—the prevalent one these days—is that the managements of all other institutions are altruistic, whereas business is profit-seeking and therefore out for itself and materialistic. But even if it is accepted that for many people nonprofit is virtuous, and profit dubious, if not outright sinful, the explanation that profit undermines the legitimacy of business management is hardly adequate. In all Western countries the legitimacy of owners, that is, of real capitalists, and their profits is generally accepted without much question. That of a professional management is not, yet professional management obtains profits for other people rather than for itself—and its main beneficiaries today are the pension funds of employees.

And then there is the situation in Japan. In no other country, not even in France or in Sweden, was the intellectual climate of the postwar period as hostile to ''profit'' as in Japan, at least until 1975 or so. The left-wing intelligentsia of Japan in the universities or the newspapers might have wanted to nationalize Japan's big businesses. But it never occurred even to the purest Marxist among them to question the necessity of management or its legitimacy.

The explanation clearly lies in the image which Japanese management has of itself and which it presents to its society. In Japanese law, as in American and European law, management is the servant of the stockholders. But this the Japanese treat as pure fiction. The reality which is seen as guiding the behavior of Japanese big-business management (even in companies that are family-owned and family-managed like Toyota) is management as an organ of the business itself.

Management is the servant of the going concern, which brings together in a common interest a number of constituencies: employees first, then customers, then creditors, and finally suppliers. Stockholders are only a special group of creditors, rather than "the owners" for whose sake the enterprise exists. As their performance shows, Japanese businesses are not run as philanthropies and know how to obtain economic results. In fact, the Japanese banks, which are the real powers in the Japanese economy, watch economic performance closely and move in on a poorly performing or lackluster top management much faster than do the boards of Western publicly held companies. But the Japanese have institutionalized the going concern and its values through lifetime employment, under which the employees' claim to job and income comes first—unless the survival of the enterprise itself is endangered.

The Japanese formulation presents very real problems, especially at a time of rapid structural change in technology and economy when labor mobility is badly needed. Still, the Japanese example indicates why management legitimacy is a problem in the West. Business management in the West (and in particular business management in the United States) has not yet faced up to the fact that our society has become a society of organizations of which management is the critical organ.

Thirty years ago or so, when the serious study of management began, Ralph Cordiner, then CEO of the General Electric Company, tried to reformulate the responsibility of corporate top management. He spoke of its being the "trustee for the balanced best interest of stockholders, employees, customers, suppliers and plant communities"—the groups which would now be called *stakeholders* or *constituencies*. As a slogan this caught on fast. Countless other American companies wrote it into their Corporate Philosophy statement. But neither Mr. Cordiner nor any of the other chairmen and presidents who embraced his rhetoric did what the Japanese have done: institutionalize their professions. They did not think through what the best-balanced interest of these different stakeholders would mean, how to judge performance against such an objective, and how to create accountability for it. The statement remained good intentions. And good intentions are not enough to make power legitimate. In fact, good intentions as the grounds for power characterize the "enlightened despot." And enlightened despotism never works.

The term *enlightened despot* was coined in the eighteenth century—with Voltaire probably its greatest and most enthusiastic exponent—when the divine right of princes was no longer generally accepted as a ground of legitimate power. The prince with the best intentions among eighteenth-century enlightened despots and the very model of the progressive, the enlightened liberal, was the Austrian emperor Joseph II (reigned 1765–90). Every one of the reforms that he pioneered was a step in the right direction—the abolition of torture; religious toleration for Protestants, Jews, and even atheists; universal free education and public hospitals in every county; abolition of serfdom; codification of the laws; and so on. Yet his subjects, and especially his subjects in the most advanced parts of his empire, the Austrian Netherlands, rose against him in revolt. And when, a few years later, the French Revolution broke out, the enlightened despots of Europe toppled like ninepins. They had no constituency to support them.

Because Ralph Cordiner and his contemporaries never even tried to ground management power in institutional arrangements, their assertion very rapidly became enlightened despotism. In the 1950s and 1960s it became *corporate capitalism,* in which an enlightened "professional" management has absolute power within its corporation, controlled only by itself and irremovable except in the event of catastrophe. "Stock ownership," it was argued, had come to be so widely dispersed that shareholders no longer could interfere, let alone exercise control.

But this is *hubris:* arrogance and sinful pride, which always rides before a fall. Within ten years after it had announced the independence of management in the large, publicly owned corporation, "corporate capitalism" began to collapse. For one, stock ownership came to be concentrated again, in the hands of the pension funds.

And then inflation distorted values, as it always does, so that stock prices, which are based on earnings expectations, came to appear far lower than book values and liquidation values. The result was the wave of hostile takeovers that has been inundating the American economy these last years and is spilling over into Europe now. Underlying it is the assertion that the business enterprise exists, and solely, for the sake of stockholder profits, and short-run, immediate profits at that.

By now it has become accepted widely—except on Wall Street and among Wall Street lawyers—that the hostile takeover is deleterious and in fact one of the major causes of the loss of America's competitive position in the world economy. One way or another, the hostile takeover will be stopped. It may be through a "crash;" speculative booms always collapse in the end. It may be through such changes as switching to different classes of common stock, with the shares owned by the outside public having a fraction of the voting power of the insiders' shares, or by giving up voting rights for publicly held common shares altogether. (I owe this suggestion to Mr. Walter Wriston, the chairman emeritus of New York's Citibank.)

No matter how the hostile takeover boom is finally stopped, it will have made certain that the problem of management legitimacy has to be tackled. We know some of the specifications for the solution. There have to be proper safeguards of the economic performance of a business: its market standing, the quality of its products or services, and its performance as an innovator. There has to be emphasis on, and control of, financial performance. If the takeover boom has taught us one thing, it is that management must not be allowed substandard financial performance.

But somehow the various "stakeholders" also have to be brought into the management process (for example, through the company's pension plan as a representative of the company's employees for whom the pension plan is the trustee). And somehow the maintenance of the wealth-producing and the job-producing capacity of the enterprise, that is, the maintenance of the going concern, needs to be built into our legal and institutional arrangements. It should not be too difficult. After all, we built the preservation of the going concern into our bankruptcy laws all of ninety years ago when we gave it priority over all other claims, including the claims of the creditors. But whatever the specifics, business management has to attain legitimacy; its power has to be grounded in a justification outside and beyond it and has to be given the "constitutional" sanction it still largely lacks.

Closely connected to the problem of the legitimacy of management is management's compensation.

Management, to be legitimate, must be accepted as "professional." Professionals have always been paid well and deserve to be paid well. But it has always been considered unprofessional to put money ahead of professional responsibility and professional standards. This means that there have to be limitations on managerial incomes. It is surely not professional for a chief executive officer to give himself a bonus of several millions at the very time at which the pay of the company's other employees is cut by 30 percent, as the chief executive officer of Chrysler did a few years ago. It is surely not professional altogether for people who are employees and not "owners" to pay themselves salaries and bonuses greatly in excess of what their own colleagues, that is, other members of management, receive. And it is not professional to pay oneself salaries and bonuses that are so far above the norm as to create social tension, envy, and resentment. Indeed there is no economic justification for very large executive incomes. German and Japanese top managers surely do as good a job as American top managers—perhaps, judging by results, an even better one. Yet their incomes are, at the most, half of what American chief executives of companies in similar industries and of similar size are sometimes being paid.

But there is also work to be done on the preparation, testing, and selection of, and on the succession to, the top-management jobs in the large business enterprises; on the structure of top management; and on performance standards for top management and the institutional arrangements for monitoring and enforcing them.

Business management is not yet fully accepted as legitimate in the West because it has not yet realized the full implications of its success. Individual executives, even those of the biggest company, are largely anonymous. They only make asses of themselves if they try to behave as if they were aristocrats. They are hired hands like the rest of us. On the day on which they retire and move out of the executive suite they become "nonpersons" even in their old company. But while in office they represent; individually almost faceless, collectively they constitute a governing group. As such their behavior is seen as representative. What is private peccadillo for ordinary mortals becomes reprehensible misconduct and indeed betrayal if done by a leader. For not only is the leader visible; it is his duty to set an example.

But then there is also the big question of what is now being called the "social responsibility" of management. It is not, despite all rhetoric to the contrary, a social responsibility of business but of all institutions—otherwise we would hardly have all the malpractice suits against American hospitals or all the suits alleging discrimination against American colleges and universities. But business is surely one of the key institutions of a society of organizations and as such needs to determine what its social responsibilities are—and what they are not.

Surely business, like anyone else, is responsible for its impacts: responsibility for one's impacts is, after all, one of the oldest tenets of the law. And surely, business, like anyone else, is in violation of its responsibilities if it allows itself impacts beyond those necessary to, and implicit in, its social purpose, for example, producing goods and services. To overstep these limits constitutes a *tort*, that is, a violation.

But what about problems that do not result from an impact or any other activity of business and yet constitute grave social ills? Clearly it is not a responsibility of business, or of any organization, to act where it lacks competence; to do so is not responsibility but irresponsibility. Thus when a former mayor of New York City in the 1960s called for "General Electric and the other big corporations of New York City to help solve the problem of the Black Ghetto by making sure that there is a man and father in the home of every Black Welfare Mother," he was not only ridiculous. He demanded irresponsibility.

But also management must not accept "responsibility" if by doing so it harms and impedes what is its first duty: the economic performance of the enterprise. This is equally irresponsible.

But beyond these caveats there is a no-man's-land where we do not even fully understand what the right questions are. The problems of New York, for instance, are in no way caused by business. They were largely caused by public policies business had warned against and fought against: primarily by rent control, which, as it always does, destroys the very housing the poor need, that is, decent, well-maintained older housing; by demagogic welfare policies; and by equally demagogic labor-relations policies. And yet when New York City was on the verge of self-destruction, in the late 1960s and early 1970s, a small group of senior executives of major New York business enterprises mobilized the business community to reverse the downward slide and to renew New York City—people like Austin Tobin of the Port of New York Authority; David Rockefeller of the Chase Manhattan Bank; Walter Wriston and William Spencer of Citibank; Felix Rohatyn of Lazard Frères, the private bankers; the top management of Pfizer, a pharmaceutical company; and several others. They did this not by "taking responsibility" for things they lacked competence in, for example, the problems of the black ghetto. They did it by doing what they were highly competent to do: they started and led the most dramatic architectural development of any major city since Napoleon III had created a new Paris and Francis Joseph a new Vienna a hundred years earlier. The black ghetto is still there, and so are all the ills associated with it, for example, crime on the streets. But the city has been revitalized.

And this did not happen because these businesses and their managements needed the city; excepting only the Port of New York Authority, they could all have moved out, as a good many of their colleagues—IBM, for instance, or General Electric, or Union Carbide—were doing. These businesses and their top managements acted because the city needed them, though, of course, they benefited in the end if only because a business—and any other institution—does better in a healthy rather than a diseased social environment.

Is there a lesson in this? There surely is a challenge.

Altogether, for management of the big business to attain full legitimacy, it will have to accept that to remain "private" it has to accept that it discharges a social, and that means a "public," function.

The Job as Property Right

When, in 1985, a fair-size Japanese company found itself suddenly threatened by a hostile takeover bid made by a group of American and British "raiders"—the first such bid in recent Japanese history—the company's management asserted that the real owners of the business, and the only ones who could possibly sell it, were not the stockholders, but the employees. This was considerable exaggeration, to be sure. The real owners of a major Japanese company are the banks, as has already been said. But it is true that the rights of the employees to their jobs are the first and overriding claim in a large Japanese company, except when the business faces a crisis so severe that its very survival is at stake.

To Western ears the Japanese company statement sounded very strange. But actually the United States—and the West in general—may be as far along in making the employees the dominant interest in business enterprise, and not only in the large one as in Japan. All along, of course, the employees' share of the revenues of a business, almost regardless of size, exceeds what the "owners" can possibly hope to get: ranging from being four times as large (that is, 7 percent for after-tax profits, as against 25 percent for wages and salaries) to being twelve times as large (that is, 5 percent for profits versus 60 percent of revenues for wages and salaries). The pension fund not only greatly increased the share of the revenues that go into the "wage fund," to the point that in poor years the pension fund may claim the entire profit and more. American law now also gives the pension fund priority over the stockholders and their property rights in a company's liquidation, way beyond anything Japanese law and Japanese custom give to the Japanese worker.

Above all, the West, with the United States in the lead, is rapidly converting the individual employee's job into a new property right and, paradoxically, at the very time at which the absolute primacy of stockholder short-term rights is being asserted in and by the hostile takeover.

The vehicle for this transformation in the United States is not the union contract or laws mandating severance pay as in many European countries. The vehicle is the lawsuit. First came the suit alleging discrimination, whether in hiring an employee, in firing, in promotion, in pay, or in job assignment—discrimination on grounds of race or sex or age or handicap. But increasingly these suits do not even allege discrimination, but violation of "due process." They claim that the employer has to treat the employee's job, including the employee's expectations for pay and promotion, as something the enjoyment of which and of its fruits can be diminished or taken away only on the basis of preset and objective standards and through an established process which includes an impartial review and the right to appeal. But these are the features that characterize "property" in the history of the law. In fact, they are the *only* features a right must possess to be called property in the Western legal tradition.

And as few managements yet seem to realize, in practically every such suit the plaintiff wins and the employer loses.

This development was predictable. Indeed, it was inevitable. And it is irreversible. It is also not "novel" or "radical." What gives access to a society's productive resources—gives access thereby to a livelihood and to social function and status and constitutes a major, if not the major, avenue to economic independence however modest—has always become a "property right" in Western society. And this is what the job has become, and especially the knowledge worker's job as a manager or a professional.

We still call land "real" property. For until quite recently it was land alone that gave to the great majority of mankind—95 percent or more—what "property" gives: access to, and control over, society's productive resources; access to a livelihood and to social status and function; and finally a chance at an *estate* (the term itself meant, at first, a landholding) and with it economic independence.

In today's developed societies, however, the overwhelming majority—all but 5 or 10 percent of the population—find access to and control over productive resources and access to a livelihood and to social status and function through being employees of organizations, that is, through their jobs. For highly educated people the job is practically the only access route. Ninety-five percent, or more, of all people with college degrees will spend their entire working lives as employees of an organization. Modern organization is the first, and so far the only, place where we can put large numbers of highly educated people to productive work and pay them for applying knowledge.

For the great majority of Americans, moreover, the pension fund at their place of employment is their only access to an "estate," that is to a little economic independence. By the time the main breadwinner in the American family, white collar or blue collar, is forty-five years old, the claim to the pension fund is likely to be the family's largest asset, far exceeding in value the equity in the home or the family's personal belongings, for example, their automobiles.

Thus the job had to become a property right—the only question is in what form and how fast.

Working things like this out through lawsuits may be "as American as apple pie," but is hardly as wholesome. There is still a chance for management to take the initiative in this development and to shape the new property rights in the job so that they equally serve the employee, the company, and the economy. We need to maintain flexibility of employment. We need to make it possible for a company to hire new people and to increase its employment. And this means that we must avoid the noose the Europeans have put around their neck: the severance pay which the law of so many Continental countries mandates makes it so expensive to lay off anybody that companies simply do not hire people. That Belgium and Holland have such extraordinary high unemployment is almost entirely the result of these countries' severance pay laws. But whichever way we structure the new property rights which the job embodies, there will be several requirements which every employer, that is, every organization, will have to satisfy. First, there must be objective and equal performance standards for everyone performing a given job, regardless of race, color, sex, or age. Secondly, to satisfy the requirements of due process, the appraisal against these standards of performance has to be reviewed by somebody who is truly disinterested. Finally, due process demands a right of appeal—something, which by the way, as "authoritarian" a company as IBM has had for more than half a century.

The evolution of the job into a "property right" changes the position of the individual within the organization. It will change equally, if not more, the position of the organization in society. For it will make clear what at present is still nebulous: organized and managed institutions have increasingly become the organs of opportunity, of achievement, and of fulfillment for the individual in the society of organizations.

Conclusion

There is still important work ahead—and a great deal of it—in areas that are conventionally considered "management" in the schools of management, in management journals, and by practicing managers themselves. But the major challenges are new ones, and well beyond the *field of management* as we commonly define it. Indeed, it will be argued that the challenges I have been discussing are not management at all, but belong in political and social theory and public law.

Precisely. The success of management has not changed the *work* of management. But it has greatly changed management's *meaning*. Its success has made management the general, the pervasive function, and the distinct organ of our society of organizations. As such, management inevitably has become "affected with the public interest." To work out what this means for management theory and management practice will constitute the "management problems" of the next fifty years.

13

Social Innovation:
Management's New Dimension

Are we overemphasizing science and technology as this century's change agents? *Social innovations*—few of them owing anything to science or technology—may have had even profounder impacts on society and economy, and indeed profound impacts even on science and technology themselves. And management is increasingly becoming the agent of social innovation.

Here are five examples—five among many:

* the research lab;
* Eurodollar and commercial paper;
* mass and mass movement;
* the farm agent; and
* management itself as an organized function and discipline.

The Research Lab

The research lab as we now know it dates back to 1905. It was conceived and built for the General Electric Company in Schenectady, New York, by one of the earliest "research managers," the German-American physicist Charles Proteus Steinmetz. Steinmetz had two clear objectives from the start: to organize science and scientific work

Written for and first published in my essay volume *The Frontier of Management*, 1986.

for purposeful technological invention and to build continuous self-renewal through innovation into that new social phenomenon—the big corporation.

Steinmetz took two of the features of his new lab from nineteenth-century predecessors. From the German engineer, Hefner-Alteneck, he took the idea of setting up within a company a separate group of scientifically and technically trained people to devote themselves exclusively to scientific and technical work—something Hefner-Alteneck had pioneered in 1872 in the Siemens Company in Berlin five years after he had joined it as the first college-trained engineer to be hired anywhere by an industrial company. From Thomas Alva Edison, Steinmetz took the *research project*: the systematic organization of research, beginning with a clear definition of the expected end result and identification of the steps in the process and of their sequence.

But Steinmetz then added three features of his own. First, his researchers were to work in teams. Hefner-Alteneck's "designers"—the term *researcher* came much later—had worked the way scientists worked in the nineteenth-century university, each in his own lab with a helper or two who ran errands for the "boss," looked up things for him, or, at most, carried out experiments the boss had specified. In Steinmetz's lab there were seniors and juniors rather than bosses and helpers. They worked as colleagues, each making his own contribution to a common effort. Steinmetz's teams thus required a research director to assign researchers to projects and projects to researchers.

Second, Steinmetz brought together on the same team people of diverse skills and disciplines—engineers, physicists, mathematicians, chemists, even biologists. This was brand-new and heretical. Indeed, it violated the most sacred principle of nineteenth-century scientific organization: the principle of maximum specialization. But the first Nobel Prize awarded to a scientist in industrial research was awarded in 1932 to a chemist, Irvin Langmuir, who worked in Steinmetz's electrotechnical lab.

Finally, Steinmetz's lab radically redefined the relationship between science and technology in research. In setting the goals of his project, Steinmetz identified the new theoretical science needed to accomplish the desired technological results and then organized the appropriate "pure" research to obtain the needed new knowledge. Steinmetz himself was originally a theoretical physicist; on a recent U.S. postage stamp he is being honored for his "contributions to electrical theory." But every one of his "contributions" was the result of research he had planned and specified as part of a project to design and to develop a new product line, for example, fractional horsepower motors. Technology, traditional wisdom held and still widely holds, is "applied science." In Steinmetz's lab science—including the purest of "pure research"—is *technology-driven*, that is, a means to a technological end.

Ten years after Steinmetz completed the General Electric lab, the famed Bell Labs were established on the same pattern. A little later Du Pont followed suit, and then IBM. In developing what eventually became nylon, Du Pont worked out a good deal of the pure science for polymer chemistry. In the 1930s when IBM started to develop what eventually became the computer, it included from the beginning research in switching theory, solid-state physics, and computer logic in its engineering project.

Steinmetz's innovation also led to the "lab without walls," which is America's specific, and major, contribution to very large scientific and technological programs. The first of these, conceived and managed by President Franklin D. Roosevelt's former law partner, Basil O'Connor, was the National Foundation for Infantile Paralysis (March of Dimes), which tackled polio in the early 1930s. This project continued for more than twenty-five years and brought together in a planned, step-by-step effort a large number of scientists from half a dozen disciplines, in a dozen different locations across the country, each working on his own project but within a central strategy and under overall direction. This then established the pattern for the great World War II projects: the RADAR lab, the Lincoln Laboratory and, most massive of them all, the Manhattan Project for atomic energy. Similarly, NASA organized a "research lab without walls" when this country decided, after Sputnik, to put a man on the moon. Steinmetz's technology-driven science is still highly controversial, is indeed anathema to many academic scientists. Still, it is the organization we immediately reach for whenever a new scientific problem emerges, for example, when AIDS suddenly became a major medical problem in 1984–85.

Eurodollar and Commercial Paper

In fewer than twenty years, the financial systems of the world have changed more perhaps than in the preceding two hundred. The change agents were two social innovations: the Eurodollar and the use of commercial paper as a new form of "commercial loan." The first created a new world economy, dominated by the "symbol" economy of capital flows, exchange rates, and credits. The second triggered the "financial revolution" in the United States. It has replaced the old, and seemingly perennial, segmentation of financial institutions into insurance companies, savings banks, commercial banks, stock brokers, and so on, by "financial supermarkets," each focused on whatever financial services the market needs rather than on specific financial products. And this financial revolution is now spreading worldwide.

Neither the Eurodollar nor commercial paper was designed as "revolutionary." The Eurodollar was invented by the Soviet Union's State Bank when General Eisenhower was elected president of the United States in 1952, in the middle of the Korean War. The Russians feared that the new president would embargo their dollar deposits in American banks to force them to stop supporting the North Koreans. They thus hurriedly withdrew these deposits from American banks. Yet they wanted to keep their money in American dollars. The solution they found was the *Eurodollar*: a deposit denominated in U.S. currency but kept in a bank outside the United States. And this then created, within twenty years, a new supranational money and capital market. It is outside and beyond the control of national central banks, and indeed totally unregulated. Yet in its totality—and there are now Euroyen and Euro-Swiss-francs and Euromark in addition to Eurodollars—it is larger in both deposits and turnover than the deposits and turnover of the banking and credit systems of all major trading nations taken together. Indeed, without this innovation on the part of the overseas executives of the Soviet State Bank—every one undoubtedly a good Communist—capitalism might not have survived. It made possible the enormous expansion of world trade, which has been the engine of economic growth and prosperity in the developed free enterprise countries these last thirty years.

At about the same time, perhaps a little later, two American financial institutions—one a brokerage house, Goldman Sachs, the other a finance company, General Electric Credit Corporation (founded to provide credit to the buyers of General Electric's electrical machinery)—hit on the idea of using an old but totally obscure financial instrument, the "commercial paper," as a new form of *commercial loan*, that is, as a substitute for bank credit. Neither institution is allowed under American financial regulations to make commercial loans—only banks are. But *commercial paper*, essentially simply a promise to pay a certain amount at a certain date, is not considered a loan in American law but a security, and this, in turn, banks are not permitted to issue. Economically, however, there is not the slightest difference between the two—something which nobody had, however, seen earlier. By exploiting this legal technicality these two firms, and dozens of others following them in short order, managed to outflank the seemingly impregnable lending monopoly of the commercial banks, especially as credit based on commercial paper can be provided at substantially lower interest rates than banks can lend money against customers' deposits. The banks at first dismissed commercial paper as a mere gimmick. But within fifteen years it had abolished most, if not all, of the demarcation lines and barriers between all kinds of credits and investments in the American economy to the point that, today, practically every financial institution and every financial instrument competes with every other institution and every other financial instrument.

For almost two hundred years economists have considered the financial and credit system to be the central core of an economy, and its most important feature. In every country it is hedged in by laws, rules, and regulations, all designed to preserve the system and to prevent any changes in it. And nowhere was the financial system more highly structured and more carefully guarded than in the United States. Commercial paper—little but a change in a term and almost insignificant as an innovation—has broken through all safeguards of law, regulation, and custom and has subverted the American financial system. We still rank the country's banks. But although New York's Citibank is surely the country's largest bank—and altogether the country's largest "financial institution"—the "number-two bank" is probably not a bank at all, but General Electric Credit Corporation. And Walter Wriston, the long-time chairman of Citibank, points out that

Citibank's biggest competitor in banking and finance is not a financial institution at all but Sears, Roebuck, the country's largest department store chain, which now gives more consumer credit than any credit institution.

Mass and Mass Movement

A third social innovation of this century are mass and mass movements. The *mass* is a collective. It has a behavior of its own and an identity of its own. It is not irrational; on the contrary, it is highly predictable. But its dynamics are what in an individual we would call "the subconscious."

The essence of the *mass movement* is concentration. Its individual "molecules," the individuals who comprise it, are what a chemist calls highly organized and highly charged. They all point in the same direction, all carry the same charge. In the nuclear physicist's terms, the mass is a *critical mass*, that is, the smallest fraction big enough to alter the nature and behavior of the whole.

The mass was first invented—for it was an invention and not just a discovery—in the closing years of the nineteenth century when, exploiting the then brand-new literacy, two Americans, Joseph Pulitzer and William Randolph Hearst, created the first mass medium, the cheap, mass-circulation newspaper. Until then a newspaper was meant to be "written by gentlemen for gentlemen," as the masthead of one of the London newspapers proudly proclaimed for many years. Pulitzer's and Hearst's "yellow press" by contrast was sneered at as "being written by pimps for guttersnipes." But it created a mass readership and a mass following.

These two men and their newspapers then created and led the first modern political mass movement, the campaign to force the United States into the Spanish-American War of 1898. The tactics that these two men developed have since become standard for all mass movements. They did not even try to win over the majority as earlier American movements—the abolitionists or the Free Soilers, for instance—had done. On the contrary, they tried to organize a minority of true believers: they probably never attracted more than 10 percent of the electorate. But they organized this following as a disciplined "shock troop" around the single cause of fighting Spain. They urged their readers in every issue to clip out and mail a postcard to their congressman demanding that America declare war on Spain. And they made a candidate's willingness to commit himself on the war issue the *sole* criterion for endorsing or opposing him regardless of his position on any other issue. Thus, their small minority had the "swing vote" and could claim control of the political future of the candidates. In the end it imposed its will on the great majority, even though almost every opinion maker in the country was opposed.

A mass movement is powerful precisely because the majority has a diversity of interests all over the lot and is thus lukewarm in regard to all of them and zealous in respect to none. The single cause gives the mass movement its discipline and its willingness to follow a leader. It thus makes it stand out and appear much bigger than it really is. It enables a single cause to dominate the news and, indeed, largely to determine what is news. And because it makes its support of parties and candidates totally dependent on their willingness or unwillingness to commit themselves to the single cause, it may cast the deciding vote.

The first to apply what Pulitzer and Hearst had invented to a permanent "crusade" was the temperance movement. For almost a century such temperance groups as the Anti-Saloon League and the Women's Christian Temperance Union had struggled and campaigned without much success. Around 1900 their support was probably at its lowest level since the Civil War. And then they adopted the new tactics of the mass movement. The Women's Christian Temperance Union even hired several of Pulitzer's and Hearst's editors. The "true believers" in Prohibition never numbered more than 5 or 10 percent of the electorate. But in less than twenty years they had Prohibition written into the Constitution.

Since then, single causes—the environment, automobile safety, nuclear disarmament, gay rights, the Moral Majority—have become commonplace. But we are only now beginning to realize how profoundly the single-cause mass movement has changed the politics of all democratic countries.

And outside of the United States the social innovation of the mass has had even greater impacts. The great tyrannies of this century—Lenin's and Stalin's Bolsheviks, Mussolini's Fascism, Hitler's Nazism, even Maoism—are all applications of the mass movement, the highly disciplined single-cause minority of true believers, to the ultimate political goal of gaining and holding power.

Surely no discovery or invention of this century has had greater impact than the social innovation of mass and mass movement. Yet none is less understood.

Indeed, in respect to the mass we are today pretty much where we were in respect to the psychodynamics of the individual a hundred years ago. Of course we knew of the "passions." But they were something one could only explain away as part of "animal nature." They lay outside the *rational*, that is, outside prediction, analysis, and understanding. All one could do was to suppress them. And then, beginning a hundred years ago, Freud showed that the passions have their reasons, that indeed, in Pascal's famous phrase, "the heart has its reasons of which Reason knows nothing." Freud showed that the subconscious is as strictly rational as the conscious, that it has its own logic and its own mechanisms. And although not all psychologists today—indeed, not even the majority—accept the specific causative factors of Freudian psychoanalysis, all accept Freud's psychodynamics of the individual.

But so far we still lack a Sigmund Freud of the mass.

The Farm Agent

The single, most important *economic* event of this century is surely the almost exponential rise in farm production and farm productivity worldwide (except, of course, in the *Soviet Union*). It was brought about primarily through a social innovation of the early years of this century: the farm agent.

Karl Marx, a hundred years ago, had good reason for his dismissal of the "peasant" as hopelessly ignorant and hopelessly unproductive. Indeed, practically every nineteenth-century observer shared Marx's contempt. By 1880, serious, systematic scientific work on agricultural methods and agricultural technology had been going on for two hundred years. Even the systematic training of farmers and agronomists in an "agricultural university" had been started one hundred years earlier. Yet only a very small number of large landowners were paying any attention. The vast majority of farms—practically all American farmers, for instance—did not, in 1880, farm any differently, farm any better, or produce any more than their ancestors had done for centuries. And twenty years later, around 1900, things still had not changed.

And then, suddenly, around the time of World War I—maybe a little later—things changed drastically. The change began in the United States. By now it has spread everywhere; indeed, the surge in farm production and farm productivity has become most pronounced in Third World countries such as India.

What happened was not that farmers suddenly changed their spots. What happened was a social innovation that put the new agricultural knowledge within farmers' reach. Julius Rosenwald, the chief executive of a mail-order company, Sears, Roebuck, himself a Chicago clothing merchant and the purest of "city slickers," invented the farm agent (and for ten years paid farm agents out of his own pocket until the U.S. government took over the Farm Extension Service). He did not do this out of philanthropy alone, but primarily to create purchasing power among his company's customers, the American farmers. The farm agent provided what had hitherto been lacking: a conduit from the steadily increasing pool of agricultural knowledge and information to the practitioners on the farm. And within a few short years the "ignorant, reactionary, tradition-steeped peasant" of Marx's time had become the "farm technologist" of the "scientific revolution on the farm."

Management

My last example of a social innovation is management. "Managers," of course, have been around a long time. The term itself is, however, of twentieth-century coinage. And it is only in this century, and largely within the last fifty years, that management has emerged as a generic function of society, as a distinct kind of work, and as a discipline. A century ago most major tasks, including the economic task we call business, were done mainly in and by the family or by family-run enterprises such as the artisan's small workshop. By now all of them have become organized in institutions: government agency and university, hospital, business enterprise, Red Cross, labor union, and so on. And all of them have to be managed. *Management* is thus the specific function of today's "society of organizations." It is the specific practice that converts a mob into an effective, purposeful, and productive group.

Management and organization have become global rather than Western or capitalist. The Japanese introduced management as an organized discipline in the early 1950s, and it became the foundation for their spectacular economic and social successes. Management is a very hot topic in the Soviet Union. And one of the first moves of the Chinese after the retreat from Maoism was to set up an Enterprise Management Agency in the prime minister's office and import an American-style management school.

The essence of modern organization is to make individual strengths and knowledge productive and to make individual weaknesses irrelevant. In traditional organizations—the ones that built the pyramids and the Gothic cathedrals, or in the armies of the eighteenth and nineteenth centuries—everybody did exactly the same unskilled jobs in which brute strength was the main contribution. Such knowledge as existed was concentrated at the top and in very few heads.

In modern organizations everybody has specialized and fairly advanced knowledge and skill. In the modern organization there are the metallurgist and the Red Cross disaster specialist, the trainer and the tool designer, the fund raiser and the physical therapist, the budget analyst and the computer programmer, all doing their work, all contributing their knowledge, but all working toward a joint end. The little each knows matters; the infinity each doesn't know, does not.

The two cultures today may not be those of the humanist and the scientist as C. P. Snow, the English physicist turned novelist, proclaimed thirty years ago. They may be the cultures of what might be called the *literati* and the *managers*: The one sees reality as ideas and symbols; the other sees reality as performance and people.

Management and organization are still quite primitive. As is common in a rapidly evolving discipline—as was true, for instance, in medicine until fairly recently—the gap between the leading practitioners and the great majority is enormously wide and is closing but slowly. Far too few, even of the most accomplished of today's managers in all organizations, realize that management is defined by responsibility and not by power. Far too few fight the debilitating disease of bureaucracy: the belief that big budgets and a huge staff are accomplishments rather than incompetence.

Still, the impact has been enormous. Management and its emergence have, for instance, rendered equally obsolete both social theories that dominated the nineteenth century and its political rhetoric: the Jeffersonian creed that sees society moving toward a crystalline structure of independent small owners—the yeoman on his forty acres, the artisan in his workshop, the shopkeeper who owns his own store, the independent professional—and the Marxist theorem of a society inexorably turning into an amorphous gas of equally impoverished, equally disenfranchised proletarians. Instead, organization has created an employee society. In the employee society, blue-collar workers are a steadily shrinking minority. Kowledge workers are the new and growing majority—both the main cost and the main resource of all developed societies. And although knowledge workers are employees, they are not proletarians but, through their pension funds, the only capitalists and, collectively, the owners of the means of production.

It is management which in large measure accounts for this century's most extraordinary social phenomenon: the educational explosion. The more highly schooled people are, the more dependent they then become on organizations. Practically all people with schooling beyond high school, in all developed countries—in the United States the figure is 90 percent plus—will spend all their working lives as employees of managed organizations and could not make their living without them. Neither could their teachers.

Conclusion

If this were a history of social innovation in the twentieth century, I would have to cite and to discuss scores of additional examples. But this is not the purpose of this essay. The purpose is not even to show the importance of social innovation. It is, above all, to show that social innovation in the twentieth century has largely become the task of the manager.

This was not always the case; on the contrary, it is quite new.

The act that, so to speak, ushered in the nineteenth century was an innovation: the American Constitution. Constitutions were, of course, nothing new; they go back to ancient Greece. But the American Constitution was different: It first provided expessly a process for its own change. Every earlier constitution had been presumed unchangeable and an "eternal verity." And then the Americans created in the Supreme Court a mechanism to adapt the Constitution to new conditions and demands. These two innovations explain why the American Constitution has survived where all earlier ones perished after a short life of total frustration.

A hundred years later, Prince Bismarck in Germany created, without any precedent, the social innovations we now call *Social Security*—health insurance, old-age pensions, and workmen's compensation insurance against industrial accidents, which were followed a little later by unemployment compensation. Bismarck knew exactly what he was trying to do: defuse a "class war" that threatened to tear asunder the very fabric of society. And he succeeded. Within fifteen years, socialism in Western and Northern Europe had ceased to be "revolutionary" and had become "revisionist."

Outside of the West, the nineteenth century produced the tremendous social innovations of the Meiji Restoration in Japan, which enabled the least Western and most isolated of all contemporary countries both to become a thoroughly "modern" state and nation and to maintain its social and cultural identity.

The nineteenth century was thus a period of very great social innovation. But, with only a few exceptions, social innovtions in the nineteenth century were made by governments. *Invention*, that is, technical discovery, the nineteenth century left to the private sector. Social innovation was governmental and a political act.

Somehow, in this century, government seems to have lost its ability to do effective social innovation. It could still do it in America's New Deal in the 1930s, although the only New Deal innovations that worked were things that had been designed and thoroughly tested earlier, often before World War I, in the large-scale "pilot experiments" conducted by individual states such as Wisconsin, New York, and California. Since then, however, we have had very few governmental innovations in any developed country that have produced the results for which they were designed—very few indeed that have produced any results except massive costs.

Instead, social innovation has been taken over by the private, nongovernmental sector. From being a political act it has become a "managerial task." We still have little methodology for it, though we now do possess a "discipline of innovation." Few social innovators in management yet know what they intend to accomplish the way the American Founding Fathers, Bismarck, and the statesmen of Meiji did—though the examples of Rosenwald's farm agent would indicate that it can be done. Still, social innovation has clearly become management's new dimension.

Part Four

Business as a Social Institution

Introduction to Part Four

There has been a veritable explosion of books on business ethics, sermons on business ethics, college courses on business ethics. And there can be little doubt that ethics are badly needed in business—though no more so than in governments, in police departments, in universities, and in all other places where there is money, power, or sex. But is there such a thing as "Business Ethics?" Can there be such a thing? This essay—it reflects the fact that I once taught philosophy and ethics—was highly unpopular when it was published in 1981 and is just as unpopular today, for it answers the question with a flat "No." There either is—the answer given by Western philosophy—only one ethic, and it is that of the individual; everything else is casuistry and must, as all history proves, rapidly degenerate into finding excuses for even the most unethical and immoral behavior of the powerful. Or—the answer which Eastern and especially Confucian philosophy would give—there can be only the same ethics for every kind of organization, that is for any human group in which people work or live together in a variety of relationships.

This essay, in other words, does not deal with business as "business." It deals with business as one of society's institutions. And so do the other essays in this part. Indeed this has been my approach to business altogether. I am widely known as a "business writer," but, in truth, I never had great interest in business. Having worked with businesses for almost fifty years now I have come to know a good deal about them. But I know just as much about universities, hospitals, churches, government agencies—and for the same reason: I have worked with them for many years. And only one of my twenty-five books is a "business book": *Managing for Results*, 1964—the first book written on what we now call "strategy." It deals exclusively with obtaining economic results in a business. That I began to study organization and management by studying the management of business enterprise—in the early 1940s—was simply because no other institution then was either as visible or as accessible. No other institution would even now invite an outsider to make a close study of its structure, its policies, its values and beliefs. Even as old and well-known an institution as the Catholic Church has only lately permitted an outsider to look at its central institution, the archdiocese, from the inside—and then the student was a priest; I doubt that a layman would have been admitted. But in 1943 a business, General Motors, invited me in; and thus the study of management began as a study of business management, for my 1946 book, *Concept of the Corporation* (to be re-issued soon as a Transaction book), was the first one to study the management of an organization, and the only one for a long time.

Still, business is a separate institution. It has its own purpose and mission and they are different from the purpose and mission of a hospital or university or church or the Red Cross. It has its own values and its own definition of "results." It is, like all our modern institutions, a special-purpose, indeed a single-purpose institution. It exists, however, to perform a *social* function: the provision of goods and services which we call the "economic" function. And as such it has specific social opportunities, specific social challenges. One such is the challenge of enabling individuals to achieve—the challenge of making people productive in work. Another one is the challenge of organizing the productive process so as to obtain maximum yield from society's scarce resources. These social challenges are being discussed in the second and third essay of this part.

14

Can There Be "Business Ethics"?

"Business ethics" is rapidly becoming the "in" subject, replacing yesterday's "social responsibilities." Business ethics is now being taught in departments of philosophy, business schools, and theological seminaries. There are countless seminars on it, speeches, articles, conferences, and books, not to mention the many earnest attempts to write business ethics into the law. But what precisely is business ethics? And what could, or should, it be? Is it just another fad, and only the latest round in the hoary American blood sport of business baiting? Is there more to business ethics than the revivalist preacher's call to the sinner to repent? And if there is indeed something that one could call business ethics and could take seriously, what could it be?

Ethics is, after all, not a recent discovery. Over the centuries philosophers in their struggle with human behavior have developed different approaches to ethics, each leading to different conclusions, indeed to conflicting rules of behavior. Where does business ethics fit in—or does it fit in anywhere at all?

The confusion is so great—and the noise level even greater—that perhaps an attempt might be in order to sort out what business ethics might be, and what it might not be, in the light of the major approaches which philosophers have taken throughout the ages (though my only qualification for making this attempt is that I once, many years before anybody even thought of business ethics, taught philosophy and reli-

Published first in *The Public Interest*, 1981.

gion, and then worked arduously on the tangled questions of "political ethics").

Business Ethics and the Western Tradition

To the moralist of the Western tradition business ethics would make no sense. Indeed, the very term would to him be most objectionable, and reeking of moral laxity. The authorities on ethics disagreed, of course, on what constitutes the grounds of morality—whether they be divine, human nature, or the needs of society. They equally disagreed on the specific rules of ethical behavior; that sternest of moral rules, the Ten Commandments, for instance, thunders "Thou shalt not covet thy neighbor's . . . maidservant." But it says nothing about sexual harassment of one's own women employees, though it was surely just as common then as now.

All authorities of the Western tradition—from the Old Testament prophets all the way to Spinoza in the seventeenth century, to Kant in the eighteenth century, Kierkegaard in the nineteenth century, and, in this century, the Englishman F. H. Bradley (*Ethical Studies*) or the American Edmond Cahn (*The Moral Decision*)—are, however, in complete agreement on one point: There is only one ethics, one set of rules of morality, one code, that of *individual* behavior in which the same rules apply to everyone alike.

A pagan could say, "*Quod licet Jovi non licet bovi.*" He could thus hold that different rules of behavior apply to Jupiter from those that apply to the ox. A Jew or a Christian would have to reject such differentiation in ethics—and precisely because all experience shows that it always leads to exempting the "Jupiters," the great, powerful, and rich, from the rules which "the ox," the humble and poor, has to abide by.

The moralist of the Western tradition accepts "extenuating" and "aggravating" circumstances. He accepts that the poor widow who steals bread to feed her starving children deserves clemency and that it is a more heinous offense for the bishop to have a concubine than for the poor curate in the village. But before there can be extenuating or aggravating circumstances, there has to be an offense. And the offense is the same for rich and poor, for high and low alike—theft is theft, concubinage is concubinage. The reason for this insistence on a code that considers only the individual, and not his status in life or society, is precisely that otherwise the mighty, the powerful, the successful will gain exemption from the laws of ethics and morality.

The only differences between what is ethically right and ethically wrong behavior which traditional moralists, almost without exception, would accept—would indeed insist on—are differences grounded in social or cultural mores, and then only in respect to venial offenses, that is, the way things are done rather than the substance of behavior. Even in the most licentious society, fidelity to the marriage vow is meritorious, all moralists would agree; but the sexual license of an extremely permissive society, say seventeenth century Restoration England or late twentieth century America, might be considered an extenuating circumstance for the sexual transgressor. And even the sternest moralist has always insisted that, excepting only true matters of conscience, practices that are of questionable morality in one place and culture might be perfectly acceptable—and indeed might be quite ethical—in another cultural surrounding. Nepotism may be considered of dubious morality in one culture, in today's United States, for instance. In other cultures, a traditional Chinese one, for example, it may be the very essence of ethical behavior, both by satisfying the moral obligation to one's family and by making disinterested service to the public a little more likely.

But—and this is the crucial point—these are qualifications to the fundamental axiom on which the Western tradition of ethics has always been based: There is only one code of ethics, that of individual behavior, for prince and pauper, for rich and poor, for the mighty and the meek alike. Ethics, in the Judaeo-Christian tradition, is the affirmation that all men and women are alike creatures—whether the Creator be called God, Nature, or Society.

And this fundamental axiom business ethics denies. Viewed from the mainstream of traditional ethics, business ethics is not ethics at all, whatever else it may be. For it asserts that acts that are not immoral or illegal if done by ordinary folk become immoral or illegal if done by business.

One blatant example is the treatment of extortion in the current discussions of business ethics. No one ever has had a good word to say for extortion, or has advocated paying it. But if you and I are found to have paid extortion money under threat of physical or material harm, we are not considered to have behaved immorally or illegally. The extortioner is both immoral and a criminal. If a business submits to extortion, however, current business ethics considers it to have acted unethically. There is no speech, article, book, or conference on business ethics, for instance, which does not point an accusing finger in great indignation at Lockheed for giving in to a Japanese airline company, which extorted money as a prerequisite to considering the purchase of Lockheed's faltering L-1011 jet plane. There was very little difference between Lockheed's paying the Japanese and the pedestrian in New York's Central Park handing his wallet over to a mugger. Yet no one would consider the pedestrian to have acted unethically.

Similarly, in Senate confirmation hearings, one of President Reagan's cabinet appointees was accused of "unethical practices" and investigated for weeks because his New Jersey construction company was alleged to have paid money to union goons under the threat of their beating up the employees, sabotaging the trucks, and vandalizing the building sites. The accusers were self-confessed labor racketeers; no one seemed to have worried about their ethics.

One can argue that both Lockheed and the New Jersey builder were stupid to pay the holdup men. But as the old saying has it: "Stupidity is not a court martial offense." Under the new business ethics, it does become exactly that, however. And this is not compatible with what ethics always was supposed to be.

The new business ethics also denies to business the adaptation to cultural mores which has always been considered a moral duty in the traditional approach to ethics. It is now considered grossly unethical—indeed it may even be a questionable practice if not criminal offense—for an American business operating in Japan to retain as a counselor the distinguished civil servant who retires from his official position in the Japanese government. Yet the business that does not do this is considered in Japan to behave antisocially and to violate its clear ethical duties. Business taking care of retired senior civil servants, the Japanese hold, makes possible two practices they consider essential to the public interest: that a civil servant past age forty-five must retire as soon as he is outranked by anyone younger than he; and that governmental salaries and retirement pensions—and with them the burden of the bureaucracy on the taxpayers—be kept low, with the difference between what a first-rate man gets in government service and what he might earn in private employment made up after his retirement through his counselor's fees. The Japanese maintain that the expectation of later on being a counselor encourages a civil servant to remain incorruptible, impartial, and objective, and thus to serve only the public good; his counselorships are obtained for him by his former ministry and its recommendation depends on his rating by his colleagues as a public servant. The Germans, who follow a somewhat similar practice—with senior civil servants expected to be taken care of through appointment as industry-association executives—share this conviction. Yet, despite the fact that both the Japanese and the German systems seem to serve their respective societies well and indeed honorably, and even despite the fact that it is considered perfectly ethical for American civil servants of equal rank and caliber to move into well-paid executive jobs in business and foundations and into even more lucrative law practices, the American company in Japan that abides by a practice the Japanese consider the very essence of social responsibility, is pilloried in the present discussion of business ethics as a horrible example of unethical practices.

Surely business ethics assumes that for some reason the ordinary rules of ethics do not apply to business. Business ethics, in other words, is not ethics at all, as the term has commonly been used by Western philosophers and Western theologians. What is it then?

Casuistry: The Ethics of Social Responsibility

"It's casuistry," the historian of Western philosophy would answer. Casuistry asserted that rulers, because of their responsibility, have to strike a balance between the ordinary demands of ethics which apply to them as individuals and their social responsibility to their subjects, their kingdom—or their company.

Casuistry was first propounded in Calvin's *Institutes*, then taken over by the Catholic theologians of the Counter-Reformation (Bellarmin, for instance, or St. Charles Borromeus) and developed into a political ethics by their Jesuit disciples in the seventeenth century.

Casuistry was the first attempt to think through social responsibility and to embed it in a set of special ethics for those in power. In this respect, business ethics tries to do exactly what the casuists did 300 years ago. And it must end the same way. If business ethics continues to be casuistry its speedy demise in a cloud of ill-repute can be confidently predicted.

To the casuist the social responsibility inherent in being a ruler—that is, someone whose actions have impact on others—is by itself an ethical imperative. As such, the ruler has a duty, as Calvin first laid down, to subordinate his individual behavior and his individual conscience to the demands of his social responsibility.

The *locus classicus* of casuistry is Henry VIII and his first marriage to Catherine of Aragon. A consummated marriage—and Catherine of Aragon had a daughter by Henry, the future "Bloody Mary"—could not be dissolved except by death, both Catholic and Protestant theologians agreed. In casuistry, however, as both Catholics and Protestants agreed, Henry VIII had an ethical duty to seek annulment of the marriage. Until his father, well within living memory, had snatched the Crown by force of arms, England had suffered a century of bloody and destructive civil war because of the lack of a legitimate male heir. Without annulment of his marriage, Henry VIII, in other words, exposed his country and its people to mortal danger, well beyond anything he could in conscience justify. The one point on which Protestants and Catholics disagreed was whether the Pope also had a social, and thereby an ethical, responsibility to grant Henry's request. By not granting it, he drove the king and his English subjects out of the Catholic Church. But had he granted the annulment, the Catholic casuists argued, the Pope would have driven Catherine's uncle, the Holy Roman Emperor, out of the Church and into the waiting arms of an emerging Protestantism; and that would have meant that instead of assigning a few million Englishmen to heresy, perdition, and hellfire, many times more souls—all the people in all the lands controlled by the Emperor, that is, in most of Europe—could have been consigned to everlasting perdition.

This may be considered a quaint example—but only because our time judges behavior by economic rather than theological absolutes. The example illustrates what is wrong with casuistry and indeed why it must fail as an approach to ethics. In the first place casuistry must end up becoming politicized, precisely because it considers social responsibility an ethical absolute. In giving primacy to political values and goals it subordinates ethics to politics. Clearly this is the approach business ethics today is taking. Its very origin is in politics rather than in ethics. It expresses a belief that the responsibility which business and the business executive have, precisely because they have social impact, must determine ethics—and this is a political rather than an ethical imperative.

Equally important, the casuist inevitably becomes the apologist for the ruler, the powerful. Casuistry starts out with the insight that the behavior of rulers affects more than themselves and their families. It thus starts out by making demands on the ruler—the starting point for both Calvin and his Catholic disciples in the Counter-Reformation fifty years later. It then concludes that rulers must, therefore, in conscience and ethics, subordinate their interests, including their individual morality, to their social responsibility. But this implies that the rules which decide what is ethical for ordinary people do not apply equally, if at all, to those with social responsibility. Ethics for them is instead a cost-benefit calculation involving the demands of individual conscience and the demands of position—and that means that the rulers are exempt from the demands of ethics, if only their behavior can be argued to confer benefits on other people. And this is precisely how business ethics is going.

Indeed, under casuist analysis the ethical violations which to most present proponents of business ethics appear the most heinous crimes turn out to have been practically saintly.

Take Lockheed's bribe story for instance. Lockheed was led into paying extortion money to a Japanese airline by the collapse of the supplier of the engines for its wide-bodied L-1011 passenger jet, the English Rolls Royce Company. At that time Lockheed employed some 25,000 people making L-1011s, most of them in southern California which then, 1972–73, was suffering substantial unemployment from sharp cutbacks in defense orders in the aerospace industry. To safeguard the 25,000 jobs, Lockheed got a large government subsidy. But to be able to maintain these jobs, Lockheed needed at least one large L-1011 order from one major airline. The only one among the major airlines not then committed to a competitor's plane was All-Nippon Airways in Japan. The self-interest of Lockheed Corporation and of its stockholders would clearly have demanded speedy abandonment of the L-1011. It was certain that it would never make money—and it has not made a penny yet. Jettisoning the L-1011 would immediately have boosted Lockheed's earnings, maybe doubled them. It would have immediately boosted Lockheed's share price; stock market analysts and investment bankers pleaded with the firm to get rid of the albatross. If Lockheed had abandoned the L-1011, instead of paying extortion money to the Japanese for ordering a few planes and thus keeping the project alive, the company's earnings, its stock price, and the bonuses and stock options of top management, would immediately have risen sharply. Not to have paid extortion money to the Japanese would to a casuist, have been self-serving. To a casuist, paying the extortion money was a duty and social responsibility to which the self-interest of the company, its shareholders, and its executives had to be subordinated. It was the discharge of social responsibility of the ruler to keep alive the jobs of 25,000 people at a time when jobs in the aircraft industry in southern California were scarce indeed.

Similarly, the other great horror story of business ethics would, to the casuist, appear as an example of business virtue if not of unselfish business martyrdom. In the "electrical apparatus conspiracy" of the late 1950s, several high-ranking General Electric executives were sent to jail. They were found guilty of a criminal conspiracy in violation of antitrust because orders for heavy generating equipment, such as turbines, were parceled out among the three electrical apparatus manufacturers in the United States—General Electric, Westinghouse, and Allis Chalmers. But this "criminal conspiracy" only served to reduce General Electric's sales, its profits, and the bonuses and stock options of the General Electric executives who took part in the conspiracy. Since the electric apparatus cartel was destroyed by the criminal prosecution of the General Electric executives, General Electric sales and profits in the heavy apparatus field have sharply increased, as has market penetration by the company, which now has what amounts to a near-monopoly. The purpose of the cartel—which incidentally was started under federal government pressure in the Depression years to fight unemployment—was the protection of the weakest and most dependent of the companies, Allis Chalmers (which is located in Milwaukee, a depressed and declining old industrial area.) As soon as government action destroyed the cartel, Allis Chalmers had to go out of the turbine business and had to lay off several thousand people. And while there is still abundant competition in the world market for heavy electric apparatus, General Electric now enjoys such market dominance in the home market that the United States, in case of war, would not have major alternative suppliers of so critical a product as turbines.

The casuist would agree that cartels are both illegal and considered immoral in the United Stats—although not necessarily anyplace else in the world. But he would also argue that the General Electric executive who violated U.S. law had an ethical duty to do so under the higher law of social responsibility to safeguard both employment in the Milwaukee area and the defense-production base of the United States.

The only thing that is surprising about these examples is that business has not yet used them to climb on the casuist bandwagon of business ethics. For just as almost any behavior indulged in by the seventeenth century ruler could be shown to be an ethical duty by the seventeenth century disciples of Calvin, of Bellarmin, and of Borromeus, so almost any behavior of the executive in organizations today—whether in a business, a hospital, a university, or a government agency—could be shown to be his ethical duty under the casuistic cost-benefit analysis between individual ethics and the demands of social responsibility. There are indeed signs aplenty that that most apolitical of rulers, the American business executive, is waking up to the political potential of business ethics. Some of the advertisements which large companies—Mobil Oil, for example—are now running to counter the attacks made on them in the name of social responsibility and business ethics, clearly use the casuist approach to defend business, and indeed to counterattack. But if business ethics becomes a tool to defend as ethical acts on the part of executives that would be condemned if committed by anyone else, the present proponents of business ethics, like their casuist predecessors 400 years ago, will have no one to blame but themselves.

Casuistry started out as high morality. In the end, its ethics came to be summed up in two well-known pieces of cynicism: "An ambassador is an honest man, lying abroad for the good of his country," went a well-known eighteenth century pun. And a hundred years later, Bismarck said, "What a scoundrel a minister would be if, in his own private life, he did half the things he has a duty to do to be true to his oath of office."

Long before that, however, casuistry had been killed off by moral revulsion. Its most lasting memory perhaps are the reactions to it which reestablished ethics in the West as a universal system, binding the individual regardless of station, function, or social responsibility: Spinoza's *Ethics*, and the *Provincial Letters* of his contemporary, Blaise Pascal. But also—and this is a lesson that might be pondered by today's proponents of business ethics, so many of whom are clergymen—it was their embracing casuistry that made the Jesuits hated and despised, made "Jesuitical" a synonym of immoral, and led to the Jesuit order being suppressed by the Pope in the eighteenth century. And it is casuistry, more than anything else, that has caused the anticlericalism of the intellectuals in Catholic Europe.

Business ethics undoubtedly is a close parallel to casuistry. Its origin is political, as was that of casuistry. Its basic thesis, that ethics for the ruler, and especially for the business executive, has to express social responsibility is exactly the starting point of the casuist. But if business ethics is casuistry, then it will not last long—and long before it dies, it will have become a tool of the business executive to justify what for other people would be unethical behavior, rather than a tool to restrain the business executive and to impose tight ethical limits on business.

The Ethics of Prudence and Self-Development

There is one other major tradition of ethics in the West, the Ethics of Prudence. It goes all the way back to Aristotle and his enthronement of prudence as a cardinal virtue. It continued for almost 2,000 years in the popular literary tradition of the "Education of the Christian Prince," which reached its ultimate triumph and its reduction to absurdity in Machiavelli's *Prince*. Its spirit can best be summed up by the advice which then-Senator Harry Truman gave to an Army witness before his committee in the early years of World War II: "Generals should never do anything that needs to be explained to a Senate Committee—there is nothing one can explain to a Senate Committee."

"Generals," whether the organization is an army, a corporation, or a university, are highly visible. They must expect their behavior to be seen, scrutinized, analyzed, discussed, and questioned. Prudence thus demands that they shun actions that cannot easily be understood, explained, or justified. But generals, being visible, are also examples. They are leaders by their very position and visibility. Their only choice is whether their example leads others to right action or to wrong action. Their only choice is between direction and misdirection, between leadership and misleadership. They thus have an ethical obligation to give the example of right behavior and to avoid giving the example of wrong behavior.

The Ethics of Prudence does not spell out what "right" behavior is. They assume that what is wrong behavior is clear enough—and if there is any doubt, it is "questionable" and to be avoided. Prudence makes it an ethical duty for the leader to exemplify the precepts of ethics in his own behavior.

And by following prudence, everyone regardless of status becomes a leader, a superior man and will "fulfill himself," to use the contemporary idiom. One becomes the superior man by avoiding any act which would make one the kind of person one does not want to be, does not respect, does not accept as superior. "If you don't want to see a pimp when you look in the shaving mirror in the morning, don't hire call girls the night before to entertain congressmen, customers, or salesmen." On any other basis, hiring call girls may be condemned as vulgar and tasteless, and may be shunned as something fastidious people do not do. It may be frowned upon as uncouth. It may even be illegal. But only in prudence is it ethically relevant. This is what Kierkegaard, the sternest moralist of the nineteenth century, meant when he said that aesthetics is the true ethics.

The Ethics of Prudence can easily degenerate. Concern with what one can justify becomes, only too easily, concern with appearances— Machiavelli was by no means the first to point out that in a Prince, that is, in someone in authority and high visibility, appearances may matter more than substance. The Ethics of Prudence thus easily decays into the hypocrisy of public relations. Leadership through right example easily degenerates into the sham of charisma and into a cloak for misdirection and misleadership—it is always the Hitlers and the Stalins who are the great charismatic leaders. And fulfilment through self-development into a superior person—what Kierkegaard called "becoming a Christian"—may turn either into the smugness of the Pharisee who thanks God that he is not like other people, or into self-indulgence instead of self-discipline, moral sloth instead of self-respect, and into saying "I like," rather than "I know."

Yet, despite these degenerative tendencies, the Ethics of Prudence is surely appropriate to a society of organizations. Of course, it will not be business ethics—it makes absolutely no difference in the Ethics of

Prudence whether the executive is a general in the Army, a bureau chief in the Treasury Department in Washington, a senator, a judge, a senior vice president in a bank, or a hospital administrator. But a society of organizations is a society in which an extraordinarily large number of people are in positions of high visibility, if only within one organization. They enjoy this visibility not, like the Christian Prince, by virtue of birth, nor by virtue of wealth—that is, not because they are personages. They are functionaries and important only through their responsibility to take right action. But this is exactly what the Ethics of Prudence is all about.

Similarly, executives set examples, whatever the organization. They set the tone, create the spirit, decide the values for an organization and for the people in it. They lead or mislead, in other words. And they have no choice but to do one or the other. Above all, the ethics or aesthetics of self-development would seem to be tailor-made for the specific dilemma of the executive in the modern organization. By himself he is a nobody and indeed anonymous. A week after he has retired and has left that big corner office on the twenty-sixth floor of his company's skyscraper or the Secretary's six-room corner suite on Constitution Avenue, no one in the building even recognizes him anymore. And his neighbors in the pleasant suburb in which he lives in a comfortable middle-class house—very different from anything one might call a palace—only know that "Joe works someplace on Park Avenue" or "does something in the government." Yet collectively these anonymous executives are the leaders in a modern society. Their function demands the self-discipline and the self-respect of the superior man. To live up to the performance expectations society makes upon them, they have to strive for self-fulfilment rather than be content with lackadaisical mediocrity. Yet at the pinnacle of their career and success, they are still cogs in an organization and easily replaceable. And this is exactly what self-fulfilment in ethics, the Kierkegaardian "becoming a Christian," concerns itself with: how to become the superior man, important and autonomous, without being a big shot let alone a Prince.

One would therefore expect the discussion of business ethics to focus on the Ethics of Prudence. Some of the words, such as to "fulfill oneself," indeed sound the same, though they mean something quite different. But by and large, the discussion of business ethics, even if more sensibly concerning itself with the ethics of organization, will have nothing to do with prudence.

The reason is clearly that the Ethics of Prudence is the ethics of authority. And while today's discussion of business ethics (or of the ethics of university administration, of hospital administration, or of government) clamors for responsibility, it rejects out of hand any authority and, of course, particularly any authority of the business executive. Authority is not legitimate; it is "elitism." But there can be no responsibility where authority is denied. To deny it is not anarchism nor radicalism, let alone socialism. In a child, it is called a temper tantrum.

The Ethics of Interdependence

Casuistry was so thoroughly discredited that the only mention of it to be found in most textbooks on the history of philosophy is in connection with its ultimate adversaries—Spinoza and Pascal. Indeed, only ten or fifteen years ago, few if any philosophers would have thought it possible for anything like "business ethics" to emerge. "Particularist ethics," a set of ethics that postulates that this or that group is different in its ethical responsibilities from everyone else, would have been considered doomed forever by the failure of casuistry. Ethics, almost anyone in the West would have considered axiomatic, would surely always be ethics of the individual and independent of rank and station.

But there is another, non-Western ethics that is situational. It is the most successful and most durable ethics of them all: the Confucian ethics of interdependence.

Confucian ethics elegantly sidesteps the trap into which the casuists fell; it is a universal ethics, in which the same rules and imperatives of behavior hold for every individual. There is no social responsibility overriding individual conscience, no cost-benefit calculation, no greater good or higher measure than the individual and his behavior, and altogether no casuistry. In the Confucian ethics, the rules are the same for all. But there are different general rules, according to the five basic relationships of interdependence, which for the Confucian embrace the totality of individual interactions in civil society: superior and subordinate (or master and servant); father and child; husband and wife; oldest brother and sibling; friend and friend. Right behavior—what in the English translation of Confucian ethics is usually called "sincerity"[1]— is that individual behavior which is truly appropriate to the specific relationship of mutual dependence because it optimizes benefits for both parties. Other behavior is insincere and therefore wrong behavior and unethical. It creates dissonance instead of harmony, exploitation instead of benefits, manipulation instead of trust.

An example of the Confucian approach to the ethical problems discussed under the heading of business ethics would be sexual harassment. To the Confucian it is clearly unethical behavior because it injects power into a relationship that is based on function. This makes it exploitation. That this insincere—that is, grossly unethical—behavior on the part of a superior takes place within a business or any other kind of organization, is basically irrelevant. The master/servant or superior/subordinate relationship is one between individuals. Hence, the Confucian would make no distinction between a general manager forcing his secretary into sexual intercourse and Mr. Samuel Pepys, England's famous seventeenth-century diarist, forcing his wife's maids to submit to his amorous advances. It would not even make much difference to the Confucian that today's secretary can, as a rule, quit without suffering more than inconvenience if she does not want to submit, whereas the poor wretches in Mrs. Pepys' employ ended up as prostitutes, either because they did not submit and were fired and out on the street, or because they did submit and were fired when they got pregnant. Nor would the Confucian see much difference between a corporation vice president engaging in sexual harassment and a college professor seducing coeds with implied promises to raise their grades.

And, finally, it would be immaterial to the Confucian that the particular insincerity involves sexual relations. The superior would be equally guilty of grossly unethical behavior and violation of fundamental rules of conduct if, as a good many of the proponents of business ethics ardently advocate, he were to set himself up as a mental therapist for his subordinates and help them to adjust. No matter how benevolent his intentions, this is equally incompatible with the integrity of the superior/subordinate relationship. It equally abuses rank based on function and imposes power. It is therefore exploitation whether done because of lust for power or manipulation or done out of benevolence—either way it is unethical and destructive. Both sexual relations and the healer/patient relationship must be free of rank to be effective, harmonious, and ethically correct. They are constructive only as friend to friend or as husband to wife relations, in which differences in function confer no rank whatever.

This example makes it clear, I would say, that virtually all the concerns of business ethics, indeed almost everything business ethics considers a problem, have to do with relationships of interdependence, whether that between the organization and the employee, the manufacturer and the customer, the hospital and the patient, the university and the student, and so on.

Looking at the ethics of interdependence immediately resolves the conundrum which confounds the present discussion of business ethics: What difference does it make whether a certain act or behavior takes place in a business, in a nonprofit organization, or outside any organization at all? The answer is clear: none at all. Indeed the questions that are so hotly debated in today's discussion of business ethics, such as whether changing a hospital from "nonprofit" to "proprietary and for profit" will affect either its behavior or the ethics pertaining to it, the most cursory exposure to the ethics of interdependence reveals as sophistry and as nonquestions.

The ethics of interdependence thus does address itself to the question which business ethics tries to tackle. But today's discussion, explicitly or implicitly, denies the basic insight from which the ethics of interdependence starts and to which it owes its strength and durability: It denies *interdependence*.

The ethics of interdependence, as Confucian philosophers first codified it shortly after their Master's death in 479 B.C., considers illegitimate and unethical the injection of power into human relationships. It asserts that interdependence demands equality of obligations. Children owe obedience and respect to their parents. Parents, in turn, owe affection, sustenance and, yes, respect, to their children. For every paragon of filial piety in Confucian hagiology, such as the dutiful daughter, there is a paragon of parental sacrifice, such as the loving father who sacrificed his brilliant career at the court to the care of his five children and their demands on his time and attention. For every minister who risks his job, if not his life, by fearlessly correcting an Emperor guilty of violating harmony, there is an Emperor laying down his life rather than throw a loyal minister to the political wolves.

In the ethics of interdependence there are only obligations, and all obligations are mutual obligations. Harmony and trust—that is, interdependence—require that each side be obligated to provide what the other side needs to achieve its goals and to fulfill itself.

But in today's American—and European—discussion of business ethics, ethics means that one side has obligations and the other side has rights, if not entitlements. This is not compatible with the ethics of interdependence and indeed with any ethics at all. It is the politics of power, and indeed the politics of naked exploitation and repression. And within the context of interdependence the exploiters and the oppressors are not the bosses, but the ones who assert their rights rather than accept mutual obligation, and with it, equality. To redress the balance in a relationship of interdependence—or at least so the ethics of interdependence would insist—demands not pitting power against power or right against right, but matching obligation to obligation.

To illustrate: Today's ethics of organization debate pays great attention to the duty to be a "whistle-blower" and to the protection of the whistle-blower against retaliation or suppression by his boss or by his organization. This sounds high-minded. Surely, the subordinate has a right, if not indeed a duty, to bring to public attention and remedial action his superior's misdeeds, let alone violation of the law on the part of a superior or of his employing organization. But in the context of the ethics of interdependence, whistle-blowing is ethically quite ambiguous. To be sure, there are misdeeds of the superior or of the employing organizations which so grossly violate propriety and laws that the subordinate (or the friend, or the child, or even the wife) cannot remain silent. This is, after all, what the word "felony" implies; one becomes a partner to a felony and criminally liable by not reporting, and thus compounding it. But otherwise? It is not primarily that to encourage whistle-blowing corrodes that bond of trust that ties the superior to the subordinate. Encouraging the whistle-blower must make the subordinate lose his trust in the superior's willingness and ability to protect his people. They simply are no longer his people and become potential enemies or political pawns. And in the end, encouraging and indeed even permitting whistle-blowers always makes the weaker one—that is, the subordinate—powerless against the unscrupulous superior, simply because the superior no longer can recognize or meet his obligation to the subordinate.

Whistle-blowing, after all, is simply another word for informing. And perhaps it is not quite irrelevant that the only societies in Western history that encouraged informers were bloody and infamous tyrannies—Tiberius and Nero in Rome, the Inquisition in the Spain of Philip II, the French Terror, and Stalin. It may also be no accident that Mao, when he tried to establish a dictatorship in China, organized whistle-blowing on a massive scale. For under whistle-blowing, under the regime of the informer, no mutual trust, no interdependencies, and no ethics are possible. And Mao only followed history's first totalitarians, the "Legalists" of the third century B.C., who suppressed Confucius and burned his books because he had taught ethics and had rejected the absolutism of political power.

The limits of mutual obligation are indeed a central and difficult issue in the ethics of interdependencies. But to start out, as the advocates of whistle-blowing do, with the assumption that there are only rights on one side, makes any ethics impossible. And if the fundamental problem of ethics is the behavior in relations of interdependence, then obligations have to be mutual and have to be equal for both sides. Indeed, in a relationship of interdependence it is the mutuality of obligation that creates true equality, regardless of differences in rank, wealth, or power.

Today's discussion of business ethics stridently denies this.It tends to assert that in relations of interdependence one side has all the duties and the other one all the rights. But this is the assertion of the Legalist, the assertion of the totalitarians who shortly end up by denying all ethics. It must also mean that ethics becomes the tool of the powerful. If a set of ethics is one-sided, then the rules are written by those that have the position, the power, the wealth. If interdependence is not equality of obligations, it becomes domination.

Looking at business ethics as an ethics of interdependence reveals an additional and equally serious problem—indeed a *more* serious problem.

Can an ethics of interdependence be anything more than ethics for individuals? The Confucians say no—a main reason why Mao outlawed them. For the Confucian—but also for the philosopher of the

Western tradition—only *law* can handle the rights and objections of collectives. *Ethics* is always a matter of the person.

But is this adequate for a society of organizations such as ours? This may be the central question for the philosopher of modern society, in which access to livelihood, career, and achievement exist primarily in and through organizations—and especially for the highly educated person for whom opportunities outside of organization are very scarce indeed. In such a society, both the society and the individual increasingly depend on the performance, as well as the sincerity, of organizations.

But in today's discussion of business ethics it is not even seen that there is a problem.

"Ethical Chic" or Ethics

Business ethics, this discussion should have made clear, is to ethics what soft porn is to the Platonic Eros; soft porn too talks of something it calls "love." And insofar as business ethics comes even close to ethics, it comes close to casuistry and will, predictably, end up as a fig leaf for the shameless and as special pleading for the powerful and the wealthy.

Clearly, one major element of the peculiar stew that goes by the name of business ethics is plain old-fashioned hostility to business and to economic activity altogether—one of the oldest of American traditions and perhaps the only still-potent ingredient in the Puritan heritage. Otherwise, we would not even talk of business ethics. There is no warrant in any ethics to consider one major sphere of activity as having its own ethical problems, let alone its own ethics. Business or economic activity may have special political or legal dimensions as in "business and government," to cite the title of a once-popular college course, or as in the antitrust laws. And business ethics may be good politics or good electioneering. But that is all. For ethics deals with the right actions of individuals. And then it surely makes no difference whether the setting is a community hospital, with the actors a nursing supervisor and the consumer a patient, or whether the setting is National Universal General Corporation, the actors a quality control manager, and the consumer the buyer of a bicycle.

But one explanation for the popularity of business ethics is surely also the human frailty of which Pascal accused the casuists of his day: the lust for power and prominence of a clerisy sworn to humility. Business ethics is fashionable, and provides speeches at conferences, lecture fees, consulting assignments, and lots of publicity. And surely business ethics, with its tales of wrongdoing in high places, caters also to the age-old enjoyment of society gossip and to the prurience which—it was, I believe, Rabelais who said it—makes it fornication when a peasant has a toss in the hay and romance when the prince does it.

Altogether, business ethics might well be called ethical chic rather than ethics—and indeed might be considered more a media event than philosophy or morals.

But this discussion of the major approaches to ethics and of their concerns surely also shows that ethics has as much to say to the individual in our society of organizations as they ever had to say to say to the individual in earlier societies. They are just as important and just as needed nowadays. And they surely require hard and serious work.

A society of organizations is a society of interdependence. The specific relationship which the Confucian philosopher postulated as universal and basic may not be adequate, or even appropriate, to modern society and to the ethical problems within the modern organization and between the modern organization and its clients, customers, and constituents. But the fundamental concepts surely are. Indeed, if there ever is a viable ethics of organization, it will almost certainly have to adopt the key concepts which have made Confucian ethics both durable and effective:

—clear definition of the fundamental relationships;

—universal and general rules of conduct—that is, rules that are binding on any one person or organization, according to its rules, function, and relationships;

—focus on right behavior rather than on avoiding wrongdoing, and on behavior rather than on motives or intentions. And finally,

—an effective organization ethic, indeed an organization ethic that deserves to be seriously considered as ethics, will have to define right behavior as the behavior which optimizes each party's benefits and thus makes the relationship harmonious, constructive, and mutually beneficial.

But a society of organizations is also a society in which a great many people are unimportant and indeed anonymous by themselves, yet are highly visible, and matter as leaders in society. And thus it is a society that must stress the Ethics of Prudence and self-development. It must expect its managers, executives, and professionals to demand of themselves that they shun behavior they would not respect in others, and instead practice behavior appropriate to the sort of person they would want to see in the mirror in the morning.

Note

1. No word has caused more misunderstanding in East/West relations than "sincerity." To a Westerner, sincerity means "words that are true to conviction and feelings"; to an Easterner, sincerity means "actions that are appropriate to a specific relationship and make it harmonious and of optimum mutual benefit." For the Westerner, sincerity has to do with intentions, that is, with morality; to the Easterner, sincerity has to do with behavior, that is, with ethics.

15

The New Productivity Challenge

The productivity of the newly dominant groups in the work force, knowledge workers and service workers, will be the biggest and toughest challenge facing managers in the developed countries for decades to come. And serious work on this daunting task has only begun.

Productivity in making and moving things—in manufacturing, farming, mining, construction, and transportation—has increased at an annual rate of 3 to 4 percent compound for the last 125 years—for a 45-fold expansion in overall productivity in the developed countries. On this productivity explosion rests all the increases in these countries in both the standard of living and the quality of life. It provided the vast increase in disposable incomes and purchasing power. But between a third and a half of its fruits were taken in the form of leisure—something known only to aristocrats or to the idle rich before 1914, when everybody else worked at least 3,000 hours a year. (Now even the Japanese work no more than 2,000 hours a year, Americans around 1,800, West Germans 1,650.) The productivity explosion also paid for the ten-fold expansion of education and for the even greater expansion of health care. Productivity has become the "wealth of nations."

Rising productivity was something so totally unprecedented that there was no term for it in any language. For Karl Marx—as for all nineteenth-century economists—it was axiomatic that worker output could be increased only by working harder or longer—all of Marx is

First published in *The Harvard Business Review*, 1991.

based on this belief. But though Frederic Winslow Taylor (1856–1915) disproved the axiom by starting to work on productivity in the early 1880s—his first substantial results came in 1883, the very year Marx died—he himself never knew the term. It did not come into general use until World War II—and then at first only in the United States. The 1950 edition of the most authoritative English dictionary, the *Concise Oxford*, did not yet list *productivity* in its present meaning. By now it is commonplace that production is the true competitive advantage.

The productivity explosion was arguably the most important social event of the past hundred years, and one that had no precedent in history. There have always been rich people and poor people. But as recently as 1850 the poor in China were not measurably worse off than those in the slums of London or Glasgow. And the average income in the richest country of 1910 was at most three times the average income of the then poorest countries—now it is twenty to forty times as large, even without counting leisure, education, or health care. Before the productivity explosion it took at least fifty years for a country to become "developed." South Korea—as late as 1955 one of the world's truly "backward" countries—did it in twenty years. This radical reversal of what has been the norm since time immemorial is in its entirety the result of the productivity revolution that started in the United States around 1870 or 1880.

Productivity in making and moving things is still going up at the same annual rate. It is going up—contrary to popular belief—fully as much in the United States as in Japan or West Germany. Indeed, the current productivity increase in U.S. farming—4½ to 5 percent a year—is far and away the biggest increase recorded anywhere at any time. And the U.S. productivity increase in manufacturing during the 1980s—3.9 percent a year—was in absolute terms actually larger than the corresponding annual increase in Japanese and German manufacturing, for the U.S. base is still quite a bit higher.

But in the developed countries the productivity revolution is over. There are simply not enough people employed making and moving things for their productivity to be decisive. They now account for no more than a fifth of the work force in a developed economy—only thirty years ago they were still a near-majority. And the productivity of the people who do make the difference—knowledge workers and service workers—is not going up. In some areas it is actually going down. Salespeople in the department stores of all developed countries now sell, adjusted for inflation, no more than two thirds of what they sold in 1929. And few people would argue, I submit, that the teacher of 1991 is more productive than the teacher of 1901.

Knowledge workers and service workers range from research scientist and cardiac surgeon through draftsman and store manager to sixteen-year-olds who work as car hops in the fast-food drive-in for a few Saturday afternoon hours. They even include large numbers of people who actually work as machine operators—washing dishes in a restaurant, polishing floors in a hospital, pushing computer keys in a claims-settling department of a insurance company. Yet knowledge workers and service workers, for all their diversity, are remarkably alike in what *does not work* in raising productivity. But there are also important similarities in what *does work* for these workers, no matter how much they might otherwise differ in knowledge, skill responsibility, social status, and pay.

2

The first thing we have learned—and it came as a rude shock—is that capital cannot be substituted for labor (i.e., for people) in knowledge and service work. Nor does new technology by itself generate higher productivity in such work. In making and moving things, capital and technology are *factors of production*, to use the economist's term. In knowledge and service work, they are *tools of production*. Whether they help productivity or harm it depends on what people do with them, on the purpose to which they are being put, for instance, or on the skill of the user. Thirty years ago we were sure that the computer would result in a massive reduction in clerical and office forces. The investment in data-processing equipment now rivals that in materials-processing technology—that is, in conventional machinery —with the great bulk of it in services. Yet office and clerical forces have grown at a much faster rate since the introduction of information technology than ever before. And there has been virtually no increase in the productivity of service work.

The most telling example are hospitals. When I first began to work with them—in the late 1940s—they were entirely labor-intensive, with little capital investment except in bricks, mortar, and beds. A good many perfectly respectable hospitals then had not yet invested in available and fairly old technologies; they had neither X-ray department nor clinical laboratory nor physical therapy. Today's hospitals are the most capital-intensive facilities around, with enormous sums invested in ultra sound, body scanners, nuclear magnetic imagers, blood and tissue analyzers, clean rooms, and a dozen more new technologies. Each of these brought with it the need for additional and expensive people without reducing by a single person the hospital's existing staff. In fact, the world-wide escalation of health-care costs is the result, in large measure, of the hospital's having become an economic monstrosity. Being both highly labor-intensive and highly capital-intensive, it is, by any economist's definition, simply not viable economically. But the hospital has at least significantly increased its performance capacity. In other areas of knowledge or service work there are only higher costs, more investment, and more people.

Only massive increases in hospital productivity can stem the health-care cost explosion. And these increases can only come from "Working Smarter."

Neither economist nor technologist gives star billing to Working Smarter as the key to the productivity explosion; the former features capital investment, the latter, technology. But Working Smarter, whether called *scientific management, industrial engineering, human relations, efficiency engineering*, or *task study* (the modest term Frederic W. Taylor himself favored)—has been the main force behind the productivity explosion. Capital investment and technology were just as copious in the then developed countries in the first hundred years of the Industrial Revolution—that is, in the century before Taylor—as they have been in the century since. But only when Working Smarter began to have an impact did productivity in making and moving things take off on its meteoric rise. Still, in making and moving things Working Smarter is but one key to increased productivity. In knowledge and service work it is *the* key.

Working Smarter means, however, very different things in knowledge and service work from what it means in making and moving things.

3

When Frederic Taylor started what later became Scientific Management by studying the shoveling of sand, it never occurred to him to ask: "*What* is the task? *Why* do it?" All he asked was "*How* is it done?" Almost fifty years later, Harvard's Elton Mayo (1880–1949) set out to demolish Scientific Management and to replace it with what later came to be called Human Relations. But, like Taylor, he never asked, "*What* is the task? *Why* do it?" In his famous experiments at the Hawthorne Works of Western Electric, he only asked, "*How* can wiring telephone equipment best be done?" In making and moving things the task is always taken for granted.

But the first question in increasing productivity in knowledge and service work has to be: "*What* is the task? *What* do we try to accomplish? *Why* do it at all? The easiest—but perhaps also the greatest—increases in productivity in such work come from redefining the task, and especially from eliminating what needs not be done.

The oldest example is still the best one—mail-order processing at the early Sears, Roebuck. Between 1906 and 1908 Sears eliminated the time-consuming counting of money in incoming mail orders. In those days there was neither paper money nor checks but only coins. Hence, the money envelopes in the incoming order were weighed automatically: if the weight tallied with the amount of the order within fairly narrow limits, the envelopes were not even opened. Similarly, Sears eliminated the even more time-consuming detailed recording of each incoming order. It scheduled order handling and shipping according to the weight of the incoming mail, assuming forty orders for each pound of mail. These two steps increased the productivity of the entire mail-order operation ten-fold within two years.[1]

A major insurance company has recently increased the productivity of claims settlement about five-fold—from an average of fifteen minutes to three minutes per claim—by eliminating detailed checking for all but claims for very large sums. Instead of verifying thirty items—as had always been done—only five are being checked now: whether the policy is still in force; matching its face amount with the amount of the claim; matching the name of the policy holder with the name on the death certificate; matching the name of the beneficiary on the policy with the name of the claimant. What led to this productivity increase was asking the question: "What is the task?" And then the answer came fairly easy: "It's to pay death claims as cheaply and as fast as possible"—and all that is now needed to control the process is to work through a small sample, that is, through every fiftieth claim, the traditional way.

A few hospitals have eliminated most of the laborious and expensive admissions process. They now admit all patients the way they used to admit emergency patients who are brought in unconscious or bleeding and unable to fill out lengthy forms. They asked, "What is the task?"—and the answer was to identify the patient's name, sex, age, address, and how to bill. This information is found, however, on the insurance identification practically all patients carry.

Another example: a well-known private college has been able to cut its financial-aid staff from eleven full-time clerks to one or two people who work in financial aid for only a few weeks a year. Like other institutions of its kind, the college admits qualified applicants without regard to their ability to pay, with the financial-aid office then determining what tuition reduction an applicant can be given. This was done—and in most colleges is still being done—by laboriously working through a long and detailed form submitted by each applicant. But for ninety-five out of each 100 applicants financial aid is actually determined by very few factors. Being told the family income; the value of the family home; what, if any, additional sources of income there are—for example, from a trust fund—and whether there are siblings presently paying college tuition, the computer figures out in a few seconds how much financial aid will be offered. The two part-timers are needed only to winnow out the five percent of unusual cases—the applicant who is a star athlete or a National Scholarship winner—which are then easily dealt with by the dean and a small faculty committee in a few afternoon hours.

These are all examples of service work. In knowledge work, however, defining the task and getting rid of what needs not be done are even more necessary and produce even greater results.

I know one significant example: the way a major multi-national company redefined its strategic planning. For many years a planning staff of forty-five brilliant people carefully prepared "strategic scenarios" down to minute details. It was first-class work, and stimulating reading, everybody admitted. But it had minimum operational impact.

A new CEO asked, "What is the task?" His answer: "It isn't to predict the future. It is to give our businesses direction and goals and the strategy to attain these goals." It took four years of hard work and several false starts. But now the planning people—still about the same number—work through only three questions for each of the company's businesses: What market standing does it need to maintain leadership? What innovative performance does it need to support the needed market standing? What rate of return is the minimum needed to earn the cost of capital? And then the planning people together with the operating executives in each business work out broad strategy guidelines to attain these goals under different assumptions regarding economic conditions. The results are far simpler and far less pretentious than the old-style plans were, and far less elegant. But they have become the "flight plans" that guide the company's businesses and its senior executives.

Except for this company I have not heard, however, of any case where the questions *What* is the task? and *Why* do it? have so far been asked in respect to knowledge work.

4

In making and moving things people do one task at a time. Taylor's laborer shoveled sand; he did not also stoke the furnace. Mayo's wiring-room women soldered, they did not also test finished telephones on the side. The Iowa farmer planting corn does not get off his tractor between rows to attend a meeting. Concentration is not unknown in knowledge and service work. The surgeon does not take telephone calls in the operating room; nor should the lawyer while in consultation with a client. But in organizing—and that's where most knowledge and service people work—there is growing splintering. The people at the very top can sometimes concentrate themselves, though far too few even try. But the people who actually do most of the knowledge and service work in organizations—engineers, teachers, salespeople, nurses, middle managers in general—carry a steadily growing load of busy work, additional activities that contribute little or no value and that have little or nothing to do with what these people are qualified and paid for.

The worst case may be the nurse in the American hospital. We hear a great deal about the shortage of nurses. But how could we possibly have one? The number of graduating nurses entering the profession has gone up steadily for a good many years. At the same time the number of bed patients has been going down sharply. The explanation of the paradox: nurses now spend only half their time doing what they have learned and are being paid for—that is, nursing. The other half of their shift is taken by activities that do not require the nurse's skill and knowledge, add neither health-care nor economic value, and have little or nothing to do with patient care and patient well-being—primarily, of course, the ever swelling avalanche of paperwork for Medicare, Medicaid, insurers, the billing office, and the prevention of malpractice suits.

The situation in higher education is not much different. Every study reports that faculty in colleges and universities spend steadily increasing hours in committee meetings rather than in the classroom, in advising students, or in research. But few of these committees would ever be missed. And they would do a better job and in less time if they had three instead of seven members.

Salespeople are just as splintered. In the department store they now spend so much time serving the computer that they have little time serving the customer—the main reason, perhaps, for the steady decline in their productivity as producers of sales and revenues. Field sales representatives spend up to one third of their time filling out reports of all kinds rather than calling on customers. And engineers sit through meeting after meeting when they should be busy at their work stations.

This is not job enrichment; it is job impoverishment. It destroys productivity. It saps motivation and morale. Nurses, every attitude survey shows, bitterly resent not being allowed to do what they went into nursing for and are trained to do—to give patient care at the patient bed. They also—understandably—feel that they are grossly underpaid for what they are capable of doing while the hospital administrator—equally understandably—feels that they are grossly overpaid for the unskilled clerical work they are actually doing.

The cure is fairly easy, as a rule. A few hospitals have taken the paperwork out of the nurses' job and given it to a floor clerk who also answers the telephone calls from friends and relatives of patients and arranges the flowers they send in. All of a sudden, these hospitals had a surplus of nurses. The level of patient care and the hours nurses devote to it went up sharply. Yet the hospitals could cut the number of nurses needed by a quarter or a third, and could thus raise nurses' salaries without incurring a higher nursing payroll.

To do this requires that we ask in respect to every knowledge and service job: "What do we pay for?" "What value is this job supposed to add?" The answer is not always obvious or uncontroversial. One department store that raised the question in respect to its salespeople on the floor answered: "Sales." Another one in the same metropolitan area and with pretty much the same kind of clientele answered: "Customer service." Each answer led to a different restructuring of the jobs on the sales floor. But each store achieved, and fairly fast, substantial increases in the revenues generated by each salesperson and each department, that is, in both productivity and profitability.

5

For all their tremendous worldwide impacts Frederic Taylor and Scientific Management have had a bad press, especially in academia. One reason, perhaps the main one, is the unrelenting campaign America's labor unions waged against both in the early years of this century. The unions actually succeeded in banning any kind of work study in army arsenals and naval shipyards, where in those years practically all defense production was done in this country.

The unions of 1911 did not oppose Taylor because they thought him pro-management or anti-labor (he was neither). His unforgivable sin was his assertion that there is no such thing as skill in making and moving things. All such work, Taylor asserted, was the same. All could be analyzed step by step as a series of unskilled operations that then could be put together into any kind of job. Anyone willing to learn these operations would be a "first-class man," deserving "first-class pay." He could do the most highly skilled work and do it to perfection.

But the unions of Taylor's time—and especially the highly respected and extremely powerful unions in arsenals and shipyards—were craft monopolies. Their power base was their control of an apprenticeship of five or seven years to which, as a rule, only relatives of members were admitted. They considered their craft a "mystery," the secrets of which no member was allowed to divulge. The skilled workers in the arsenals and navy yards in particular were paid extremely well—more than most physicians of those times and triple what Taylor's "first-class man" could then expect to get. No wonder that Taylor's denial of the mystery of craft and skill infuriated these "aristocrats of labor" as subversion and pestilential heresy.

Most contemporaries, eighty years ago, agreed with the unions. Even thirty years later the belief in the mystery of craft and skill persisted, and also in the long years of apprenticeship needed to acquire either. Hitler, for instance, was convinced that it would take the United States at least five years to train optical craftsmen, and modern war requires precision optics. It would therefore take many years, Hitler was sure, before America could field an effective army and air force in Europe—the conviction that made him declare war on America when Japan attacked Pearl Harbor.

We now know that Taylor was right. The United States did indeed have almost no optical craftsmen in 1941. And modern war does indeed require precision optics, and in large quantities. But by applying Taylor's Scientific Management the U.S. trained in a few months semiskilled workers to turn out more highly advanced optics than the Germans with their craftsmen ever did, and on the assembly line to boot. And by that time Taylor's first-class men with their increased productivity also made a great deal more money than any craftsman of 1911 could ever have dreamed of.

Eventually knowledge work and service work may turn out to be like work making and moving things—that is, "just work," to use an old Scientific Management slogan. At least this is the position of the more radical proponents of Artificial Intelligence, Taylor's true children or grandchildren. But for the time being, knowledge and service jobs must not be treated as just work. They cannot be assumed to be homogeneous. They must be treated as falling into a number of distinct categories—probably three. Each requires different analysis and different organization. In making and moving things the focus in increasing productivity is on *work*. In knowledge and service work it has to be on *performance*.

To be specific: for some jobs in knowledge and service work performance means quality. One example is the research lab, in which quantity—that is, the number of results—is quite secondary to their quality. One new drug generating annual sales of $500 million and dominating the market for a decade is infinitely more valuable than twenty "me too" drugs each with annual sales of $20 or $30 million. The same holds for basic policy or for strategic decisions. But it also applies to much less grandiose work—the physician's diagnosis, for instance, or packaging design, or editing a magazine.

Then there is a wide range of knowledge and service jobs in which quality and quantity together constitute performance. The salesperson's performance on the department-store floor is one example. A "satisfied customer" is a qualitative statement, and indeed not so easy to define. But it is as important as the amounts on the sales tickets, or the quantity of output. In architectural design quality largely defines performance. In the draftsman's work quality is an integral part of performance. But so is quantity. And the same applies to the engineer; to the sales rep in the local stockbroker's office; to the medical technologist; to the branch manager of the local bank; to the reporter, to the nurse; to the claims adjuster for the automotive insurer—in fact, to a vast range of knowledge and service jobs. Performance in them is always both, quantity and quality. To increase productivity in these jobs therefore always requires work on both.

Finally, there are a good many jobs—filing, handling death claims in the life insurance office, making beds in the hospital—in which performance is similar to performance in making and moving things. Quality is a condition and a restraint. It is external rather than in itself performance. It has to be built into the process. But once this has been done, performance is largely defined by quantity—for example, the number of minutes it takes to make a hospital bed the prescribed way. These jobs are, in effect, "production jobs" even though they do not result in making and moving things.

Thus increasing productivity in knowledge and service work requires thinking through into which category of performance a given job belongs. Only then do we know what we should be working on. Only then can we decide what needs to be analyzed, what needs to be improved, what needs to be changed. For only then do we know what productivity means in a specific knowledge or service job.

6

There is more to increasing productivity in knowledge work and service work than defining the task, concentrating on the task, and defining performance. We do not yet know how to analyze the process in jobs in which performance predominantly means quality. We need to ask instead, "What works?" For jobs in which performance means both quality and quantity, we need to do both: ask what works and analyze the process step by step and operation by operation. In production work we need to define the quality standards and build them into the process, but the actual productivity improvement then comes through fairly conventional industrial engineering, that is, through task analysis followed by putting together the individual simple operations into a complete "job."

But the three steps outlined above will by themselves produce substantial productivity increases—perhaps most of what can be attained at any one time. They need to be worked through again and again—maybe every three or five years, and certainly whenever we change work or its organization. But then, according to all the experience we have, the resulting productivity increases will equal, if not exceed, whatever Industrial Engineering, Scientific Management, or Human Relations ever achieved in making and moving things. In other words, they should by themselves give us the "productivity revolution" we need in knowledge and service work.

But on one condition only: that we actually apply what we have learned since World War II about increasing productivity in making and moving things: the work has to be done in partnership with the people who hold the knowledge and service jobs, the people who are to become more productive. The goal has to be to build responsibility for productivity and performance into every knowledge and service job regardless of level, difficulty, or skill.

Frederick Taylor has often been criticized for never once asking the workers whose jobs he studied; he told them. Nor did Elton Mayo ever ask—he also told them. But there is also no record of Sigmund Freud's ever asking patients what they thought might be their problem. Neither Marx nor Lenin ever thought of asking the masses. And it did not occur to any High Command in World War I or World War II to ask junior officers or enlisted men in the front lines about weapons, uniforms, or even food (in the American armed forces this became the custom only during Vietnam). Taylor simply shared the belief of his age in the wisdom of the expert. He thought both workers and managers to be "dumb oxen." Mayo, forty years later, had high respect for managers, but workers, he thought, were "immature" and "maladjusted" and needed the expert guidance of the psychologist.

When World War II came, however, we had no choice; we had to ask the workers. In the plants we had neither engineers nor psychologists nor foremen—they were all in uniform. And when we asked the workers, we found—to our immense surprise, as I still recollect—that the workers were neither dumb oxen nor immature and maladjusted. They knew a great deal about the work they were doing, its logic and rhythm, the tools, the quality and so on. Asking them was the way to get started on productivity and quality.[2] At first only a few businesses accepted this novel proposition—IBM was perhaps the first one, and for a long time also the only one. Then in the late 1950s and early 1960s it was picked up by the Japanese, whose earlier attempts to return to prewar autocracy in the plant had collapsed in bloody strikes and near civil war. Nowadays, while still far from being widely practiced, it is at least generally accepted in theory that the workers' knowledge of their job is the starting point for improving productivity, quality, and performance altogether.

In making and moving things, partnership with the responsible worker is, however, only the *best* way—after all, Taylor's telling them worked, too, and quite well. In knowledge and service work, partnership with the responsible worker is the *only* way; nothing else will work at all.

Two more lessons that neither Taylor nor Mayo knew; increased productivity needs continuous learning. It is not enough to redesign the job and then to train the worker in the new way of doing it—which is what Taylor did and taught. That's when learning begins, and it never ends. Indeed, as the Japanese can teach us—it came out of their ancient tradition of Zen learning—the greatest benefit of training is not in learning the new. It is to do better what we already do well. And equally important, an insight of the last few years: knowledge people and service people learn the most when they teach. The best way to improve the productivity of the star salesperson is for him or her to present "the secrets of my success" at a sales convention. The best way for the surgeon to improve his or her performance is to give a talk about it at the county medical society. The best way for a nurse to improve her performance is to teach her fellow nurses. It is often being said that in the information age every enterprise has to become a learning institution. It also has to become a teaching institution.

Conclusion

Developed economies face economic stagnation if they do not raise the productivity of knowledge and service work. Even Japan—still heavily manufacturing-intensive—can no longer expect increased productivity in making and moving things to sustain economic growth. Even there the great majority of working people are knowledge workers and service workers with productivities as low as those in any other developed country. And when farmers are down to a mere 3 percent of the employed population, as they are in the U.S. and Japan—and in most of Western Europe as well—even record increases in their productivity such as the 4 to 5 percent the United States boasts of add virtually nothing to the country's overall productivity, its wealth, its competitiveness.

Raising the productivity of knowledge and service work must therefore be an *economic* priority for developed countries. Whichever country first succeeds in satisfying it will economically dominate the twenty-first century. And the key is raising the productivity of *knowledge* work, on all levels.

But the need to raise the productivity of *service* work may be even greater. It is a *social* priority in developed countries. Unless it is met, the developed work faces increasing social tensions, increasing polarization, increasing radicalization. It may ultimately face a new class war.

In the knowledge society, access to opportunities for careers and advancement is becoming limited to people of advanced schooling, people qualified for knowledge work. But such people will always be a minority. They will always be outnumbered by people who lack the qualifications for anything but fairly low-skilled service work. In their social position such people are comparable to the proletarians of years ago: the poorly educated, unskilled masses who thronged the exploding industrial cities and streamed into their factories.

When Frederick Taylor started his work on the productivity of making and moving things in the early 1880s, class war between industrial proletarian and "bourgeois"—its reality but even more the fear of it—obsessed every developed country. Fear of it motivated Taylor to start his work. And the belief in the inevitability of class war was by no means confined to the Left. A generation before Taylor, Benjamin Disraeli, the greatest of the nineteenth-century conservatives, had predicted it. And Henry James, the chronicler of American wealth and European aristocracy, was so frightened by it that he made it the central theme of one of his most haunting novels, *The Princess Casamassima*—it appeared in 1885, two years after Marx's death and four years after Taylor had begun studying the productivity of shoveling sand.

Marx has been proven wrong in his prophesy of the inevitable "immiseration" of the proletariat leading inevitably to a revolution. But when he made these prophesies they seemed eminently reasonable—indeed, almost self-evident—to well-informed and highly intelligent contemporaries. What defeated Marx and Marxism in the end was the rising productivity of making and moving things—that is, in essence, the work Taylor started. It gave the proletarians the productivity that allowed their being paid a middle-class income and to achieve middle-class status despite lack of skill, wealth, and education. By the time of the Great Depression—when, according to Marx and the Marxists the "Proletarian Revolution" should surely have become triumphant—the proletarian had become a bourgeois.

Unless the productivity of service work is rapidly improved, both the social and the economic position of that large class—as large a group as people making and moving things ever were at their peak—must steadily go down. Real incomes cannot for any length of time be higher than productivity. The service workers may use their numerical strength to get higher wages than their economic contribution justifies. But this only impoverishes all of society with everybody's real income going down and unemployment going up. Or the incomes of the unskilled service workers are allowed to go down in relation to the steadily rising wages of affluent knowledge workers, with an increasing gulf between the two groups, an increasing polarization into classes. In either case service workers must become alienated, increasingly bitter, increasingly see themselves as *a class apart*. And the escape hatch—the productive and therefore well-paid jobs for poorly educated and poorly skilled people in making and moving things—is closing rapidly. By the end of this century the number of such jobs in every developed country will be at most two fifths of what it was at their peak only forty years ago.

We are in a much better position than our ancestors were a century ago. We know what Marx and his contemporaries did not know: *productivity can be raised*. We also know *how to raise it*. And we know this best for the work where the social need is most urgent: unskilled and semiskilled service work—the jobs in maintenance, whether of factories, schools, hospitals, or offices; in restaurants and in supermarkets; in a host of clerical jobs. This, as has already been stated, is production work—and what we have learned during the past hundred years about increasing productivity applies to such work with a minimum of adaptation. Indeed, in such work substantial productivity increases have already been achieved. Some multinational maintenance companies—both in the United States and in Europe—have systematically applied to low-skilled service jobs the approaches this article discusses. They have defined the task; concentrated work on it; defined performance; made the employee a partner in productivity improvement and the first source of ideas for it; and built continuous learning and continuous teaching into the job of every employee and of every work team. They have substantially raised productivity—in some cases, doubled it. This then has allowed them to raise wages. But it has also greatly raised self-respect and pride. It is, incidentally, by no means coincidence that these increases were achieved by outside contractors rather than by the organization (e.g., the hospitals) in which the service people actually do their work. To obtain major productivity increases in production-type service work usually requires contracting out such work to a firm that has no other business, understands this work, respects it, and offers opportunities for advancement for low-skill service workers—for example, to become its local or regional manager. The organizations in which this work is being done—for example, the hospital in which the people work who make the beds, or the college whose students they feed—neither understand such work nor respect it enough to devote to it the time and hard work needed to make it productive, no matter how much they pay for it.

The task is known and doable, but the urgency is great. To raise the productivity of service work cannot be done by governmental action or by politics altogether. It is the task of managers and executives in businesses and nonprofit organizations. It is, in fact, the *first social responsibility* of management in the knowledge society.

Notes

1. Boris Emmet and John E. Jeucks. *Catalogues and Counters: A History of Sears, Roebuck* (University of Chicago Press, 1950).
2. I was the first to draw this conclusion in my two early books, *The Future of Industrial Man* (1942) and *The New Society* (1949), in which I argued for the "responsible worker" as "part of management." As a result of their wartime experiences, Edwards Deming and Joseph Juran each developed what we now call "quality circles" and "total quality management." Finally, the idea was forcefully presented by Douglas McGregor in his 1960 book *The Human Side of Enterprise* with its "Theory X" and "Theory Y."

16

The Emerging Theory of Manufacturing

We cannot build it yet. But already we can specify the "post-modern" factory of 1999. Its essence will not be mechanical, though there will be plenty of machines. Its essence will be conceptual—the product of four principles and practices that together constitute a new approach to manufacturing.

Each of these concepts is being developed separately, by different people with different starting points and different agendas. Each concept has its own objectives and its own kinds of impact. Statistical Quality Control is changing the social organization of the factory. The new manufacturing accounting lets us make production decisions as business decisions. The "flotilla," or module, organization of the manufacturing process promises to combine the advantages of standardization and flexibility. Finally, the systems approach embeds the physical process of making things, that is, manufacturing, in the economic process of business, that is, the business of creating value.

As these four concepts develop, they are transforming how we think about manufacturing and how we manage it. Most manufacturing people in the United States now know we need a new theory of manufacturing. We know that patching up old theories has not worked and that further patching will only push us further behind. Together these concepts give us the foundation for the new theory we so badly need.

First published in *The Harvard Business Review*, 1990.

The most widely publicized of these concepts, Statistical Quality Control (SQC), is actually not new at all. It rests on statistical theory formulated seventy years ago by Sir Ronald Fisher. Walter Shewhart, a Bell Laboratories physicist, designed the original version of SQC in the 1930s for the zero-defects mass production of complex telephone exchanges and telephone sets. During World War II, W. Edwards Deming and Joseph Juran, both former members of Shewhart's circle, separately developed the versions used today.

The Japanese owe their leadership in manufacturing quality largely to their embrace of Deming's precepts in the 1950s and 1960s. Juran too had great impact in Japan. But U.S. industry ignored their contributions for forty years and is only now converting to SQC, with companies such as Ford, General Motors, and Xerox among the new disciples. Western Europe also has largely ignored the concept. More important, even SQC's most successful practitioners do not thoroughly understand what it really does. Generally it is considered a production tool. Actually, its greatest impact is on the factory's social organization.

By now, everyone with an interest in manufacturing knows that SQC is a rigorous, scientific method of identifying the quality and productivity that can be expected from a given production process in its current form so that control of both attributes can be built into the process itself. In addition, SQC can instantly spot malfunctions and show where they occur—a worn tool, a dirty spray gun, an overheating furnace. And because it can do this with a small sample, malfunctions are reported almost immediately, allowing machine operators to correct problems in real time. Further, SQC quickly identifies the impact of any change on the performance of the entire process. (Indeed, in some applications developed by Deming's Japanese disciples, computers can simulate the effects of a proposed change in advance.) Finally, SQC identifies where, and often how, the quality and productivity of the entire process can be continuously improved. This used to be called the "Shewhart Cycle" and then the "Deming Cycle"; now it is *kaizen,* the Japanese term for continuous improvement.

But these engineering characteristics explain only a fraction of SQC's results. Above all, they do not explain the productivity gap between Japanese and U.S. factories. Even after adjusting for their far greater reliance on outside suppliers, Toyota, Honda, and Nissan turn out two or three times more cars per worker than comparable U.S. or European plants do. Building quality into the process accounts for no more than one-third of this difference. Japan's major productivity gains are the result of social changes brought about by SQC.

The Japanese employ proportionately more machine operators in direct production work than Ford or GM. In fact, the introduction of SQC almost always increases the number of machine operators. But this increase is offset many times over by the sharp drop in the number of nonoperators: inspectors, above all, but also the people who do not *do* but *fix,* like repair crews and "fire fighters" of all kinds.

In U.S. factories, especially mass-production plants, such non-operating, blue-collar employees substantially outnumber operators. In some plants, the ratio is two to one. Few of these workers are needed under SQC. Moreover, first-line supervisors also are gradually eliminated, with only a handful of trainers taking their place. In other words, not only does SQC make it possible for machine operators to be in control of their work, it makes such control almost mandatory. No one else has the hands-on knowledge needed to act effectively on the information that SQC constantly feeds back.

By aligning information with accountability, SQC resolves a heretofore irresolvable conflict. For more than a century, two basic approaches to manufacturing have prevailed, especially in the United States. One is the engineering approach pioneered by Frederick Winslow Taylor's "scientific management." The other is the "human relations" (or "human resources") approach developed before World War I by Andrew Carnegie, Julius Rosenwald of Sears Roebuck, and Hugo Münsterberg, a Harvard psychologist. The two approaches have always been considered antitheses, indeed, mutually exclusive. In SQC, they come together.

Taylor and his disciples were just as determined as Deming to build quality and productivity into the manufacturing process. Taylor asserted that his "one right way" guaranteed zero-defects quality; he was as vehemently opposed to inspectors as Deming is today. So was Henry Ford, who claimed that his assembly line built quality and productivity into the process (though he was otherwise untouched by Taylor's scientific management and probably did not even know about it). But

minutely, accounted for. The remaining costs—and that can mean 80 percent to 90 percent—are allocated by ratios that everyone knows are purely arbitrary and totally misleading: in direct proportion to a product's labor cost, for example, or to its dollar volume.

Second, the benefits of a change in process or in method are primarily defined in terms of labor cost savings. If other savings are considered at all, it is usually on the basis of the same arbitrary allocation by which costs other than direct labor are accounted for.

Even more serious is the third limitation, one built into the traditional cost accounting system. Like a sundial, which shows the hours when the sun shines but gives no information on a cloudy day or at night, traditional cost accounting measures only the costs of producing. It ignores the costs of nonproducing, whether they result from machine downtime or from quality defects that require scrapping or reworking a product or part.

Standard cost accounting assumes that the manufacturing process turns out good products 80 percent of the time. But we now know that even with the best SQC, nonproducing time consumes far more than 20 percent of total production time. In some plants, it accounts for 50 percent. And nonproducing time costs as much as producing time does—in wages, heat, lighting, interest, salaries, even raw materials. Yet the traditional system measures none of this.

Finally, manufacturing cost accounting assumes the factory is an isolated entity. Cost savings in the factory are "real." The rest is "speculation"—for example, the impact of a manufacturing process change on a product's acceptance in the market or on service quality. GM's plight since the 1970s illustrates the problem with this assumption. Marketing people were unhappy with top management's decision to build all car models, from Chevrolet to Cadillac, from the same small number of bodies, frames, and engines. But the cost accounting model showed that such commonality would produce substantial labor cost savings. And so marketing's argument that GM cars would lose customer appeal as they looked more and more alike was brushed aside as speculation. In effect, traditional cost accounting can hardly justify a product *improvement,* let alone a product or process *innovation.* Automation, for instance, shows up as a cost but almost never as a benefit.

All this we have known for close to forty years. And for thirty years, accounting scholars, government accountants, industry accountants, and accounting firms have worked hard to reform the system. They have made substantial improvements. But since the reform attempts tried to build on the traditional system, the original limitations remain.

What triggered the change to the new manufacturing accounting was the frustration of factory-automation equipment makers. The potential users, the people in the plants, badly wanted the new equipment. But top management could not be persuaded to spend the money on numerically controlled machine tools or robots that could rapidly change tools, fixtures, and molds. The benefits of automated equipment, we now know, lie primarily in the reduction of nonproducing time by improving quality (that is, getting it right the first time) and by sharply curtailing machine downtime in changing over from one model or product to another. But these gains cost accounting does not document.

Out of this frustration came Computer-Aided Manufacturing International, or CAM-I, a cooperative effort by automation producers, multinational manufacturers, and accountants to develop a new cost accounting system. Started in 1986, CAM-I is just beginning to influence manufacturing practice. But already it has unleashed an intellectual revolution. The most exciting and innovative work in management today is found in accounting theory, with new concepts, new approaches, new methodology—even what might be called new economic philosophy—rapidly taking shape. And while there is enormous controversy over specifics, the lineaments of the new manufacturing accounting are becoming clearer every day.

As soon as CAM-I began its work, it became apparent that the traditional accounting system could not be reformed. It had to be replaced. Labor costs are clearly the wrong unit of measurement in manufacturing. But—and this is a new insight—so are all the other elements of production. The new measurement unit has to be time. The costs for a given period of time must be assumed to be fixed; there are no ''variable'' costs. Even material costs are more fixed than variable, since defective output uses as much material as good output does. The only thing that is both variable and controllable is how much time a given process takes. And ''benefit'' is whatever reduces that time. In one fell swoop, this insight eliminates the first three of cost accounting's four traditional limitations.

But the new cost concepts go even further by redefining what costs and benefits really are. For example, in the traditional cost accounting system, finished-goods inventory costs nothing because it does not absorb any direct labor. It is treated as an "asset." In the new manufacturing accounting, however, inventory of finished goods is a "sunk cost" (an economist's, not an accountant's, term). Stuff that sits in inventory does not earn anything. In fact, it ties down expensive money and absorbs time. As a result, its time costs are high. The new accounting measures these time costs against the benefits of finished-goods inventory (quicker customer service, for instance).

Yet manufacturing accounting still faces the challenge of eliminating the fourth limitation of traditional cost accounting: its inability to bring into the measurement of factory performance the impact of manufacturing changes on the total business—the return in the marketplace of an investment in automation, for instance, or the risk in not making an investment that would speed up production changeovers. The in-plant costs and benefits of such decisions can now be worked out with considerable accuracy. But the business consequences are indeed speculative. One can only say, "Surely, this should help us get more sales," or "If we don't do this, we risk falling behind in customer service." But how do you quantify such opinions?

Cost accounting's strength has always been that it confines itself to the measurable and thus gives objective answers. But if intangibles are brought into its equations, cost accounting will only raise more questions. How to proceed is thus hotly debated, and with good reason. Still, everyone agrees that these business impacts have to be integrated into the measurement of factory performance, that is, into manufacturing accounting. One way or another, the new accounting will force managers, both inside and outside the plant, to make manufacturing decisions as *business* decisions.

Henry Ford's epigram, "The customer can have any color as long as it's black," has entered American folklore. But few people realize what Ford meant: flexibility costs time and money, and the customer won't pay for it. Even fewer people realize that in the mid-1920s, the "new" cost accounting made it possible for GM to beat Ford by giving customers both colors and annual model changes at no additional cost.

By now, most manufacturers can do what GM learned to do roughly seventy years ago. Indeed, many go quite a bit further in combining standardization with flexibility. They can, for example, build a variety of end products from a fairly small number of standardized parts. Still, manufacturing people tend to think like Henry Ford: you can have either standardization at low cost or flexibility at high cost, but not both.

The factory of 1999, however, will be based on the premise that you not only *can* have both but also *must* have both—and at low cost. But to achieve this, the factory will have to be structured quite differently.

Today's factory is a battleship. The plant of 1999 will be a "flotilla," consisting of modules centered either around a stage in the production process or around a number of closely related operations. Though overall command and control will still exist, each module will have its own command and control. And each, like the ships in a flotilla, will be maneuverable, both in terms of its position in the entire process and its relationship to other modules. This organization will give each module the benefits of standardization and, at the same time, give the whole process greater flexibility. Thus it will allow rapid changes in design and product, rapid response to market demands, and low-cost production of "options" or "specials" in fairly small batches.

No such plant exists today. No one can yet build it. But many manufacturers, large and small, are moving toward the flotilla structure: among them are some of Westinghouse's U.S. plants, Asea Brown Boveri's robotics plant in Sweden, and several large printing plants, especially in Japan.

The biggest impetus for this development probably came from GM's failure to get a return on its massive (at least $30 billion and perhaps $40 billion) investment in automation. GM, it seems, used the new machines to improve its existing process, that is, to make the assembly line more efficient. But the process instead became less flexible and less able to accomplish rapid change.

Meanwhile, Japanese automakers and Ford were spending less and attaining more flexibility. In these plants, the line still exists, but it is discontinuous rather than tightly tied together. The new equipment is being used to speed changes, for example, automating changeovers of jigs, tools, and fixtures. So the line has acquired a good bit of the flexibility of traditional batch production without losing its standardization. Standardization and flexibility are thus no longer an either-or proposition. They are—as indeed they must be—melded together.

This means a different balance between standardization and flexibility, however, for different parts of the manufacturing process. An "average" balance across the plant will do nothing very well. If imposed throughout the line, it will simply result in high rigidity and big costs for the entire process, which is apparently what happened at GM. What is required is a reorganization of the process into modules, each with its own optimal balance.

Moreover, the relationships between these modules may have to change whenever product, process, or distribution change. Switching from selling heavy equipment to leasing it, for instance, may drastically change the ratio between finished-product output and spare-parts output. Or a fairly minor model change may alter the sequence in which major parts are assembled into the finished product. There is nothing very new in this, of course. But under the traditional line structure, such changes are ignored, or they take forever to accomplish. With competition intensifying and product life cycles shortening all the time, such changes cannot be ignored, and they have to be done fast. Hence the flotilla's modular organization.

But this organization requires more than a fairly drastic change in the factory's physical structure. It requires, above all, different communication and information. In the traditional plant, each sector and department reports separately upstairs. And it reports what upstairs has asked for. In the factory of 1999, sectors and departments will have to think through what information they owe to whom and what information they need from whom. A good deal of this information will flow sideways and across department lines, not upstairs. The factory of 1999 will be an information network.

Consequently, all the managers in a plant will have to know and understand the entire process, just as the destroyer commander has to know and understand the tactical plan of the entire flotilla. In the factory of 1999, managers will have to think and act as team members, mindful of the performance of the whole. Above all, they will have to ask: What do the people running the other modules need to know about the characteristics, the capacity, the plans, and the performance of *my* unit? And what, in turn, do we in my module need to know about theirs?

The last of the new concepts transforming manufacturing is systems design, in which the whole of manufacturing is seen as an integrated process that converts materials into goods, that is, into economic satisfactions.

Marks & Spencer, the British retail chain, designed the first such system in the 1930s. Marks & Spencer designs and tests the goods (whether textiles or foods) it has decided to sell. It designates one manufacturer to make each product under contract. It works with the manufacturer to produce the right merchandise with the right quality at the right price. Finally, it organizes just-in-time delivery of the finished products to its stores. The entire process is governed by a meticulous forecast as to when the goods will move off store shelves and into customers' shopping bags. In the last ten years or so, such systems management has become common in retailing.

Though systems organization is still rare in manufacturing, it was actually first attempted there. In the early 1920s, when the Model T was in its full glory, Henry Ford decided to control the entire process of making and moving all the supplies and parts needed by his new plant, the gigantic River Rouge. He built his own steel mill and glass plant. He founded plantations in Brazil to grow rubber for tires. He bought the railroad that brought supplies to River Rouge and carried away the finished cars. He even toyed with the idea of building his own service centers nationwide and staffing them with mechanics trained in Ford-owned schools. But Ford conceived of all this as a financial edifice held together by ownership. Instead of building a system, he built a conglomerate, an unwieldy monster that was expensive, unmanageable, and horrendously unprofitable.

In contrast, the new manufacturing system is not "controlled" at all. Most of its parts are independent—independent suppliers at one end, customers at the other. Nor it is plant centered, as Ford's organization was. The new system sees the plant as little more than a wide place in the manufacturing stream. Planning and scheduling start with shipment to the final customer, just as they do at Marks & Spencer. Delays, halts, and redundancies have to be designed into the system— a warehouse here, an extra supply of parts and tools there, a stock of old products that are no longer being made but are still occasionally demanded by the market. These are necessary imperfections in a continuous flow that is governed and directed by information.

What has pushed American manufacturers into such systems design is the trouble they encountered when they copied Japan's just-in-time methods for supplying plants with materials and parts. The trouble could have been predicted, for the Japanese scheme is founded in social and logistic conditions unique to that country and unknown in the United States. Yet the shift seemed to American manufacturers a matter of procedure, indeed, almost trivial. Company after company found, however, that just-in-time delivery of supplies and parts created turbulence throughout their plants. And while no one could figure out what the problem was, the one thing that became clear was that with just-in-time deliveries, the plant no longer functions as a step-by-step process that begins at the receiving dock and ends when finished goods move into the shipping room. Instead, the plant must be redesigned from the end backwards and managed as an integrated flow.

Manufacturing experts, executives, and professors had urged such an approach for two or three decades now. And some industries, such as petroleum refining and large-scale construction, do practice it. But by and large, American and European manufacturing plants are neither systems designed nor systems managed. In fact, few companies have enough knowledge about what goes on in their plants to run them as systems. Just-in-time delivery, however, forces managers to ask systems questions: Where in the plant do we need redundancy? Where should we place the burden of adjustments? What costs should we incur in one place to minimize delay, risk, and vulnerability in another?

A few companies are even beginning to extend the systems concept of manufacturing beyond the plant and into the marketplace. Caterpillar, for instance, organizes its manufacturing to supply any replacement part anywhere in the world within forty-eight hours. But companies like this are still exceptions; they must become the rule. As soon as we define manufacturing as the process that converts things into economic satisfactions, it becomes clear that producing does not stop when the product leaves the factory. Physical distribution and product service are still part of the production process and should be integrated with it, coordinated with it, managed together with it. It is already widely recognized that servicing the product must be a major consideration during its design and production. By 1999, systems manufacturing will have an increasing influence on how we design and remodel plants and on how we manage manufacturing businesses.

Traditionally, manufacturing businesses have been organized "in series," with functions such as engineering, manufacturing, and marketing as successive steps. These days, that system is often complemented by a parallel team organization (Procter & Gamble's product management teams are a well-known example), which brings various functions together from the inception of a new product or process project. If manufacturing is a system, however, every decision in a manufacturing business becomes a manufacturing decision. Every decision should meet manufacturing's requirements and needs and in turn should exploit the strengths and capabilities of a company's particular manufacturing system.

When Honda decided six or seven years ago to make a new, upscale car for the U.S. market, the most heated strategic debate was not about design, performance, or price. It was about whether to distribute the Acura through Honda's well-established dealer network or to create a new market segment by building separate Acura dealerships at high cost and risk. This was a marketing issue, of course. But the decision was made by a team of design, engineering, manufacturing, and marketing people. And what tilted the balance toward the separate dealer network was a manufacturing consideration: the design for which independent distribution and service made most sense was the design that best utilized Honda's manufacturing capabilities.

Full realization of the systems concept in manufacturing is years away. It may not require a new Henry Ford. But it will certainly require very different management and very different managers. Every manager in tomorrow's manufacturing business will have to know and understand the manufacturing system. We might well adopt the Japanese custom of starting all new management people in the plant and in manufacturing jobs for the first few years of their careers. Indeed, we might go even further and require managers throughout the company to rotate into factory assignments throughout their careers—just as army officers return regularly to troop duty.

In the new manufacturing business, manufacturing is the integrator that ties everything together. It creates the economic value that pays for everything and everybody. Thus the greatest impact of the manufacturing systems concept will not be on the production process. As with SQC, its greatest impact will be on social and human concerns—on career ladders, for instance, or more important, on the transformation of *functional* managers into *business* managers, each with a specific role, but all members of the same production and the same cast. And surely, the manufacturing businesses of tomorrow will not be run by financial executives, marketers, or lawyers inexperienced in manufacturing, as to many U.S. companies are today.

There are important differences among these four concepts. Consider, for instance, what each means by "the factory." In SQC, the factory is a place where people work. In management accounting and the flotilla concept of flexible manufacturing, it is a place where work is being done—it makes no difference whether by people, by white mice, or by robots. In the systems concept, the factory is not a place at all; it is a stage in a process that adds economic value to materials. In theory, at least, the factory cannot and certainly should not be designed, let alone built, until the entire process of "making"—all the way to the final customer—is understood. Thus defining the factory is much more than a theoretical or semantic exercise. It has immediate practical consequences on plant design, location, and size; on what activities are to be brought together in one manufacturing complex; even on how much and in what to invest.

Similarly, each of these concepts reflects a particular mind-set. To apply SQC, you don't have to think, you have to do. Management accounting concentrates on technical analysis, while the flotilla focuses on organization design and work flow. In the systems concept, there is great temptation to keep on thinking and never get to the doing. Each concept has its own tools, its own language, and addresses different people.

Nevertheless, what these four concepts have in common is far more important than their differences. Nowhere is this more apparent than in their assumption that the manufacturing process is a configuration, a whole that is greater than the sum of its parts. Traditional approaches all see the factory as a collection of individual machines and individual operations. The nineteenth-century factory was an assemblage of machines. Taylor's scientific management broke up each job into individual operations and then put those operations together into new and different jobs. "Modern" twentieth-century concepts—the assembly line and cost accounting—define performance as the sum of lowest cost operations. But none of the new concepts is much concerned with performance of the parts. Indeed, the parts as such can only underperform. The process produces results.

Management also will reflect this new perspective. SQC is the most nearly conventional in its implications for managers, since it does not so much change their job as shift much of it to the work force. But even managers with no business responsibility (and under SQC, plant people have none) will have to manage with an awareness of business considerations well beyond the plant. And every manufacturing manager will be responsible for integrating people, materials, machines, and time. Thus every manufacturing manager ten years hence will have to learn and practice a discipline that integrates engineering, management of people, and business economics into the manufacturing process. Quite a few manufacturing people are doing this already, of course—though usually unaware that they are doing something new and different. Yet such a discipline has not been systematized and is still not taught in engineering schools or business schools.

These four concepts are synergistic in the best sense of this much-abused term. Together—but only together—they tackle the conflicts that have most troubled traditional, twentieth-century mass-production plants: the conflicts between people and machines, time and money, standardization and flexibility, and functions and systems. The key is that every one of these concepts defines performance as productivity and conceives of manufacturing as the physical process that adds economic value to materials. Each tries to provide economic value in a different way. But they share the same theory of manufacturing.

17

The Hostile Takeover and Its Discontents

Almost every week these last few years there has been a report of another "hostile takeover bid," another stock-market maneuver to take over, merge, or split up an existing publicly held company against determined opposition by the company's board of directors and management. No such wave of stock-market speculation has hit the United States since the "bears" and the "bulls" of the 1870s, when the Goulds and the Drews and the Vanderbilts battled each other for control of American railroads. The new wave of hostile takeovers has already profoundly altered the contours and landmarks of the American economy. It has become a dominant force—many would say *the* dominant force—in the behavior and actions of American management, and, almost certainly, a major factor in the erosion of American competitive and technological leadership. Yet the papers usually report it only on the financial page. And very few people, outside of business, really quite know what goes on or, indeed, what a hostile takeover really is.

The hostile takeover usually begins with a *raider*—a company or an individual who is legally incorporated and works through a corporation—buying a small percentage of the target company's share capital on the open market, usually with money borrowed expressly for this purpose. When, and as the raider expects, the target's board of directors and its management spurn his takeover bid, the raider borrows more money—sometimes several billion dollars—buys more of

First published in *The Public Interest,* 1986.

the target's shares on the market, and goes directly to the target's stockholders, offering them substantially more than the current share price on the stock exchange. If enough of the target's shareholders accept to give the raider complete control, he then typically unloads the debt he has incurred in the takeover onto the company he has acquired. In a hostile takeover the victim thus ends up paying for his own execution.

The raider not only now controls a big company: he has made a tidy profit on the shares he bought at the lower market price. Even if the takeover attempt fails, the raider usually wins big. The target may only be able to escape the raider by finding a *white knight,* that is, someone who is less odious to the management of the target company and willing to pay even more for its shares, including those held by the raider. Alternatively, the target company pays ransom to the raider— which goes by the Robin Hood-like name of *greenmail*—and buys out the shares the raider acquired at a fancy price, way beyond anything its earnings and prospects could justify.

Hostile takeovers were virtually unknown before 1980. Harold Geneen, who built ITT into the world's largest and most diversified conglomerate in the 1960s and 1970s, made literally hundreds of acquisitions—perhaps as many as a thousand. But he never made an offer to a company unless its management had first invited him to do so. Indeed, in a good many of Geneen's acquisitions the original initiative came from the company to be acquired; it offered itself for sale. In those days it would have been impossible to finance hostile takeovers: no bank would have lent money for such a purpose. But since 1980 they have become increasingly easy to finance.

At first, hostile takeovers were launched by large companies intent on rapid growth or rapid diversification. This phase reached a climax in 1982 with a months-long battle of three giants: Bendix (defense and automotive), Martin-Marietta (defense, aerospace, and cement), and Allied (chemicals). Bendix began the fight with a hostile takeover bid for Martin-Marietta, which promptly counterattacked with a hostile takeover bid for Bendix. When these two, like two scorpions in a bottle, had finished each other off, Allied joined the fray, paid ransom to an exhausted Martin-Marietta, took over Bendix, and in the process ousted the Bendix management that had started the battle.

Since then, raiders increasingly are individual stock-market operators whose business is the hostile takeover. Some, like Carl Icahn, range over the lot, attacking all kinds of business. T. Boone Pickens, originally a small, independent oil producer, specializes in large petroleum companies—his targets have included such major companies as Gulf Oil, Phillips Petroleum, and Union Oil. Ted Turner of Atlanta specializes in the media and was embroiled in a hostile takeover bid for the smallest of the three television networks, CBS. But there are dozens of smaller raiders abroad, many of them looking for fast-growing medium-size companies, especially companies in such currently "sexy" fields as electronics, computers, or biotechnology. Others primarily raid financial institutions. Practically all of them do so on money borrowed at high interest rates.

Why the Raider Succeeds

How many hostile takeover bids there have been, no one quite knows. Conservative estimates run to four hundred or five hundred, with at least one-half ending in the disappearance of the target company either because the raider succeeds or because the target finds a white knight. Such a massive phenomenon—whether considered destructive or constructive—surely bespeaks fundamental changes in the underlying economic structure and the environment of American business and the American economy. Yet to my knowledge there has so far been practically no discussion of what might explain the takeover phenomenon, of its meaning, and of the policy questions it raises.

What, for instance, explains the *vulnerability* of companies, among them a good many big, strong, well-established ones? Few of the raiders have much financial strength of their own. Most have little managerial or business achievement behind them. In the 1960s and early 1970s, the managements of big, publicly owned companies were widely believed to be impregnable; nothing short of the company's bankruptcy could threaten, let alone dislodge, them. It was then considered almost a "self-evident truth" in highly popular books (those of John Kenneth Galbraith, for instance) that we had moved into "corporate capitalism" as a new and distinct "stage," one in which professional managers perpetuated themselves and ran the country's big business autonomously, ruling without much interference from any of their supposed "constituencies." But in the last few years, any number of companies, and particularly large companies doing well by any yardstick, have been swallowed up by hitherto unknown and obscure newcomers despite the most vigorous defense by their management.

These raiders often have no capital of their own, but have to borrow every penny they need to buy a small percentage of the company's stock and then to make their takeover bid. By now, to bar a hostile takeover bid even giants like General Motors are forced into expensive and complicated subterfuges such as splitting their shares into a number of different issues, each with different voting rights. What has happened to corporate capitalism and to the absolute control by professional autonomous management, seemingly so firmly established only a little while ago?

Fundamentally, there are three explanations for this extreme vulnerability of established companies to the hostile takeover.

One explanation is inflation.

Then there are structural changes within the economy that make a good many of yesterday's most successful companies no longer appropriate to today's economic realities.

Finally, *corporate capitalism*—that is, the establishment of a management accountable only to itself—has made managements and companies exceedingly vulnerable. They have no constituencies to come to their succor when attacked.

Inflation distorts: It distorts values. It distorts relationships. It creates glaring discrepancies between economic assumptions and economic realities. The fifteen years of inflation in America which began during Lyndon Johnson's presidency and continued into the early 1980s were no exception. And the most predictable, indeed the most typical, distortion of inflation is between the value of assets and their earning power. *In any inflation the cost of capital goods tends to rise much faster than the price of the goods they produce. It thus becomes economical to buy already existing capital assets rather than to invest in new facilities and new machines.* So any company that is rich in fixed assets is worth more when dismembered—that is, when its capital assets are being sold as pieces of real estate, as factories, as machinery and equipment—than it is worth on a realistic price/earnings ratio based on the value of its output. This is one of the distortions that the raiders exploit.

The stock market values companies on the basis of their earnings. It values them, in other words, as "going concerns." It does not value them on their liquidation value. As a result, a company heavy with fixed assets—and especially a company that also has a lot of cash with which the raider, after the takeover, can repay himself (and make a sizable profit to boot)—becomes a most inviting target. This situation accounts for one-quarter, perhaps even one-third, of all the successful hostile takeovers.

Equally important are the tremendous structural changes in the American and world economies in the last fifteen years. They have made inappropriate a good many of the traditional forms of economic integration. The best example is probably the large, integrated petroleum company. One need not sympathize with Mr. T. Boone Pickens, the raider who successfully forced one of the country's largest and proudest petroleum companies, Gulf Oil, into a shotgun marriage with a white knight and almost succeeded in taking over two other old and well-established petroleum companies, Union Oil and Phillips. But Pickens has a point. He has forced the petroleum companies at which he leveled his guns to do something sensible: that is, to split the company into two parts, one making and selling petroleum products, one keeping reserves of crude oil in the ground.

Large integrated petroleum companies have performed poorly since 1980 or 1981. Their earnings basically reflect a crude oil price of around $12 or $15 a barrel, which is what the market price would have been all along had there been no OPEC cartel. But all of them have tried desperately, since the OPEC oil shock of 1973, to build up underground crude oil reserves for the future. And these reserves were priced in the market, and especially by people looking for a long-term tax shelter, on the expectation of a petroleum price many times the present price, twenty or thirty years hence. In fact, to justify what the market was paying for crude and crude oil reserves in the ground, one would have to assume a petroleum price of around $100 a barrel by the year 2015 or so; otherwise, on a discounted cash-flow basis, the present valuation of these proven underground reserves could not possibly be justified.

Whether the expectation of high petroleum prices twenty or thirty years hence is rational is not the point. (Every historical experience would indicate that the only rational expectation is for the petroleum prices thirty years hence to be lower than they are today—but this is another issue.) The fact is that it makes little sense today to be an "integrated" petroleum company. The interests of the people who want the earnings of the present petroleum company, and the interests of the people who look for a long-term tax shelter (and who, in other words, do not care much about present earnings), are not compatible. Therefore Pickens's proposal, that the integrated petroleum company split itself into two pieces, made sense.

A similar situation exists in the steel industry, and in fact in a good many of the traditional large-scale, integrated, capital-intensive materials producers. Every one of these situations invites a raider.

But perhaps the biggest single reason companies are vulnerable to the raider is "corporate capitalism" itself: that is, autonomous management, accountable to no one, controlled by no one, and without constituents. It has made management arrogant. And far from making management powerful, corporate capitalism has actually made it impotent. Management has become isolated and has lost its support base, in its own board of directors, among its own stockholders, and among its own employees.

Wherever a management threatened by a raider has been able to organize a "constituency," it has beaten off the hostile takeover. One example is Phillips Petroleum in Bartlesville, Oklahoma, which mobilized the employees and the community; this was enough to defeat Pickens. But where managements have given in to the temptation to become omnipotent they have in effect rendered themselves impotent. When they are then attacked, they have nobody to support them if someone offers a few dollars above the current market price to the company shareholders.

Where the Cash Comes From

The vulnerability of the victims does not, by itself, explain how the raiders finance their takeover. To mount a hostile takeover bid for a large company takes a very large war chest. One and a half billion dollars is just about the minimum needed to attack a big company. In some recent cases the amount went up to $4 billion. It has to be in cash, as a rule. To be sure, if the takeover bid succeeds, the target company then pays. But the money has to be available from the beginning—that is, when it is by no means sure that the takeover bid will succeed. If the takeover bid is launched by an individual, as more and more of them have been in recent years, there usually is no security whatever for the cash that the raider has to borrow. The raider himself usually has negligible assets, certainly compared to the sums needed. Even if the takeover bid is being launched by another large company, the amount needed to finance it is usually way beyond anything the company could normally raise as additional debt. *Yet the only "security" for the loan that a raider has to obtain is the promise of repayment if the raid is successful.* This is hardly what was once normally considered a "bankable" loan, yet the raiders have had no difficulty obtaining these loans. Indeed, when the financing of hostile takeover bids switched (mostly for regulatory reasons) from being done by means of bank loans to being done by means of bonds, the market promptly named the bonds *junk bonds,* and with good reason. Nevertheless, there is no difficulty in getting such bonds underwritten and issued, with commercial banks being avid buyers.

Bank loans—or junk bonds—to finance hostile takeovers are available for the same reason that enabled countries like Brazil, Zaire, or Argentina, in the early 1980s, to obtain loans from Western banks in amounts that were clearly beyond their capacity to pay interest on (let alone to repay the loan itself), and for the same reason that large money-center banks, such as Continental Illinois in Chicago, were willing, indeed eager, to snap up highly speculative and sometimes fraudulent loans often to nonexistent oil and gas speculators. *The American commercial bank is pinched by the shrinkage of its traditional sources of income and almost desperate to find new ones, and especially to find borrowers willing to pay very high interest rates.* And the raider who makes a hostile takeover bid is, of course, perfectly willing to promise very high interest rates; after all, he will not pay them—the company he is aiming to take over will, after it has succumbed.

Commercial banks, as every textbook states, make their living as *liquidity arbitrageurs:* They obtain their money from "demand deposits" which have perfect liquidity—that is, the right to withdraw the money at any time. The bank then lends out that money for longer periods of time (from ninety days to three years is the ordinary time span of a commercial loan); so the amounts owed to the bank have far less liquidity than the amounts it owes. This, then, justifies the bank's charging a substantially higher interest rate, with the difference between the interest rate on what the bank lends out and the interest rate on what the bank borrows being the bank's income.

Increasingly, this does not work anymore, for the bank either is not able to be the liquidity arbitrageur or does not get paid for it. One reason is, of course, the zero-interest demand deposits, once prescribed by the regulatory authorities, have all but disappeared. Historically, businesses have provided the bulk of the demand deposits. But few businesses these days keep large cash supplies, and the typical checking account of individuals now pays 5.5 percent interest. Adding to this the costs of administration, acquisition, and so on, the bank probably ends up paying 8 or 9 percent for the money on deposit in customers' checking accounts—which means that even demand deposits no longer provide a substantial "interest spread." And most American consumers today keep in their checking account only a minimum balance. The rest is in accounts that pay much higher interest, such as money-market accounts, which still allow high liquidity.

On the demand side, too, the liquidity arbitrage has become far less profitable. Increasingly, American businesses do not finance themselves through commercial loans, but through *commercial paper*—the business version of an installment loan. This, however, bypasses the banking system. The company with a temporary cash surplus directly buys commercial paper issued by another company with a temporary cash need. But the "spread" on commercial paper between what the borrower pays and what the lender gets is much lower than that between the traditional noninterest on demand deposits and the bank's lending rate on commercial loans. That spread may be 1.5 percent as against 4 or 5 percent previously.

By now, most U.S. banks, especially the larger ones, know that they cannot hope to continue to build their business on the "spread" of interest rates between what they pay for their money and what they charge for it. They will have to shift their income base to fees and commissions. But even those few banks which accepted this ten years ago and which since then have been working hard on shifting their source of income from being paid for money to being paid for information and service—Citibank in New York was probably the first, and is by far the leader—still have a very long way to go. And in the meantime the banks are hurting for sources of income. Hence the pressure on them to look for borrowers willing to pay higher interest— or at least to promise they will—whether they be Oklahoma wildcatters, military governments engulfed by inflation in their own country (such as Brazil and Argentina), or takeover raiders.

The Lure of Easy Money

That the raiders can obtain money they need still does not explain why the shareholders team up with the raider to take over, merge, or liquidate the company they own.

They do not do so, it is abundantly clear, because they believe the takeover will be advantageous to the company. On the contrary, the shareholders clearly know that the takeover bid is usually a disaster for the company. Increasingly, they sell their shares to the raider only if they get cash, that is, if they get out of the company and have nothing to do with it anymore. Or, if they take securities in exchange, they immediately sell them. And yet, again and again, the shareholders, in their great majority, either accept the bid of the raider or turn it down only if a white knight offers more than the raider does. But then they also immediately sell whatever the white knight has given them in exchange against their holdings in the company that has disappeared.

The shareholders who control our large, publicly held companies simply have no choice but to accept the raider's offer. They are forced, perhaps even legally forced, to accept the raider's bid if it is higher than the currently quoted price for the stock. This is essentially a result of the shift of share ownership from individuals to institutions who are "trustees," and especially to pension funds. Pension funds (increasingly also mutual funds) are the legal "owners" of the publicly owned companies of America, with their holdings amounting to about 50 percent of the common shares. The percentage is even higher for large companies because the institutional holders concentrate on them. The people who manage these large and growing aggregations of capital, especially the pension fund managers, are trustees rather than owners. They are trustees both for the management of the companies which owe the pensions to their employees and for the ultimate beneficiaries, the employees and future pensioners. As trustees they have, however, little choice about whether they want to sell their shares if someone bids substantially above what the same shares fetch at the market price. They have to accept. If they were to say no, they would lay themselves open to an enormous and uninsurable liability. The trustees could be sued by both the management of the company and the ultimate beneficiaries, the employees, were those shares, six months later, quoted below the price the raider had offered. Trustees do not have the right to superimpose *their* judgment on what a "prudent man" would do. And a prudent man surely will take the bird in the hand, especially if there is no reason to believe that there is a bird in the bush.

Pension fund managers are also under tremendous pressure to show better-than-average returns—and yet are unable to do so as a rule. The pension funds of most American businesses are *defined-benefit* plans: The company promises to pay the employee upon retirement a fixed proportion of the employee's salary, usually 60 percent or so of the last five years' earnings. What the employee is to receive is fixed, or rather will be fixed by the time the employee reaches retirement. What the company contributes, however, is flexible. The contribution payable by the company goes down if the pension fund increases in value—if, for instance, it shows a high return or profit from its investments. If the pension fund does not show earnings, or shows earnings lower than anticipated, the company's contribution goes up.

This is in sharp contrast to plans based on a *defined contribution*, under which what the company pays in each year is fixed, with the employee upon retirement receiving either a fixed stipend or a variable one which depends upon what the earnings of the pension fund have been.

In the defined-benefit plan, therefore, management constantly pushes the pension fund manager to show profits, especially from investments, so that the company's contribution can be minimized. But this is a total delusion. It is in fact an impossibility. The pension funds by now *are* the American stock market. And if one is the market, one cannot possibly beat it. The performance record of the pension funds bears this out. It has been abysmal, almost without exception. In fact, the desire to "beat the market" is in itself the reason that most pension funds have performed substantially worse than the market. As a result, the pension funds waste their substance by supporting a huge stock market that only fritters away in commissions the money that should go to the future beneficiaries. In the long and checkered history of investment and finance, there is probably no more uniformly dismal record than that of American pension fund management in the last twenty years.

And yet company managements still believe that their fund can "beat the odds"—the way each slot machine player in Las Vegas believes that he can beat them. And the pension fund manager who does not operate short term, and who refuses to speculate, to trade, and to show "results" over the next three months, is likely to lose the account quickly. There is probably no more competitive industry around. This makes irresistible an offer by the raider to pay $55 for a share that is quoted at $40 on the market.

Pension fund managers know that the raider's bid is deleterious to the company whose stock they own. But they cannot consider the welfare and interests of their "property." They are not owners. They are of necessity speculators, even though they are legally vested with the owner's power. And so they behave as speculators. They have to accept the raider's bid unless a white knight makes a better offer.

The Dangers of Defensiveness

The wave of hostile takeovers is a result of profound structural changes in the American economy. But it is in itself a serious disorder. There is a great deal of discussion about whether hostile takeovers are good or bad for the shareholders. There can be absolutely no doubt, however, that they are exceedingly bad for the economy. They force management into operating short term. More and more of our businesses, large, medium-size, and small, are not being run for business results but for protection against the hostile takeover. This means that more and more of our businesses are forced to concentrate on results in the next three months. They are being run so as to encourage the institutional investors, on which all publicly traded companies today depend for their supply of capital, to hold onto the company shares rather than to toss them overboard the moment the first hostile takeover bid appears.

But worse still, companies are being forced to do stupid things to prevent themselves from being raided. It is, for instance, becoming dangerous for any company to be liquid. Liquidity can only attract the raider who can expect to repay himself, and the debt he incurs in bidding for the company, out of the company's own cash. And thus companies who find themselves in a liquid position, no matter how much cash they may need only a few months further on, hasten to squander the cash—for instance, in buying up something that is totally alien to their own business and has only one advantage: it absorbs a lot of money. Even worse, companies increasingly cut back on expenses for the future, such as research and development. One of the most ominous developments for the future of America is the speed with which the Japanese are taking over the markets of the rapidly industrializing countries: Brazil, for instance, or India. They do so because they can invest in the distribution system in these countries in anticipation of the future market. American company managements are perfectly aware of this. But when asked why they do not do likewise, they tend to say, "We cannot afford to set aside this money and invest it in tomorrow. We need it to make a good showing in next month's or next quarter's profits."

The fear of the raider is undoubtedly the largest single cause for the increasing tendency of American companies to manage for the short term and let the future go hang. The fear of the raider demoralizes and paralyzes. The impact on the morale of management people and of professional people in the company can hardly be overestimated. And worse still, after the successful takeover, the morale in a company is destroyed, often forever. The people who can leave, do. The others do their minimum. "What's the point in my trying to do a good job if the rug will be pulled out from under me tomorrow?" is a frequent comment. Add to this that the raiders, in order to reimburse themselves, usually start out by selling off the company's most promising businesses. Hence the impact of a takeover on morale is total catastrophe.

Altogether, the record is poor for all companies that have been merged, especially into a conglomerate or into a business with which they had little in common: for example, the typical financial conglomerate. Only three out of every ten such acquiring companies do as well two years later as they did before the merger. But the record of companies that have been acquired in a hostile takeover is uniformly dismal.

Clearly the hostile takeover cannot be justified as leading to a more efficient allocation of resources. Most of them have no aim except to enrich the raider. To achieve this end, he offers the stockholders more money for their shares than they would get on the market, which is to say, he bribes them. And to be able to pay the bribe he loads a heavy debt on the company that is being taken over, which by itself severely impairs the company's potential for economic performance. The fact that, almost without exception, the result of the hostile takeover is also a demoralization and severe impairment of the human organization disproves the argument that the hostile takeover results in a more efficient allocation of resources. Actually, all it proves is that "resources" in the modern business enterprise are not primarily bricks and mortar—or even oil in the ground. They are the human organization.

There are indeed cases where a human organization becomes more productive by being dissociated from its former enterprise, by being set up separately—in fact, a good many of today's large organizations, and especially the conglomerates, would greatly increase their productivity by being split into smaller units, or by establishing parts as separate businesses. But this is not what the hostile takeover accomplishes. On the contrary, the most valuable parts of the acquired business are invariably put on the block after a hostile takeover so as to raise money to pay off some of the debt. And this impairs both their productivity and that of the remaining assets.

There are serious questions about resource allocation in the American economy. But the hostile takeover is clearly not the right tool to bring about a more efficient allocation. It does severe damage to the true productive resource, the human organization, its spirit, its dedication, its morale, its confidence in its management, and its identification with the enterprise that employs its people.

Even if hostile takeovers are "good for the shareholders"—and they are "good" only for the very shortest time—they are surely not good for the economy. They are indeed so bad that we will be forced to put an end to them, one way or another.

One way to do so might be to emulate the British and create a "takeover panel" with the power to stop takeover bids considered to be contrary to the best long-term interest of the enterprise and the economy. Whether such a panel could be set up in this country— or whether it would just become another troublesome government agency—is very debatable. The way the British are doing it would immediately run afoul of our antitrust laws.

It is therefore more probable that we will put an end to the hostile takeover—or at least put serious obstacles in its path—by abandoning the concept of "one share, one vote" and go to shares that, although participating equally in profits (and in the proceeds of a liquidation), have unequal voting power, at least as long as the company shows adequate financial results. General Motors has already gone this way, and quite a few smaller firms are following. This would not be a radical departure. The British, for many years, had the *private limited company,* in which management held the voting power as long as it showed specified results. Similarly, the Germans, for well over a hundred years, have had the *Kommanditgesellschaft auf Aktien,* in which management holds the majority of the voting power even though it has a very small minority of the share ownership—again, as long as there is adequate performance and results. In other words, a shift to a system in which different classes of shares have differential voting power—with Class A shares, for instance, having one hundred times the votes of Class B shares—would only need a few fairly simple safeguards to be functional: First, vesting the Class A shares, with their superior voting power, in a truly independent and strong board of directors rather than in management, a board on which independent outside directors have a clear majority (which is what the Germans require, by the way). Second, making the extra voting strength of the Class A shares conditional on the company's showing specified results and adequate performance. Thus a two-class system of shares would control the hostile takeover and yet give protection against managerial malperformance and even against managerial mediocrity.

But perhaps the takeover wave will come to an end with a whimper rather than a bang by losing its financial underpinnings. It would not take much to bring this about. One default on one big loan to finance a hostile takeover, one "billion-dollar-scandal" similar to the Penn Square collapse in Oklahoma that brought down Chicago's mighty Continental Illinois Bank—and there would be no more money available to finance hostile takeovers. And in the long history of finance, every single scheme that lured lenders into advancing money for economically nonproductive purposes by promising them returns substantially above the going market rate has come to a bad end sooner or later—and usually sooner.

Even if we control the hostile takeover, however, there will remain the underlying structural problems of which the hostile takeover is only a symptom. It clearly raises basic questions about the following: the role, functions, and governance of pension funds; the legitimacy of management; and finally, the purpose of business enterprise, especially large enterprise. Are the stockholders the *only* constituents to whom all other interests, including that of the enterprise itself as a going concern, must be totally subordinated?

Where Wall Street Meets Las Vegas

Abatement of the boom in hostile takeovers can, paradoxically, only aggravate the pension fund problem. It would eliminate the windfall profits pension funds now reap by accepting the inflated prices the raider offers. These windfall profits are, by and large, the only way for pension fund managers to obtain the quick stock-market gains that their bosses, the company's managers, expect and demand of them. If they are eliminated, company managers will predictably put even greater pressure for quick results on the people who manage their company's pension fund; and those in turn will put even greater pressure for short-term results on the companies in the shares of which the fund invests—thus pushing businesses even further toward managing for the short term, which, as by now everybody agrees, is a significant factor in the steady erosion of America's competitive position in the world economy. As long as American pension funds are based on "defined benefits," no relief is in sight.

It might have made sense for the labor unions to push for defined benefits when pension funds first became widespread some thirty years ago. For under that system, the employer shoulders all future risks. Actually, in the negotiation of the contract that set the pattern for today's pension fund, the General Motors contract in the spring of 1950, the United Automobile Workers strongly opposed defined benefits and wanted "defined contributions" instead.[1] It was the company that chose defined benefits, even though General Motors' president at the time, Charles E. Wilson, recommended a defined-contributions plan as financially much sounder. He as well as Reuther were, however, overruled by GM's finance committee, and the rest of the country's pension plans then followed GM's lead. Yet the choice of defined benefits—as we now realize, a major blunder—was based on the same delusion that makes people in Las Vegas believe they will "make a fortune" if only they keep on feeding more quarters into the slot machine.

Under defined benefits the company commits itself to paying the future pensioner a fixed percentage of his wage or salary during his retirement years. The contribution the company makes is then dependent on the value of the fund's assets in a given years, as against the present value of the future pension obligation. The higher the fund's present value the lower the current contribution, and vice versa. And so management deluded itself into believing that an ever-rising New York stock market is a law of nature—or at least of history—and that, therefore, under a defined-benefits plan, it would be the stock market, through its preordained continuing rise, that would eventually provide the money to discharge the pension obligation rather than the company itself. Indeed, quite a few managements then promised their boards of directors that the defined-benefits plan would in the long run become a "money spinner" for the company and would produce income well above anything it would have to put into the plan.

And then the second delusion: Practically every company adopting a defined-benefits pension plan did so with the firm belief that its own pension plan, if only administered "professionally," would "beat" *even* an ever-rising market.

There is, of course, no law that prescribes an ever-rising stock market. Stock markets historically have not even been particularly effective as hedges against inflation: The American market in the last twenty years, for instance, barely kept pace with it. Indeed a large pool of money which over any long period of time grows at all, let alone as fast as the economy, is the rarest of all exceptions—neither the Medici nor the Fuggers, nor the Rothschilds nor the Morgans, succeeded in this endeavor. Similarly, no large-company pension fund, to the best of my knowledge, has over the last twenty or thirty years done as well as the stock market. The only performance that counts in the pension fund is performance over the long run, because the obligations extend over twenty-five years or longer. Indeed, some of the large defined-contributions funds have produced better results for their beneficiaries, the employees and future pensioners, and at lower cost to the employers, than the great majority of the large defined-benefits plans. This is most definitely true of the largest of the defined-contributions plans, that of the employers and employees of the American nonprofit institutions, the Teachers Insurance and Annuity Association.

The misconceptions that led American management into opting for defined benefits thus practically guaranteed from the beginning that the funds would have to become "speculators" and increasingly short term in their focus.

The choice of defined benefits also explains in large part the poor *social* performance of the American pension fund. For it was conceived, as Walter Reuther quite rightly suspected, as much to be a social as to be a financial institution. It was conceived to create a visible center of common interest between employer and employee. Indeed, what General Motors had in mind was very similar to what the Japanese, acting quite independently a few years later, achieved by "lifetime employment." But unlike lifetime employment the American pension fund has not created any community of interest in the worker's mind.

The laws that govern the private pension plans in America define their managers as "trustees" for the eventual beneficiaries, the employees. In reality the managers of defined-benefits plans are of necessity appointed by and accountable only to company management. For the one at risk is the employer—and so the fund has to be run to reduce, as much as possible, the burden on the employer. As a result, the employees feel no responsibility for the fund. It is "deferred wages" rather than a "stake in the company." The employees do not feel that it makes any difference how the fund performs. And they are right: Unless the company goes bankrupt it does not make any difference to them. They also, in a defined-benefits plan, cannot in any meaningful way be brought into the decision-making process through the pension fund, the actual owners of America's productive resources.

A defined-contributions plan is no panacea, but it minimizes the problems. The right model was easily available thirty years ago—for the Teachers Insurance and Annuity Association goes back much further, to the early 1920s. The TIAA runs on "flexible contributions" rather than on defined contributions. The contribution made each year into the plan by a university, or a nonprofit organization such as the Boy Scouts, or for a minister in a Protestant denomination, is a fixed percentage of salary. It goes up as the employee advances. It thereby also—an important point—automatically adjusts the annual contribution to inflation. And yet this annual premium is known and predictable. And because of this, the TIAA can, and does indeed, invest for the long term, which in turn explains why its results have been better than those of any of the large defined-benefits plans. At the same time, the TIAA, precisely because the employer has discharged his obligation in full once he remits a fixed percentage of the employee's salary, has been able to bring the future beneficiaries, that is, today's employees, into the government of the institution. University faculty do not consider the TIAA the "employer's pension fund"; it is "our pension fund," in which they take an active interest and which to them meaningfully symbolizes the basic identity of economic interest between their employer, that is, the university, and themselves.

Pension plans are beginning to change, albeit very slowly. Many companies, especially medium-size ones, now encourage employees to make their own pension provisions, often out of company-provided funds, for example, through an Individual Retirement Account. That at least makes possible a rational (that is, a long-term) investment policy. But for the foreseeable future the bulk of our corporate pension plans will remain committed to the defined-benefits formula. The principle legal owners of our large companies, the large pension funds, will therefore continue to be forced to act as speculators rather than as investors, let alone as owners. And thus there will, for the foreseeable future, be a need to protect the economy's wealth-producing resources—that is, its businesses, and the pension funds themselves—against the pressure to manage for the immediate, the short term, the next month, the next quarter, and above all against the takeover.

Demise of the Corporate Guardians

Corporate capitalism—the rule of autonomous managers as the "philosopher-kings" of the modern economy accountable at best to a professional code but not controlled by shareholders or by any other constituency—was first proclaimed more than fifty years ago by Adolph Berle and Gardner Means in their 1932 classic, *The Modern Corporation and Private Property.* "Control," Berle and Means argued, had become divorced from "property." Indeed, property was no longer ownership. It had become investment, concerned only with dividends and capital gains but not with the welfare or the governance of the property itself.

From the beginning, anyone with any knowledge of political theory or political history could have predicted that this would not work. Management, one could confidently say, would not last any longer as philosopher-king than any earlier philosopher-kings had, which was never very long. Management has power. Indeed, to do its job, it *has* to have power. But power does not last, regardless of its performance, its knowledge, and its good intentions, unless it be grounded in some sanction outside and beyond itself, some "grounds of legitimacy," whether divine institution, or election, or the consent of the governed. Otherwise, power is not legitimate. It may be well-meaning, it may perform well, it may even test out as "highly popular" in surveys and polls. Yet illegitimate power always succumbs to the first challenger. It may have no enemies, but no one believes in it, either, and no one owes it allegiance.

This should have been obvious to American management fifty years ago when Berle and Means first pointed out that there was no more real ownership in the American corporation. After all, that the philosopher-king—that is, power grounded in performance rather than in legitimacy—would not last had been known since Aristotle's dismissal of Plato's philosopher-king, all of twenty-three hundred years ago. But American management did exactly what all earlier philosopher-kings have done—for example, the "enlightened despots" of eighteenth-century Europe. It gloried in its good intentions. And American managements busily engaged in removing what they considered as the last obstacle to their enlightened rule, an independent and powerful board of directors. And then, when the investors of Berle and Means became the speculators of the pension fund, management found itself powerless against the first challenger, the raider. The hostile takeover bid is thus the final failure of corporate capitalism.

But we do need management. The business enterprise needs a government, and it needs a government that has power, has continuity, and can perform. In other words, it needs a government that has legitimacy. How can legitimacy be restored to the management of America's large, publicly owned companies?

One step, surely the first one, is to restore an independent and strong board of directors. Indeed, as has already been pointed out, where such a board exists, raiders, by and large, have been beaten off. Even shareholders whose only interest is the quick buck are likely to listen to a strong and independent board that has standing and respect in the community and is not dismissed as management's puppet. The hostile takeover may thus finally succeed in bringing about the reform and restoration of the strong, independent board of directors which a great many voices within the business community have been demanding for many years.

But such a board would not and could not be a representative of the shareholders alone. The board member who commands enough respect to be listened to is likely to be an *independent director,* that is, somebody who does not represent any constituency, including the nominal owners, but rather the integrity and the interest of the enterprise itself. The hostile takeover is thus almost certain to speed up a development that is already under way: the emergence of professionally qualified men and women who serve on a very small number of boards, maybe no more than four at a time; who have independent respect and standing in a broader community based on their achievements and known integrity; and who take seriously their responsibilities, including the responsibility to set performance goals for top management and to police them, to monitor the behavior and ethics of top management, and to remove even the proudest chief executive officer who does not live up to the standards set by the board in the interest of the enterprise.

But is this not simply replacing one set of philosopher-kings by another group of technocrats or wise men? To be sure, independent outside board members, unlike a company president, do not fight for their own jobs when they resist a takeover. But they still do not represent any identifiable constituency, neither do they have any grounds of legitimacy other than disinterested performance and knowledge. Will the large public corporation in America have to learn to mobilize new constituents, to bring in other "interests" to balance the former owners now become speculators, and to create new bonds of allegiance?

Reportedly, would-be raiders refrain from making a hostile takeover bid for a company in which employees hold a substantial portion of the shares. They know that employee-owners are not likely to accept the raider's offer. Most employees stand to lose more, of course, if their jobs are endangered than they can possibly gain by getting more for their stock. Above all employees identify with the company and are personally and emotionally attached to its remaining independent. And the most spectacular defeat of a hostile takeover bid was not achieved by a management with a strong performance record. It was the previously mentioned defeat of the bid for Phillips Petroleum in Bartlesville, Oklahoma, when the town itself rallied to the defense of its major employer.

Thirty years ago it was popular in American big business to talk of management as being the "trustee for the best-balanced interests of stockholders, employees, plant community, customers, and suppliers alike." In many cases, of course, this was pure phrase meant to clothe with respectability the managerial philosopher-king and his enlightened despotism. But even where there was more to this assertion than self-interest, nothing has been done as a rule to convert the phrase into reality. Few attempts have been made to institutionalize the relationship of these supposed "constituencies" to the enterprise and its management. Will this now have to be undertaken in earnest to safeguard both enterprise and management? And what form should such institutional relationships take?

The Challenge to Free Enterprise

The question being most hotly debated currently is whether hostile takeovers are good or bad for shareholders. But what other groups may have a legitimate stake in the fight for the control and survival of the enterprise is probably more important, though less discussed. Does the modern, publicly owned, large enterprise exist *exclusively* for the sake of the shareholders? This is, of course, what orthodox "capitalism" asserts. But the term *free enterprise* was coined forty or fifty years ago to assert that the shareholder interest, although important, is only one interest and that the enterprise has functions well beyond that of producing returns for the shareholder—functions as an employer, as a citizen of the community, as a customer, and as a supplier. The British, in establishing a "take-over panel," have expressly recognized that a decision on mergers and takeovers affects the public interest. So far in the United States, this is expressed only negatively, that is, in forbidding mergers that violate antitrust laws. Will we have to bring in considerations of the impact on other groups and on the community and economy as a whole—and in which form? That is the central question. The answers this country will give to it will largely define the future shape of the American economy.

If the answer is, however, that the speculator's interest—never mind that the speculator has legal title as an owner—is the only interest to be considered, the free-enterprise system is unlikely to survive. It will rapidly lose public support. For most people, even though they do benefit—however indirectly (that is, as ultimate beneficiaries in a pension fund)—from the speculator's game, stand to lose more from the hostile takeover as employees, whether blue-collar or managers, and as citizens of a community. And more and more people are concerned with the hostile takeover as a moral issue. It deeply offends the sense of justice of a great many Americans.

Most Americans today are employees of an organization. There is a good deal of evidence that people in an organization, and especially managerial and professional people, will accept even the most painful adjustment, such as the closing of a business or the sale of parts of it, if the rationale is economic performance of the lack thereof. But this of course is not the rationale for the purchase or sale of the human organization or of its parts in the hostile takeover. There the only rationale is to enrich somebody who has nothing to do with the performance of the enterprise and who, quite admittedly, has not the slightest interest in it. And this goes against the grain of employees who feel that the hostile takeover treats them as "chattel" and not as a "resource," let alone as human beings. "Is the hostile takeover even compatible with our laws against peonage and involuntary servitude?" the middle-level executives in my advanced-management classes have been asking me of late.

Almost a hundred years ago the United States decided that the rights of the creditor are not absolute and amended its bankruptcy laws to put maintenance and restoration of the "going concern" ahead of the rights of the creditor, which till then had ruled when a business got into difficulties. This has worked remarkably well. The protection of the going concern during reorganization has indeed proved to be in the ultimate interest of the creditor, too. Will we now do the same thing with respect to the hostile takeover and given consideration to the protection of the going concern as a resource, and to the interests of employees, whether blue-collar, white-collar, or managerial; of the community; of suppliers and customers? Actually we are already moving in this direction through extending the protection of the bankruptcy laws to nonbankrupt going concerns threatened by subjection to one single interest. The Johns-Manville Corporation—a leading producer of asbestos and other building materials—successfully invoked the bankruptcy laws to preserve itself as a going concern and to protect its shareholders and its employees against a tidal wave of asbestos-damage liability suits. Continental Airlines similarly used the bankruptcy laws successfully to preserve itself against union wage claims that had become unbearable when deregulation made airfares and airline routes hotly competitive. It is by no means inconceivable that a clever lawyer will similarly construe the bankruptcy laws to preserve the going concern against the hostile takeover—and that the courts will go along as they did in the Johns-Manville and Continental Airlines cases. But one way or another—by law, by moving to multitier stock ownership, or by judicial exegesis—we surely will find a way to protect the going concern against hostile takeovers that subordinate all other interests—employees, the enterprise's long-range growth and prosperity, and the country's competitive position in an increasingly competitive world economy—to short-term speculative gain.

Note

1. I had occasion to discuss this issue repeatedly with Walter Reuther, then head of the UAW, when the contract was under negotiation. Reuther feared, groundlessly as it turned out, that "defined benefits" would make the worker conscious of the identity of interest between himself and the employer and thus alienate him from the union.

Part Five

Work, Tools, and Society

Introduction to Part Five

The essays in this section—excepting only the short one "India and Appropriate Technology"—all came out of work on yet another book I planned but never wrote. It was to be entitled "A Short History of Work"—something no one has ever attempted to do (and I found out I couldn't do it either). Its motto was a quotation from Alfred Russell Wallace, whose theory of evolution coincided with Charles Darwin's, with which the first essay in this section opens: "Man, alone of all animals, is capable of purposeful, non-organic evolution; he makes tools." And its purpose was to write the history of technology as a history of society.

Technologists—engineers or chemists—view technology as having to do with tools: ploughs, camshafts, airplane propellers. Economists—with few exceptions (i.e., Joseph Schumpeter) see technology as a dangerous outsider—akin to hurricanes or earthquakes—an exterior force that plays hob with their nice models of equilibrium. Sociologists and anthropologists, amazingly enough, do not see technology at all. The one major exception was Thorstein Veblen—but Max Weber, for instance, hardly mentioned it.

And yet, next to the bond of family and kinship, the work bond is the strongest social bond. The organization of work, fully as much as the organization of kinship and family, shapes community and determines social order. And the organization of work is, in turn, largely determined by technology, by tools, by motive power, by materials. That is the theme of the last essay in this part, "The First Technological Revolution and Its Lessons," which deals with the technological revolution which, more than 6000 years ago, created the Irrigation City and the Irrigation Empire. The only major thinker who paid attention to this was Karl Marx. What he called the "Mode of Production'' is quite similar to what I call the "work bond." But Marx lost his insight when he decided—in a fatal mistake—to make "ownership of the means of production" the determining factor of social order. We now know that ownership of the means of production has a great deal to do with productivity. We know, in other words, that collective or government ownership is unlikely to result in efficient production—for the simple reason that governments underwrite the losses of government-owned enterprises. But we also know that ownership of the means of production has nothing, absolutely nothing, to do with social organization and community structure. The productivity of a government-owned automobile plant in Communist East Germany or Soviet Russia was a fraction of that of the worst-run privately owned and locally run automobile plant in the West, not to mention the productivity of a Japanese automobile plant. But the social organization was exactly the same and dictated by the work bond which in turn was dictated by the technology of mass production and of the assembly line. And so the History of Work was intended to view technology as a human, nay a social, phenomenon rather than as a merely "technical" one, and society as shaped by, and formed around, work and work bond.

And while, as said already, nothing came out of the planned book, the essays in this part present the basic argument and, I hope, with enough supporting evidence to make the case.

18

Work and Tools

Man, alone of all animals, is capable of purposeful nonorganic evolution; he makes tools. This observation by Alfred Russell Wallace, co-discoverer with Darwin of the theory of evolution, may seem obvious if not trite. But it is a profound insight. And though made some seventy or eighty years ago, its implications have yet to be thought through by biologists and technologists.

One such implication is that from a biologist's (or a historian's) point of view, the technologist's identification of tool with material artifact is quite arbitrary. Language, too, is a tool, and so are all abstract concepts. This does not mean that the technologist's definition should be discarded. All human disciplines rest after all on similarly arbitrary distinctions. But it does mean that technologists ought to be conscious of the artificiality of their definition and careful lest it become a barrier rather than a help to knowledge and understanding.

This is particularly relevant for the history of technology, I believe. According to the technologist's definition of "tool," the abacus and the geometer's compass are normally considered technology, but the multiplication table or table of logarithms is not. Yet this arbitrary division makes all but impossible the understanding of so important a subject as the development of the technology of mathematics. Similarly the technologist's elimination of the fine arts from his field of vision blinds the historian of technology to an understanding of the relationship between scientific knowledge and technology. (See, for

First published in *Technology and Culture*, 1959.

instance, volumes III and IV of Singer's monumental *History of Technology*.) For scientific thought and knowledge were married to the fine arts, at least in the West, long before they even got on speaking terms with the mechanical crafts: in the mathematical number theories of the designers of the Gothic cathedral,[1] in the geometric optics of Renaissance painting, or in the acoustics of the great Baroque organs. And Lynn T. White, Jr. has shown in several recent articles that to understand the history and development of the mechanical devices of the Middle Ages we must understand something so nonmechanical and nonmaterial as the new concept of the dignity and sanctity of labor which St. Benedict first introduced.

Even within the technologist's definition of technology as dealing with mechanical artifacts alone, Wallace's insight has major relevance. The subject matter of technology, according to the preface to *History of Technology*, is "how things are done or made"; and most students of technology, to my knowledge, agree with this. But the Wallace insight leads to a different definition: the subject matter of technology would be "how man does or makes." As to the meaning and end of technology, the same source, again presenting the general view, defines them as "mastery of his [man's] natural environment." Oh no, the Wallace insight would say (and in rather shocked tones): the purpose is to overcome man's own natural, that is, animal, limitations. Technology enables man, a land-bound biped, without gills, fins, or wings, to be at home in the water or in the air. It enables an animal with very poor body insulation, that is, a subtropical animal, to live in all climate zones. It enables one of the weakest and slowest of the primates to add to his own strength that of elephant or ox, and to his own speed that of the horse. It enables him to push his life span from his "natural" twenty years or so to threescore years and ten; it even enables him to forget that natural death is death from predators, disease, starvation, or accident, and to call death from natural causes that which has never been observed in wild animals: death from organic decay in old age.[2]

These developments of man have, of course, had impact on his natural environment—though I suspect that until recent days the impact has been very slight indeed. But this impact on nature outside of man is incidental. What really matters is that all these developments alter man's biological capacity—and not through the random genetic mutation of biological evolution but through the purposeful nonorganic development we call technology.

What I have called here the "Wallace insight," that is, the approach from human biology, thus leads to the conclusion that technology is not about things: tools, processes, and products. It is about work: the specifically human activity by means of which man pushes back the limitations of the iron biological law which condemns all other animals to devote all their time and energy to keeping themselves alive for the next day, if not for the next hour. The same conclusion would be reached, by the way, from any approach, for instance, from that of the anthropologist's "culture," that does not mistake technology for a phenomenon of the physical universe. We might define technology as human action on physical objects or as a set of physical objects characterized by serving human purposes. Either way the realm and subject matter of the study of technology would be human work.

For the historian of technology this line of thought might be more than a quibble over definitions. For it leads to the conclusion that the study of the development and history of technology, even in its very narrowest definition as the study of one particular mechanical artifact (either tool or product) or a particular process, would be productive only within an understanding of work and in the context of the history and development of work.

Not only must the available tools and techniques strongly influence what work can and will be done, but how it will be done. Work, its structure, organization, and concepts must in turn powerfully affect tools and techniques and their development. The influence, one would deduce, should be so great as to make it difficult to understand the development of the tool or of the technique unless its relationship to work was known and understood. Whatever evidence we have strongly supports this deduction.

Systematic attempts to study and to improve work only began some seventy-five years ago with Frederick W. Taylor. Until then work had always been taken for granted by everyone—as it is still, apparently, taken for granted by most students of technology. "Scientific Management," as Taylor's efforts were called misleadingly ("scientific work study" would have been a better term and would have avoided a great deal of confusion), was not concerned with technology. Indeed, it took tools and techniques largely as given and tried to enable the individual worker to manipulate them more economically, more systematically, and more effectively. And yet this approach resulted almost immediately in major changes and development in tools, processes, and products. The assembly line with its conveyors was an important tool change. An even greater change was the change in process that underlay the switch from building to assembling a product. Today we are beginning to see yet another powerful consequence of Taylor's work on individual operations: the change from organizing production around the doing of things to things to organizing production around the flow of things and information, the change we call "automation".

A similar, direct impact on tools and techniques is likely to result from another and even more recent approach to the study and improvement of work: the approach called variously "human engineering," "industrial psychology", or "industrial physiology". Scientific Management and its descendants study work as operation; human engineering and its allied disciplines are concerned with the relationship between technology and human anatomy, human perception, human nervous system, and human emotion. Fatigue studies were the earliest and most widely known examples; studies of sensory perception and reaction, for instance, of aircraft pilots, are among the presently most active areas of investigation, as are studies of learning. We have barely scratched the surface here; yet we know already that these studies are leading us to major changes in the theory and design of instruments of measurement and control, and into the redesign of traditional skills, traditional tools, and traditional processes.

But of course we worked on work, if only through trial and error, long before we systematized the job. The best example of Scientific Management is after all not to be found in our century: it is the alphabet. The assembly line as a concept of work was understood by those unknown geniuses who, at the very beginning of historical time, replaced the aristocratic artist of warfare (portrayed in his last moments of glory by Homer) by the army soldier with his uniform equipment, his few repetitive operations, and his regimented drill. The best example of human engineering is still the long handle that changed the sickle into the scythe, thus belatedly adjusting reaping to the evolutionary change that had much earlier changed man from crouching quadruped into upright biped. Every one of these developments in work had immediate and powerful impact on tools, process, and product, that is, on the artifacts of technology.

The aspect of work that has probably had the greatest impact on technology is the one we know the least about: the organization of work.

Work, as far as we have any record of man, has always been both individual and social. The most thoroughly collectivist society history knows, that of Inca Peru, did not succeed in completely collectivizing work; technology—in particular, the making of tools, pottery, textiles, and cult objects—remained the work of individuals. It was personally specialized rather than biologically or socially specialized, as is work in a beehive or in an ant heap. The most thoroughgoing individualist society, the perfect market model of classical economics, presupposed a tremendous amount of collective organization in respect to law, money and credit, transportation, and so on. But precisely because individual effort and collective effort must always be calibrated with one another, the organization of work is not determined. To a very considerable extent there are genuine alternatives here, genuine choices. The organization of work, in other words, is in itself one of the major means of that purposeful and non-organic evolution which is specifically human; it is in itself an important tool of man.

Only within the very last decades have we begun to look at the organization of work.[3] But we have already learned that the task, the tools, and the social organization of work are not totally independent but mutually influence and affect one another. We know, for instance, that the almost preindustrial technology of the New York women's dress industry is the result not of technological, economic, or market conditions but of the social organization of work which is traditional in that industry. The opposite has been proven, too: when we introduce certain tools into locomotive shops, for instance, the traditional organization of work, the organization of the crafts, becomes untenable; and the very skills that made men productive under the old technology now become a major obstacle to their being able to produce at all. A good case can be made for the hypothesis that modern farm implements have made the Russian collective farm socially obsolete as an organization of work, have made it yesterday's socialist solution of farm organization rather than today's, let alone tomorrow's.

This interrelationship between organization of work, tasks, and tools must always have existed. One might even speculate that the explanation for the mysterious time gap between the early introduction of the potter's wheel and the so very late introduction of the spinning wheel lies in the social organization of spinning work as a group task performed, as the Homeric epics describe it, by the mistress working with her daughters and maids. The spinning wheel with its demand for individual concentration on the machinery and its speed is hardly conducive to free social intercourse; even on a narrowly economic basis, the governmental, disciplinary, and educational yields of the spinning bee may well have appeared more valuable than faster and cleaner yarn.

If we know far too little about work and its organization scientifically, we know nothing about it historically. It is not lack of records that explains this, at least not for historic times. Great writers—Hesiod, Aristophanes, Virgil, for instance—have left detailed descriptions. For the early empires and then again for the last seven centuries, beginning with the High Middle Ages, we have an abundance of pictorial material: pottery and relief paintings, woodcuts, etchings, prints. What is lacking is attention and objective study.

The political historian or the art historian, still dominated by the prejudices of Hellenism, usually dismisses work as beneath his notice; the historian of technology is "thing-focused." As a result, we not only still repeat as fact traditions regarding the organization of work in the past which both our available sources and our knowledge of the organization of work would stamp as old wives' tales. We also deny ourselves a fuller understanding of the already existing and already collected information regarding the history and use of tools.

One example of this is the lack of attention given to materials-moving and materials-handling equipment. We know that moving things—rather than fabricating things—is the central effort in production. But we have paid little attention to the development of materials-moving and materials-handling equipment.

The Gothic cathedral is another example. H. G. Thomson in *History of Technology* (II, 384) states flatly, for instance, "there was no exact medieval equivalent of the specialized architect" in the Middle Ages; there was only "a master mason". But we have overwhelming evidence to the contrary (summarized, for instance, in Simson[4]); the specialized, scientifically trained architect actually dominated. He was sharply distinguished from the master by training and social position. Far from being anonymous, as we still commonly assert, he was a famous man, sometimes with an international practice ranging from Scotland to Poland to Sicily. Indeed, he took great pains to make sure that he would be known and remembered, not only in written records but above all by having himself portrayed in the churches he designed in his full regalia as a scientific geometer and designer—something even the best known of today's architects would hesitate to do. Similarly we still repeat early German Romanticism in the belief that the Gothic cathedral was the work of individual craftsmen. But the structural fabric of the cathedral was based on strict uniformity of parts. The men worked to moulds which were collectively held and administered as the property of the guild. Only roofing, ornaments, doors, statuary, windows, and so on, were individual artists' work. Considering both the extreme scarcity of skilled people and the heavy dependence on local, unskilled labor from the countryside to which all our sources attest, there must also have been a sharp division between the skilled men who made parts and the unskilled who assembled them under the direction of a foreman or a gang boss. There must thus have been a fairly advanced materials-handling technology which is, indeed, depicted in our sources but neglected by the historians of technology with their uncritical Romanticist bias. And while the moulds to which the craftsman worked are generally mentioned, no one, to my knowledge, has yet investigated so remarkable a tool, and one that so completely contradicts all we otherwise believe we know about medieval work and technology.

I do not mean to suggest that we drop the historical study of tools, processes, and products. We quite obviously need to know much more. I am saying that the history of work is in itself a big, rewarding, and challenging area which students of technology should be particularly well equipped to tackle. I am saying also that we need work on work if the history of technology is truly to be history and not just the engineer's antiquarianism.

One final question must be asked: without study and understanding of work, how can we hope to arrive at an understanding of technology? Singer's great *History of Technology* abandons the attempt to give a comprehensive treatment of its subject with 1850; at that time, the editors tell us, technology became so complex as to defy description, let alone understanding. But it is precisely then that technology began to be a central force and to have major impact both on man's culture and on man's natural environment. To say that we cannot encompass modern technology is very much like saying that medicine stops when the embryo issues from the womb. We need a theory that enables us to organize the variety and complexity of modern tools around some basic, unifying concept.

To a layman who is neither professional historian nor professional technologist, it would, moreover, appear that even the old technology, the technology before the great explosion of the last hundred years, makes no real sense and cannot be understood, can hardly even be described, without such basic concepts. Every writer on technology acknowledges the extraordinary number, variety, and complexity of factors that play a part in technology and are in turn influenced by it: economy and legal system, political institutions and social values, philosophical abstractions, religious beliefs, and scientific knowledge. No one can know all these, let alone handle them all in their constantly shifting relationship. Yet all of them are part of technology in one way or another, at one time or another.

The typical reaction to such a situation has of course always been to proclaim one of these factors as *the* determinant—the economy, for instance, or the religious beliefs. We know that this can only lead to complete failure to understand. These factors profoundly influence but do not determine each other; at most they may set limits to each other or create a range of opportunities for each other. Nor can we understand technology in terms of the anthropologist's concept of culture as a stable, complete, and finite balance of these factors. Such a culture may exist among small, primitive, decaying tribes, living in isolation. But this is precisely the reason why they are small, primitive, and decaying. Any viable culture is characterized by capacity for internal self-generated change in the energy-level and direction of any one of these factors and in their interrelationships.

Technology, in other words, must be considered as a system,[5] that is, a collection of interrelated and intercommunicating units and activities.

We know that we can study and understand such a system only if we have a unifying focus where the interaction of *all* the forces and factors within the system registers some discernible effect, and where in turn the complexities of the system can be resolved in one theoretical model. Tools, processes, and products are clearly incapable of providing such focus for the understanding of the complex system we call technology. It is just possible, however, that work might provide the focus, might provide the integration of all these interdependent, yet autonomous variables, might provide one unifying concept which will enable us to understand technology both in itself and in its role, its impact on and relationships with values and institutions, knowledge and beliefs, individual and society.

Such understanding would be of vital importance today. The great, perhaps the central, event of our times is the disappearance of all non-Western societies and cultures under the inundation of Western technology. Yet we have no way of analysing this process, of predicting what it will do to man, his institutions and values, let alone of controlling it, that is, of specifying with any degree of assurance what needs to be done to make this momentous change productive or at least bearable. We desperately need a real understanding, and a real theory, a real model of technology.

History has never been satisfied to be a mere inventory of what is dead and gone—that, indeed, is antiquarianism. True history always aims at helping us understand ourselves, at helping us make what shall be. Just as we look to the historian of government for a better understanding of government, and to the historian of art for a better understanding of art, so we are entitled to look to the historian of technology for a better understanding of technology. But how can he give us such an understanding unless he himself has some concept of technology and not merely a collection of individual tools and artifacts? And can he develop such a concept unless work rather than things becomes the focus of his study of technology and of its history?

Notes

1. S. B. Hamilton only expresses the prevailing view of technologists when he says (in Singer's *History of Technology*, IV, 469) in respect to the architects of the Gothic cathedral and their patrons that there is "nothing to suggest that either party was driven or pursued by any theory as to what would be beautiful". Yet we have overwhelming and easily accessible evidence to the contrary; both architect and patron were not just "driven", they were actually obsessed by rigorously mathematical theories of structure and beauty. See, for instance, Sedlmayr, *Die Entste-*

hung der Kathedrale (Zürich, 1950); von Simson, *The Gothic Cathedral* (London: Routledge and Kegan Paul, 1956); and especially the direct testimony of one of the greatest of the cathedral designers, Abbot Suger of St-Denis, in *Abbot Suger on the Abbey Church of St-Denis and its Art Treasures*, ed. Erwin Panofsky (Princeton, 1946).

2. See on this Sir P. B. Medawar, the British biologist, in "Old Age and Natural Death" in his *The Uniqueness of the Individual* (London: Methuen, 1957).

3. Among the studies ought to be mentioned the work of the late Elton Mayo, first in Australia and then at Harvard, especially his two slim books: *The Human Problems of an Industrial Civilization* (2nd edn, Boston, 1946) and *The Social Problems of an Industrial Civilization* (Boston, 1945); the studies of the French sociologist Georges Friedmann, especially his *Industrial Society* (Glencoe: Free Press, 1964); the work carried on at Yale by Charles R. Walker and his group, especially the book by him and Robert H. Guest: *The Man on the Assembly Line* (Cambridge, Mass: Harvard University Press, 1952). I understand that studies of the organization of work are also being carried out at the Polish Academy of Science, but I have not been able to obtain any of the results.

4. O. G. von Simson, *The Gothic Cathedral* (London: Routledge and Kegan Paul, 1956).

5. The word is here used as in Kenneth Boulding's "General Systems Theory—The Skeleton of Science", *Management Science*, vol. 2, no. 3 (April 1956), p.197, and in the publications of the Society for General Systems Research.

19

Technology, Science, and Culture

The standard answer to the question, "What brought about the explosive change in the human condition these last two hundred years?" is "The Progress of Science." This paper enters a demurrer. It argues that the right answer is more likely: "A fundamental change in the concept of technology." Central to this was the reordering of old technologies into systematic public disciplines with their own conceptual equipment, for example, the "differential diagnosis" of nineteenth-century medicine. In the century between 1750 and 1850 the three main technologies of man—Agriculture, the Mechanical Arts (today's Engineering), and Medicine—went in rapid succession through this process, which resulted almost immediately in an agricultural, an industrial, and a medical "revolution" respectively.

This process owed little or nothing to the new knowledge of contemporary science. In fact, in every technology the practice with its rules of thumb was far ahead of science. Technology, therefore, became the spur to science; it took, for instance, seventy-five years until Clausius and Kelvin could give a scientific formulation to the thermodynamic behavior of Watt's steam engine. Science could, indeed, have had no impact on the Technological Revolution until the transformation from craft to technological discipline had first been completed.

But technology had an immediate impact on science, which was transformed by the emergence of systematic technology. The change was the most fundamental one—a change in science's own definition

First published in *Technology and Culture*, 1961.

and image of itself. From being "natural philosophy", science became a social institution. The words in which science defined itself remained unchanged: "the systematic search for rational knowledge". But "knowledge" changed its meaning from being "understanding", that is, focused on man's mind, to being "control", that is, focused on application in and through technology. Instead of raising, as science had always done, fundamental problems of metaphysics, it came to raise, as it rarely had before, fundamental social and political problems.

It would be claiming too much to say that technology established itself as the paramount power over science. But it was technology that built the future home, took out the marriage license, and hurried a rather reluctant science through the ceremony. And it is technology that gives the union of the two its character; it is a coupling of science *to* technology, rather than a coupling of science and technology.

The evidence indicates that the key to this change lies in new basic concepts regarding technology, that is, in a genuine Technological Revolution with its own causes and its own dynamics.

Of all major technologies medicine alone has been taught systematically for any length of time. An unbroken line leads back for one thousand years, from the medical school of today to the medical schools of the Arab caliphates. The trail, though partly overgrown, goes back, another fourteen hundred years, through the School of Alexandria to Hippocrates. From the beginning, medical schools taught both theoretical knowledge and clinical practice, engaged simultaneously, therefore, in science and technology. Unlike any other technologist in the West, the medical practitioner has continuously enjoyed social esteem and position.

Yet, until very late—1850 or thereabouts—there was no organized or predictable relationship between scientific knowledge and medical practice. The one major contribution to health care which the West made in the Middle Ages was the invention of spectacles. The generally accepted date is 1286; by 1290 the use of eyeglasses is fully documented.[1] This invention was, almost certainly, based directly upon brand-new scientific knowledge, most probably on Roger Bacon's optical experiments. Yet Bacon was still alive when spectacles came in—he died in 1294. Until the nineteenth century there is no other example of such all but instantaneous translation of new scientific knowledge into technology—least of all in medicine. Yet Galen's theory of vision, which ruled out any mechanical correction, was taught in the medical schools until 1700.[2]

Four hundred years later, in the age of Galileo, medicine took another big step—Harvey's discovery of the circulation of the blood, the first major new knowledge since the Ancients. Another hundred years, the Jenner's smallpox vaccination brought both the first specific treatment and the first prevention of a major disease.

Harvey's findings disproved every single one of the theoretical assumptions that underlay the old clinical practice of bleeding. By 1700 Harvey's findings were taught in every medical school and repeated in every medical text. Yet bleeding remained the core of medical practice and a universal panacea for another hundred years, and was still applied liberally around 1850.[3] What killed it finally was not scientific knowledge—available and accepted for two hundred years—but clinical observation.

In contrast to Harvey, Jenner's achievement was essentially technological and without any basis in theory. It is perhaps the greatest feat of clinical observation. Smallpox vaccination had hard sledding—it was, after all, a foolhardy thing deliberately to give oneself the dreaded pox. But what no one seemed to pay any attention to was the complete incompatibility of Jenner's treatment with any biological or medical theory of the time, or of any time thereafter until Pasteur, one hundred years later. That no one, apparently, saw fit to try explaining vaccination or to study the phenomenon of immunity appears to us strange enough. But how can one explain that the same doctors who practised vaccination, for a century continued to teach theories which vaccination had rendered absurd?

The only explanation is that science and technology were not seen as having anything to do with one another. To us it is commonplace that scientific knowledge is being translated into technology, and vice versa. This assumption explains the violence of the arguments regarding the historical relationship between science and the "useful arts". But the assumptions of the debate are invalid: the presence of a tie proves as little as its absence—it is our age, not the past, which presumes consistency between theory and practice.

The basic difference was not in the content but in the focus of the

two areas. Science was a branch of philosophy, concerned with understanding. Its object was to elevate the human mind. It was misuse and
degradation of science to use it—Plato's famous argument. Technology, on the other hand, was focused on use. Its object was increase of
the human capacity to do. Science dealt with the most general, technologies with the most concrete. Any resemblances between the two
were 'purely coincidental'.[4] There are no hard and fast dates for a
major change in an attitude, a world view. And the Technological
Revolution was nothing less. We do know, however, that it occurred
within the half-century 1720 to 1770—the half-century that separates
Newton from Benjamin Franklin.

Few people today realize that Swift's famous encomium on the man
who makes two blades grow where one grew before, was not in praise
of the scientists. On the contrary, it was the final, crushing argument in
a biting attack on them, and especially against the august Royal
Society. It was meant to extol the sanity and benefits of nonscientific
technology against the arrogant sterility of an idle inquiry into nature
concerned with understanding; this is against Newtonian Science,
for Swift was, as always, on the unpopular side. But his basic
assumption—that science and application were radically different and
worlds apart—was clearly the prevailing one in the opening decades of
the eighteenth century. No one scientist spoke out against the weirdest
technological "projects" of the South Sea Bubble of 1720, even
though their theoretical infeasibility must have been obvious to them.
Many, Sir Isaac Newton taking the lead, invested heavily in them.[5]
And while Newton, as Master of the Royal Mint, reformed its business
practices, he did not much bother with its technology.

Fifty years later, around 1770, Dr Franklin is the "philosopher" par
excellence and the West's scientific lion. Franklin, though a first-rate
scientist, owed his fame to his achievements as a technologist—
"artisan" in eighteenth-century parlance. He was a brilliant gadgeteer,
as witness the Franklin stove and bifocals. Of his major scientific
exploits, one—the investigation of atmospheric electricity—was immediately turned into useful application: the lightning rod. Another,
his pioneering work in oceanography with its discovery of the Gulf
Stream, was undertaken for the express purpose of application, namely, to speed up the transatlantic mail service. Yet the scientists hailed
Franklin as enthusiastically as did the general public.

In the fifty years between 1720 and 1770—not a particularly distinguished period in the history of science, by the way—a fundamental change in the attitude towards technology, both of laity and of scientists, must have taken place. One indication is the change in English attitude towards patents. During the South Sea Bubble they were still unpopular and attacked as "monopolies". They were still given to political favorites rather than to an inventor. By 1775 when Watt obtained his patent, they had become the accepted means of encouraging and rewarding technological progress.

We know in detail what happened to technology in the period which includes both the Agricultural Revolution and the opening of the Industrial Revolution. Technology as we know it today, that is, systematic, organized work on the material tools of man, was born then. It was produced by collecting and organizing existing knowledge, by applying it systematically, and by publishing it. Of these steps the last one was both the most novel—craft skill was not for nothing called a "mystery"—and the most important.

The immediate effect of the emergence of technology was not only rapid technological progress: it was the establishment of technologies as systematic disciplines to be taught and learned and, finally, the reorientation of science towards feeding these new disciplines of technological application.

Agriculture[6] and the mechanical arts[7] changed at the same time, though independently.

Beginning with such men as Jethro Tull and his systematic work on horse-drawn cultivating machines in the early years of the seventeenth century and culminating towards its end in Coke of Holkham's work on balanced large-scale farming and selective livestock breeding, agriculture changed from a "way of life" into an industry. Yet this work would have had little impact but for the systematic publication of the new approach, especially by Arthur Young. This assured both rapid adoption and continuing further work. As a result, yields doubled while manpower needs were cut in half—which alone made possible that large-scale shift of labor from the land into the city and from producing food to consuming food on which the Industrial Revolution depended.

Around 1780, Albrecht Thaer in Germany, an enthusiastic follower of the English, founded the first agricultural college—a college not of "farming" but of "agriculture". This in turn, still in Thaer's lifetime, produced the first, specifically application-focused new knowledge, namely, Liebig's work on the nutrition of plants, and the first science-based industry, fertilizer.

The conversion of the mechanical arts into a technology followed the same sequence and a similar timetable. The hundred years between the 1714 offer of the famous £20,000 prize for a reliable chronometer and Eli Whitney's standardization of parts was, of course, the great age of mechanical invention—of the machine tools, of the prime movers, and of industrial organization. Technical training, though not yet in systematic form, began with the founding of the *École des Pontes et Chaussées* in 1747. Codification and publication in organized form goes back to Diderot's *Encyclopédie*, the first volume of which appeared in 1750. In 1776—that miracle year that brought the Declaration of Independence, *The Wealth of Nations*, Blackstone's *Commentaries*, and Watt's first practical steam engine—the first modern technical university opened: the *Bergakademie* (Mining Academy) in Freiberg, Saxony. Significantly enough, one of the reasons for its establishment was the need for technically trained managers created by the increasing use of the Newcomen steam engine, especially in deep-level coal mining.

In 1794, with the establishment of the *École Polytechnique* in Paris, the profession of engineer was established. And again, within a generation, we see a reorientation of the physical sciences—organic chemistry and electricity begin their scientific career, being simultaneously sciences and technologies. Liebig, Woehler, Faraday, Henry, Maxwell were great scientists whose work was quickly applied by great inventors, designers, and industrial developers.

Only medicine, of the major technologies, did not make the transition in the eighteenth century. The attempt was made—by the Dutchman Gerhard van Swieten,[8] not only a great physician but politically powerful as adviser to the Habsburg court. Van Swieten attempted to marry the clinical practice which his teacher Boerhaave had started at Leyden around 1700 with the new scientific methods of such men as the Paduan Morgagni whose *Pathological Anatomy* (1761)[9] first treated diseases as afflictions of an organ rather than as "humours". But—a lesson one should not forget—the very fact that medicine (or rather, something by that name) was already respectable and organized as an academic faculty defeated the attempt. Vienna relapsed into medical scholasticism as soon as van Swieten and his backer, the Emperor Joseph II, died.

It was only after the French Revolution had abolished all medical schools and medical societies that a real change could be effected. Then another court physician, Corvisart, Napoleon's doctor, accomplished, in Paris around 1820, what van Swieten had failed in. Even then, opposition to the scientific approach remained powerful enough to drive Semmelweis out of Vienna and into exile when he found, around 1840, that traditional medical practices were responsible for lying-in fever with its ghastly death toll. Not until 1850, with the emergence of the modern medical school in Paris, Vienna, and Wuerzburg, did medicine become a genuine technology and an organized discipline.

This, too, happened, however, without benefit of science. What was codified and organized was primarily old knowledge, acquired in practice. Immediately *after* the reorientation of the practice of medicine, the great medical scientists appeared—Claude Bernard, Pasteur, Lister, Koch. And they were all application-focused, all driven by a desire to do, rather than by a desire to know.

We know the results of the Technological Revolution, and its impacts. We know that, contrary to Malthus, food supply in the last two hundred years has risen a good deal more than an exploding human population. We know that the average life span of man a hundred and fifty years ago was still close to the "natural life span": the twenty-five years or so needed for the physical reproduction of the species. In the most highly developed and prosperous areas, it has almost tripled. And we know the transformation of our lives through the mechanical technologies, their potential, and their dangers.

Most of us also know that the Technological Revolution has resulted in something even more unprecedented: a common world civilization. It is corroding and dissolving history, tradition, culture, and values throughout the world, no matter how old, how highly developed, how deeply cherished and loved.

And underlying this is a change in the meaning and nature of knowledge and of our attitude to it. Perhaps one way of saying this is that the non-Western world does not want Western science primarily because it wants better understanding. It wants Western science be-

cause it wants technology and its fruits. It wants control, not understanding. The story of Japan's Westernization between 1867 and her
emergence as a modern nation in the Chinese War of 1894 is the
classical, as it is the earliest, example.[10]

But this means that the Technological Revolution endowed technology with a power which none of the "useful arts"—whether agricultural, mechanical, or medical—had ever had before: impact on
man's mind. Previously, the useful arts had to do only with how man
lives and dies, how he works, plays, eats, and fights. How and what he
thinks, how he sees the world and himself in it, his beliefs and values,
lay elsewhere—in religion, in philosophy, in the arts, in science. To
use technological means to affect these areas was traditionally
"magic"—considered at least evil, if not asinine to boot.

With the Technological Revolution, however, application and cognition, matter and the mind, tool and purpose, knowledge and control
have come together for better or worse.

There is only one thing we do not know about the Technological
Revolution—but it is essential: what happened to bring about the basic
change in attitudes, beliefs, and values which released it? Scientific
progress, I have tried to show, had little to do with it. But how
responsible was the great change in world outlook which, a century
earlier, had brought about the great Scientific Revolution? What part
did the rising capitalism play? And what as the part of the new,
centralized national state with its mercantilistic policies on trade and
industry and its bureaucratic obsession with written, systematic, rational procedures everywhere? (After all, the eighteenth century codified
the laws as it codified the useful or applied arts.) Or do we have to do
here with a process, the dynamics of which lie in technology? Is it the
"progress of technology" which piled up to the point when it suddenly
turned things upside down, so that the "control" which nature had
always exercised over man now became, at least potentially, control
which man exercises over nature?

This should be, I submit, a central question both for the general
historian and for the historian of technology.

For the first, the Technological Revolution marks one of the great
turning points—whether intellectually, politically, culturally, or economically. In all four areas the traditional—and always unsuccessful—

drives of systems, powers, and religions for world domination are replaced by a new and highly successful world-imperialism, that of technology. Within a hundred years, it penetrates everywhere and puts, by 1900, the symbol of its sovereignty, the steam engine, even into the Dalai Lama's palace in Lhasa.

For the historian of technology, the Technological Revolution is not only the cataclysmic event within his chosen field; it is the point at which such a field as technology emerges. Up to that point there is, of course, a long and exciting history of crafts and tools, artifacts and mechanical ingenuity, slow, painful advances and sudden, rapid diffusion. But only the historian, endowed with hindsight, sees this as technology, and as belonging together. To contemporaries, these were separate things, each belonging to its own sphere, application, and way of life.

Neither the general historian nor the historian of technology has yet, however, concerned himself much with the Technological Revolution. The first—if he sees it at all—dismisses technology as the bastard child of science. The only general historian of the first rank (excepting only that keen connoisseur of techniques and tools, Herodotus) who devotes time and attention to technology, its role and impact is, to my knowledge, Franz Schnabel.[11] That Schnabel taught history at a technical university (Karlsruhe) may explain his interest. The historians of technology, for their part, tend to be historians of materials, tools, and techniques rather than historians of technology. The rare exceptions tend to be nontechnologists such as Lewis Mumford or Roger Burlingame who, understandably, are concerned more with the impact of technology on society and culture than with the development and dynamics of technology itself.

Yet technology is important today precisely because it unites both the universe of doing and that of knowing, connects both the intellectual and the natural histories of man. How it came thus to be in the center—when it always before had been scattered around the periphery—has yet to be probed, thought through, and reported.

Notes

1. E. Rosen, "The Invention of Eyeglasses", *Journal for the History of Medicine*, vol. 11 (1956), pp. 13–46, 183–218.

2. It was among the great Boerhaave's many "firsts" to have taught the first course in ophthalmology and to examine actual eyes—in 1708 in Leyden. Newton's *Optics* was the acknowledged inspiration. (See George Sarton, "The History of Medicine versus the History of Art", *Bulletin of the History of Medicine*, vol. 10 (1941), pp. 123–35.)

3. Bleeding actually reached a peak in the 1820s, when it was touted as the universal remedy by no less an authority than Broussais, the most famous professor at the Paris Academy of Medicine. According to Henry E. Sigrist (*Great Doctors*; London: Allen and Unwin, 1933), it became so popular that in the one year, 1827, thirty-three million leeches were important into France.

4. There was, to be sure, one famous dissent, one important and highly effective approach to science as a means to doing and as a foundation for technology. Its greatest spokesman was St. Bonaventura, the thirteenth-century antiphonist to St. Thomas Aquinas (see especially St. Bonaventura's *Reduction of all Arts to Theology*). A hundred years earlier the dissenters actually dominated in the twelfth-century Platonism of the theologian-technologist schools of St. Victoire and Chartres, builders alike of mysticism and of the great cathedrals. On this see Charles Homer Haskins, *The Renaissance of the Twelfth Century* (Cambridge, Mass: Harvard University Press, 1927); Otto von Simson, *The Gothic Cathedral* (London: Routledge and Kegan Paul, 1956); and *Abbot Suger on the Abbey Church of St. Denis and its Art Treasures*, edited by Erwin Panofsky (Princeton, 1946).

The dissenters did not, of course, see material technology as the end of knowledge; rational knowledge was a means towards the knowledge of God or at least His glorification. But knowledge, once its purpose was application, immediately focused on material technology and purely worldly ends—as St. Bernard pointed out in his famous attack on Suger's 'technocracy' as early as 1127.

The dissent never died down completely. But after the Aristotelian triumph of the thirteenth century, it did not again become respectable, let alone dominant until the advent of Romantic Natural Philosophy in the early nineteenth century, well *after* the Technological Revolution and actually its first (and so far only) literary offspring. It is well known that there was the closest connection between the Romantics—with Novalis their greatest poet, and with Schelling their official philospher—and the first major discipline which, from its inception, was always both science and technology: organic chemistry. Less well known is the fact that the Romantic movement, its philosophers, writers, and statesmen came largely out of the first technical university, the Mining Academy in Freiberg (Saxony) that had been founded in 1776.

5. J. Carswell, *The South Sea Bubble* (London: Cresset Press, 1960).

6. G. E. Fussell, *The Farmer's Tools, 1500–1900* (London, 1952); A. J. Bourde, *The Influence of England on the French Agronomes* (Cambridge, 1953); A. Demolon, *L'Evolution Scientifique et l'Agriculture Française* (Paris, 1946); R. Krzymowski; *Geschichte der deutschèn Landwirtschaft* (Stuttgart, 1939).

7. A. P. Usher, *History of Mechanical Inventions* (Rev Ed, Cambridge, Mass, 1954); also the same author's 'Machines & Mechanisms' in vol. III of Singer, *et al*, *A History of Technology* (Oxford, 1957); J. W. Roe, *English and American Tool Builders* (London, 1916); K. R. Gilbreth, 'Machine Tools', in *History of Technology*, vol. IV (Oxford, 1958); on early technical education see: Franz Schnabel, *Die Anfaenge des Technischen Hochschulwesens* (Freiburg, 1925).

8. The standard biography of van Swieten is W. Mueller, *Gerhard van Swieten* (Vienna, 1883); on the organized resistance of academic medicine to the scientific approach, see G. Strakosch-Grassmann, *Geschichte des oesterreichischen Unterrichtswesens* (Vienna, 1905).

9. This is the name commonly used for the work. Its actual title was *De Sedibus et causis morborum per anatomen indigatis*; the first English translation appeared in 1769 under the title, *The Seats and Causes of Diseases Investigated by Anatomy*.

10. This is brought out most clearly in William Lockwood, *The Economic Development of Japan, 1868–1938* (Princeton: University Press, 1954).

11. Franz Schnabel, *Deutsche Geschichte im 19. Jahrhundert* (4 Vols, Freiburg iB, 1929–1937); the discussion of technology and medicine is found chiefly in vol. III.

20

India and Appropriate Technology

"The big mistake Gandhi made was to advocate the spinning wheel," said one influential Indian government economist. "It's much too efficient. With the unemployment and underemployment we have in our villages the truly appropriate technology is the hand-held spindle, the spinning whorl." Yet this is hardly how the Indian villagers define "appropriate technology" for themselves.

What struck me most when traveling in the winter of 1978–79 for six weeks through rural India was not the pervasive poverty or the palpable unemployment; I had expected both. What I had not expected, however, were the four or five brand-new bicycles standing outside every one of the miserable hovels—and not one of them chained or locked. There may still be more bullock carts in rural India than bicycles; there surely are still infinitely more bullock carts than there are small tractors. But what powers India's Green Revolution, what has given the subcontinent a food surplus for the first time in its thousands of years, is not the digging stick or the wooden plow. It is the ubiquitous gasoline pump in the tub well and the irrigation ditch in an arid land.

From every bullock cart, every camel cart, every pedicab, and every howdah on the back of an elephant there issue the strains of the transistor radio. And the most crowded stand in every one of the countless village markets is the one that sells motor scooters on the installment plan.

First published in *The Wall Street Journal*, 1979.

Much as it pains the Indian government economist and his boss, the Prime Minister, the bicycle, transistor radio, gasoline pump, and motor scooter—rather than the spinning wheel, let alone the distaff or the spinning whorl—are indeed the appropriate technology for India and for most developing countries. They create jobs and purchasing power—the distaff would destroy both.

No one in India could tell me what the economic policy of the government is. The only governmental actions are expansion of already large government enterprises, unchecked growth of an already obese bureaucracy, and more bureaucratic regulations. The cabinet cannot agree on anything and has no policy whatever. Substantial sums are being allocated to the villages but without programs, let alone goals. But there is a pervasive rhetoric of smallness and of antitechnology.

India's Prime Minister at the time of my trip (i.e., before Mrs. Gandhi's return to power), Morarji Desai, eighty-four years old but looking fifty-five (which he attributes to his eating only raw mashed vegetables and drinking his own urine), preached to me "small is beautiful," "rural development," and "appropriate [that is, preindustrial] technology." It is this rhetoric that his economic adviser echoes when he counsels a return to the spinning whorl. And pretty much the same rhetoric can now be heard in many developing countries, for example in Indonesia or from the Islamic fundamentalists in Iran. It is very much the same rhetoric that underlay the disastrous Great Leap Forward in Maoist China twenty years ago, with its emphasis on the village and on backyard steel furnaces.

As a reaction to the delusion of "the bigger the better," which enthralled earlier Indian governments, especially Nehru, Desai's emphasis on rural India was overdue. Earlier governments neglected the village, where 90 percent of India's 550 million people still live. But "small is beautiful" is just as much a delusion as "the bigger the better." What is appropriate is not what uses the most capital or the most labor; it is not what is "small" or "big," "preindustrial" or a "scientific marvel." What is appropriate is quite simply what makes the economy's resources most productive. What is appropriate in a country of huge population and rapid population growth is what multiplies productive jobs. What is development in a country which like India has sizable resources of managerial and entrepreneurial skill and at the same time huge unfulfilled consumer needs is whatever creates purchasing power.

Steel mills, those prestige investments of the 1960s into which earlier Indian governments poured very large chunks of the country's scarce capital resources, are becoming the white elephants of the 1970s and 1980s. Steel mills are inappropriate technology for a country like India. They are highly capital-intensive rather than labor-intensive. They supply a commodity which is in ample supply on the world market and available everywhere at a low price. Above all, they create practically no jobs beyond those in the mill itself.

But the automotive industry—passenger cars, motorbikes, trucks, and tractors—is probably the most efficient multiplier of jobs around. Its own plants have a fairly high ratio of labor to capital, and the industry generates about four to five secondary or tertiary jobs throughout the economy for every one in the manufacturing plant. It creates jobs in road building and road maintenance, in traffic control, in dealerships, service stations, and repair work. And it creates enormous purchasing power with these jobs.

Similarly, making transistor radios and bicycles requires both a large manufacturing base and a large dealer system; and both multiply jobs and create purchasing power. And like the automotive industry, both create human capital, that is, skills accessible to the unlearned. The same might be said for the manufacture of synthetic fertilizer or pharmaceuticals or pesticides—all require big enterprises and national distribution and service (moreover, these products, together with the gasoline pump, underlie the two great successes of India since independence: the rapid increase in food production and the rapid decrease in infant mortality).

Equally appropriate as a creator of productivity, jobs, and purchasing power is the cosmetics manufacturer, who may be quite small. I saw a highly efficient and highly successful multinational cosmetics firm in Bangalore which, with twenty employees, produces five times as much foreign exchange per rupee of investment or sales as any of the huge state-owned Indian enterprises.

The development decades of the fifties and sixties worshiped capital investment. The best testament to this superstition is *The Stages of Economic Growth*, which Walt W. Rostow, later President Johnson's foreign policy adviser, wrote in the early 1960s and which then became the bible of the developing countries. Mr. Rostow proclaimed that development is an automatic and direct function of the size of capital investment. But that is not productivity; it's waste and incompetence. Today there is a tendency to define productivity as whatever uses the most labor—especially in the developing countries, with their huge, unemployed, young population.

But that, too, is incompetence. Productivity is whatever generates the highest overall yield from an economy's resources of capital, labor, physical resources, and time. This will also give the largest number of jobs and the maximum purchasing power. It will even produce the lowest possible inequality in the distribution of incomes attainable at a given stage of economic development. And surely poor countries cannot afford to support unproductive poeple—that is, people who appear busy winding a few strands of cotton around a wooden staff. Rich countries may be able to keep unproductive people gainfully unemployed, but poor countries have no surplus to distribute.

Above all, the troubadours of "small is beautiful" forget—as does so much of official Washington—that a healthy economy and society need *both* the large and the small. Indeed, the two are interdependent in both a developed and developing country. There can be no small manufacturer in a large market—whether that of the United States or of India—unless there is a large assembler or a large retailer, an IBM, a GM, a Sears Roebuck. It is only in their products or their stores that the small man's output can reach the market. But there would also be no GM except for the existence of a multitude of small autonomous tool-and-die shops and a host of small parts suppliers, or local dealers, service stations, and repair shops.

Pharmaceutical research requires big—if not very big—enterprises. But pharmaceutical sales depend on some 200,000 drugstores and 200,000 physicians—each of necessity decentralized and indeed autonomous. Rural development in India not only means national marketing organizations for village products and national credit and banking institutions. It means huge power stations. Above all—something the advocates of "small is beautiful" always conveniently overlook—it means centralized government bureaucracies, which surely could not be called "small" whether or not they deserve to be called "beautiful."

None of these arguments, I am afraid, made much impact on the

Indian government economist—nor I realized would they have made much impact on his Prime Minister. But once Indians have the bicycle, the motor scooter, the transistor radio, and the gasoline pumps, are they really going to go back to the spinning whorl?

21

The First Technological Revolution
and Its Lessons

Aware that we are living in the midst of a technological revolution, we are becoming increasingly concerned with its meaning for the individual and its impact on freedom, on society, and on our political institutions. Side by side with messianic promises of utopia to be ushered in by technology, there are the most dire warnings of man's enslavement by technology, his alienation from himself and from society, and the destruction of all human and political values.

Tremendous though today's technological explosion is, it is hardly greater than the first great revolution technology wrought in human life seven thousand years ago when the first great civilization of man, the irrigation civilization, established itself. First in Mesopotamia, and then in Egypt and in the Indus Valley, and finally in China, there appeared a new society and a new polity: the irrigation city, which then rapidly became the irrigation empire. No other change in man's way of life and in his making a living, not even the changes under way today, so completely revolutionized human society and community. In fact, the irrigation civilizations were the beginning of history, if only because they brought writing.

The age of the irrigation civilization was preeminently an age of technological innovation. Not until a historical yesterday, the eighteenth century, did technological innovations emerge which were com-

Presidential address to the Society for the History of Technology, 1965.

parable in their scope and impact to those early changes in technology, tools, and processes. Indeed, the technology of man remained essentially unchanged until the eighteenth century in so far as its impact on human life and human society is concerned.

But the irrigation civilizations were not only one of the great ages of technology. They represent also mankind's greatest and most productive age of social and political innovation. The historian of ideas is prone to go back to Ancient Greece, to the Old Testament prophets, or to the China of the early dynasties for the sources of the beliefs that still move men to action. But our fundamental social and political institutions antedate political philosophy by several thousand years. They all were conceived and established in the early dawn of the irrigation civilizations. Anyone interested in social and governmental institutions and in social and political processes will increasingly have to go back to those early irrigation cities. And, thanks to the work of archaeologists and linguists during the last fifty years, we increasingly have the information, we increasingly know what the irrigation civilizations looked like, we increasingly can go back to them for our understanding both of antiquity and of modern society. For essentially our present-day social and political institutions, practically without exception, were then created and established. Here are a few examples.

1. The irrigation city first established government as a distinct and permanent institution. It established an impersonal government with a clear hierarchical structure in which very soon there arose a genuine bureaucracy—which is, of course, what enabled the irrigation cities to become irrigation empires.

Even more basic: the irrigation city first conceived of man as a citizen. It had to go beyond the narrow bounds of tribe and clan and had to weld people of very different origins and blood into one community. This required the first supratribal deity, the god of the city. It also required the first clear distinction between custom and law and the development of an impersonal, abstract, codified legal system. Indeed, practically all legal concepts, whether of criminal or of civil law, go back to the irrigation city. The first great code of law, that of Hammurabi, almost four thousand years ago, would still be applicable to a good deal of legal business in today's highly developed, industrial society.

The irrigation city also first developed a standing army—it had to. For the farmer was defenseless and vulnerable and, above all, immobile. The irrigation city which, thanks to its technology, produced a surplus, for the first time in human affairs, was a most attractive target for the barbarian outside the gates, the tribal nomads of steppe and desert. And with the army came specific fighting technology and fighting equipment: the war horse and the chariot, the lance and the shield, armor and the catapult.

2. It was in the irrigation city that social classes first developed. It needed people permanently engaged in producing the farm products on which all the city lived; it needed farmers. It needed soldiers to defend them. And it needed a governing class with knowledge, that is, originally a priestly class. Down to the end of the nineteenth century these three "estates" were still considered basic in society.[1]

But at the same time the irrigation city went in for specialization of labor resulting in the emergence of artisans and craftsmen: potters, weavers, metalworkers, and so on; and of professional people: scribes, lawyers, judges, physicians.

And because it produced a surplus, it first engaged in organized trade which brought with it not only the merchant but money, credit, and a law that extended beyond the city to give protection, predictability, and justice to the stranger, the trader from far away. This, by the way, also made necessary international relations and international law. In fact, there is not very much difference between a nineteenth-century trade treaty and the trade treaties of the irrigation empires of antiquity.

3. The irrigation city first had knowledge, organized it, and institutionalized it. Both because it required considerable knowledge to construct and maintain the complex engineering works that regulated the vital water supply and because it had to manage complex economic transactions stretching over many years and over hundreds of miles, the irrigation city needed records, and this, of course, meant writing. It needed astronomical data, as it depended on a calendar. It needed means of navigating across sea or desert. It, therefore, had to organize both the supply of the needed information and its processing into learnable and teachable knowledge. As a result, the irrigation city developed the first schools and the first teachers. It developed the first systematic observation of natural phenomena, indeed the first ap-

proach to nature as something outside of and different from man and governed by its own rational and independent laws.

4. Finally, the irrigation city created the individual. Outside the city, as we can still see from those tribal communities that have survived to our days, only the tribe had existence. The individual as such was neither seen nor paid attention to. In the irrigation city of antiquity, however, the individual became, of necessity, the focal point. And with this emerged not only compassion and the concept of justice; with it emerged the arts as we know them, the poets, and eventually the religions and the philosophers.

This is, of course, not even the barest sketch. All I wanted to suggest is the scope and magnitude of social and political innovation that underlay the rise of the irrigation civilizations. All I wanted to stress is that the irrigation city was essentially "modern", as we have understood the term, and that until today, history largely consisted in building on the foundations laid five thousand or more years ago. In fact, one can argue that human history, in the last five thousand years, has largely been an extension of the social and political institutions of the irrigation city to larger and larger areas, that is, to all areas on the globe where water supply is adequate for the systematic tilling of the soil. In its beginnings, the irrigation city was the oasis in a tribal, nomadic world. By 1900 it was the tribal, nomadic world that had become the exception.

The irrigation civilization was based squarely upon a technological revolution. It can with justice be called a "technological polity". All its institutions were responses to opportunities and challenges that new technology offered. All its institutions were essentially aimed at making the new technology most productive.

I hope you will allow me one diversion.

The history of the irrigation civilizations has yet to be written. There is a tremendous amount of material available now, where fifty years ago we had, at best, fragments. There are splendid discussions available of this or that irrigation civilization, for instance of Sumer. But the very big job of recreating this great achievement of man and of telling the story of his first great civilization is yet ahead of us.

This should be preeminently a job for historians of technology such as we profess to be. At the very least the job calls for a historian with high interest in, and genuine understanding of, technology. The essential theme around which this history will have to be written must be the impacts and capacities of the new technology and the opportunities and challenges which this, the first great technological revolution, presented. The social, political, cultural institutions, familiar though they are to us today—for they are in large measure the institutions we have been living with for five thousand years—were all brand new then, and were all the outgrowth of new technology and of attempts to solve the problems the new technology posed.

It is our contention in the Society for the History of Technology that the history of technology is a major, distinct strand in the web of human history. We believe that the history of mankind cannot be properly understood without relating to it the history of man's work and man's tools, that is, the history of technology. Some of our colleagues and friends—let me mention only such familiar names as Lewis Mumford, Fairfield Osborn, Joseph Needham, R. J. Forbes, Cyril Stanley Smith, and Lynn White—have in their own works brilliantly demonstrated the profound impact of technology on political, social, economic, and cultural history. But while technological change has always had impact on the way men live and work, surely at no other time has technology so literally shaped civilization and culture as during the first technological revolution, that is, during the rise of the irrigation civilizations of antiquity.

Only now, however, is it possible to tell the story. No longer can its neglect be justified. For the facts are available, as I stated before. And we now, because we live in a technological revolution ourselves, are capable of understanding what happened then—at the very dawn of history. There is a big job to be done: to show that the traditional approach to our history—the approach taught in our schools—in which "relevant" history really begins with the Greeks (or with the Chinese dynasties), is shortsighted and distorts the real "ancient civilization".

I have, however, strayed off my topic: the question I posed at the beginning, what we can learn from the first technological revolution regarding the impacts likely to result on man, his society, and his government from the new industrial revolution, the one we are living in. Does the story of the irrigation civilization show man to be determined by his technical achievements, in thrall to them, coerced by them? Or does it show him capable of using technology to his own, to human ends, and of being the master of the tools of his own devising?

The answer which the irrigation civilizations give us to this question is threefold.

1. Without a shadow of doubt, major technological change creates the need for social and political innovation. It does make obsolete existing institutional arrangements. It does require new and very different institutions of community, society, and government. To this extent there can be no doubt: technological change of a revolutionary character coerces; it *demands innovation*.

2. The second answer also implies a strong necessity. There is little doubt, one would conclude from looking at the irrigation civilizations, that specific technological changes demand equally specific social and political innovations. That the basic institutions of the irrigation cities of the Old World, despite great cultural difference, all exhibited striking similarity may not prove much. After all, there probably was a great deal of cultural diffusion (though I refuse to get into the quicksand of debating whether Mesopotamia or China was the original innovator). But the fact that the irrigation civilizations of the New World around the Lake of Mexico and in Maya Yucatan, though culturally completely independent, millennia later evolved institutions which, in fundamentals, closely resemble those of the Old World (e.g., an organized government with social classes and a permanent military, and writing) would argue strongly that the solutions to specific conditions created by new technology have to be specific and are, therefore, limited in number and scope.

In other words, one lesson to be learned from the first technological revolution is that new technology creates what a philosopher of history might call "objective reality". And objective reality has to be dealt with on *its* terms. Such a reality would, for instance, be the conversion, in the course of the first technological revolution, of human space from "habitat" into "settlement", that is, into a permanent territorial unit always to be found in the same place—unlike the migrating herds of pastoral people or the hunting grounds of primitive tribes. This alone made obsolete the tribe and demanded a permanent, impersonal, and rather powerful government.

3. But the irrigation civilizations can teach us also that the new objective reality determines only the gross parameters of the solutions.

It determines where, and in respect to what, new institutions are needed. It does not make anything "inevitable". It leaves wide open *how* the new problems are to be tackled, what the purpose and values of the new institutions are to be.

In the irrigation civilizations of the New World the individual, for instance, failed to make his appearance. Never, as far as we know, did these civilizations get around to separating law from custom nor, despite a highly developed trade, did they invent money.

Even within the Old World, where one irrigation civilization could learn from the others, there were very great differences. They were far from homogeneous even though all had similar tasks to accomplish and developed similar institutions for these tasks. The different specific answers expressed above all different views regarding man, his position in the universe, and his society—different purposes and greatly differing values.

Impersonal bureaucratic government had to arise in all these civilizations; without it they could not have functioned. But in the Near East it was seen at a very early stage that such a government could serve equally to exploit and hold down the common man and to establish justice for all and protection for the weak. From the beginning the Near East saw an ethical decision as crucial to government. In Egypt, however, this decision was never seen. The question of the purpose of government was never asked. And the central quest of government in China was not justice but harmony.

It was in Egypt that the individual first emerged, as witness the many statues, portraits, and writings of professional men, such as scribes and administrators, that have come down to us—most of them superbly aware of the uniqueness of the individual and clearly asserting his primacy. It is early Egypt, for instance, which records the names of architects who built the great pyramids. We have no names for the equally great architects of the castles and palaces of Assur or Babylon, let alone for the early architects of China. But Egypt suppressed the individual after a fairly short period during which he flowered (perhaps as part of the reaction against the dangerous heresies of Ikhnaton). There is no individual left in the records of the Middle and New Kingdoms, which perhaps explains their relative sterility.

In the other areas two entirely different basic approaches emerged. One, that of Mesopotamia and of the Taoists, we might call "personalism'', the approach that found its greatest expression later in the Hebrew prophets and in the Greek dramatists. Here the stress is on developing to the fullest the capacities of the person. In the other approach—we might call it "rationalism", taught and exemplified above all by Confucius—the aim is the moulding and shaping of the individual according to preestablished ideals of rightness and perfection. I need not tell you that both these approaches still permeate our thinking about education.

Or take the military. Organized defense was a necessity for the irrigation civilization. But three different approaches emerged: a separate military class supported through tribute by the producing class, the farmers; the citizen-army drafted from the peasantry itself; and mercenaries. There is very little doubt that from the beginning it was clearly understood that each of these three approaches had very real political consequences. It is hardly coincidence, I believe, that Egypt, originally unified by overthrowing local, petty chieftains, never developed afterwards a professional permanent military class.

Even the class structure, though it characterizes all irrigation civilizations, showed great differences from culture to culture and within the same culture at different times. It was being used to create permanent castes and complete social immobility, but it was also used with great skill to create a very high degree of social mobility and a substantial measure of opportunities for the gifted and ambitious.

Or take science. We now know that no early civilization excelled China in the quality and quantity of scientific observations. And yet we also know that early Chinese culture did not point towards anything we could call science. Perhaps because of their rationalism the Chinese refrained from generalization. And though fanciful and speculative, it is the generalizations of the Near East and the mathematics of Egypt which point the way towards systematic science. The Chinese, with their superb gift for accurate observation, could obtain an enormous amount of information about nature. But their view of the universe remained totally unaffected thereby—in sharp contrast to what we know about the Middle Eastern developments out of which Europe arose.

In brief, the history of man's first technological revolution indicates the following:

1. Technological revolutions create an objective need for social and

political innovations. They create a need also for identifying the areas in which new institutions are needed and old ones are becoming obsolete.

2. The new institutions have to be appropriate to specific new needs. There are right social and political responses to technology and wrong social and political responses. To the extent that only a right institutional response will do, society and government are largely circumscribed by new technology.

3. But the values these institutions attempt to realize, the human and social purposes to which they are applied, and, perhaps most important, the emphasis and stress laid on one purpose as against another, are largely within human control. The bony structure, the hard stuff of a society, is prescribed by the tasks it has to accomplish. But the ethos of the society is in man's hands and is largely a matter of the "how" rather than of the "what."

For the first time in thousands of years, we face again a situation that can be compared with what our remote ancestors faced at the time of the irrigation civilization. It is not only the speed of technological change that creates a revolution, it is its scope as well. Above all, today, as seven thousand years ago, technological developments from a great many areas are growing together to create a new human environment. This has not been true of any period between the first technological revolution and the technological revolution that got under way two hundred years ago and has still clearly not run its course.

We, therefore, face a big task of identifying the areas in which social and political innovations are needed. We face a big task in developing the institutions for the new tasks, institutions adequate to the new needs and to the new capacities which technological change is casting up. And, finally, we face the biggest task of them all, the task of ensuring that the new institutions embody the values we believe in, aspire to the purposes we consider right, and serve human freedom, human dignity, and human ends.

If an educated man of those days of the first technological revolution—an educated Sumerian perhaps or an educated ancient Chinese—looked at us today, he would certainly be totally stumped by our technology. But he would, I am sure, find our existing social and political institutions reasonably familiar—they are after all, by and large, not fundamentally different from the institutions he and his contemporaries first fashioned. And, I am quite certain, he would have nothing but a wry smile for both those among us who predict a technological heaven and those who predict a technological hell of "alienation", of "technological unemployment", and so on. He might well mutter to himself, "This is where I came in." But to us he might well say, "A time such as was mine and such as is yours, a time of true technological revolution, is not a time for exultation. It is not a time for despair either. It is a time for work and for responsibility."

Note

1. See the brilliant though one-sided book by Karl A. Wittvogel, *Oriental Despotism: A Comparative Study of Total Power* (New Haven, Conn., 1957).

Part Six

The Information-Based Society

Introduction to Part Six

The three essays in this short part are quite different. The first one, "Information, Communications and Understanding," is very much "theoretical" though free from the jargon of theory. It discusses the nature of information and the conditions under which information becomes communications, that is, logic, understanding, and thereby meaning. Its foundation, though never made explicit, is the ancient theory of logic and rhetoric as first expounded in two great Platonic dialogues, the *Phaedo* and the *Phaedrus*, the dialogues respectively, about logic and rhetoric, their requirements, and their limitations. But the essay then melds into these traditional concepts the findings of modern logic and perception theory and does this without using "scientific" language and in a form easily accessible to the layman and easily applicable in and to an organization. The second essay, "Information and the Future of the City" shows the impact on social and community structure of our new ability to define, move, and use information. It is particularly concerned with the impact on that proudest of nineteenth-century achievements, the modern city, based as it was on the then new technology of moving people. The last essay, "Information-Based Organization," projects information on the social organization of people at work and shows how our newly gained information ability changes both the conceptual basis of organization and the relationships within it.

I was among the first, in the early 1950s, to realize that the computer would have profound social impacts—not because of its technical capabilities but because it would force us to use information. And, as I early realized, the work in logic done since the early years of the nineteenth century (culminating in Russell and Whitehead's *Principia Mathematica* in the early years of the twentieth century) had made information something that could be defined, organized, focused, applied—and used. Yet none of these essays discusses the computer; it is treated as what it is, a tool. How to put to use the performance capacity it gives us, is the topic of these three essays.

22

Information, Communications, and Understanding

Concern with "information" and "communications" started shortly before World War I. Russell and Whitehead's *Principia Mathematica*, which appeared in 1910, is still one of the foundation books. And a long line of illustrious successors—from Ludwig Wittgenstein through Norbert Wiener and A. N. Chomsky's "mathematical linguistics" today—has continued the work on the *logic* of information. Roughly contemporaneous is the interest in the *meaning* of communication; Alfred Korzybski started on the study of "general semantics," that is, on the meaning of communications, around the turn of the century. It was World War I, however, which made the entire Western world communications-conscious. When the diplomatic documents of 1914 in the German and Russian archives were published, soon after the end of the fighting, it became appallingly clear that the catastrophe had been caused, in large measure, by communications failure despite copious and reliable information. And the war itself—especially the total failure of its one and only strategic concept, Winston Churchill's Gallipoli campaign in 1915–16—was patently a tragicomedy of noncommunications. At the same time, the period immediately following World War I—a period of industrial strife and of total noncommunication between Westerners and "revolutionary" communists (and a little later, equally revolutionary fascists)—showed both the need for, and

Paper read before the Fellows of the International Academy of Management, Tokyo, Japan, 1969.

the lack of, a valid theory or a functioning practice of communications, inside existing institutions, inside existing societies, and between various leadership groups and their various "publics."

As a result, communications suddenly became, forty to fifty years ago, a consuming interest of scholars as well as of practitioners. Above all, communications in management has this last half-century been a central concern to students and practitioners in all institutions—business, the military, public administration, hospital administration, university administration, and research administration. In no other area have intelligent men and women worked harder or with greater dedication than psychologists, human relations experts, managers, and management students have worked on improving communications in our major institutions.

We have more attempts at communications today, that is, more attempts to talk to others, and a surfeit of communications media, unimaginable to the men who, around the time of World War I, started to work on the problems of communicating. The trickle of books on communications has become a raging torrent. I recently received a bibliography prepared for a graduate seminar on communications; it ran to ninety-seven pages. A recent anthology (*The Human Dialogue*, edited by Floyd W. Matson and Ashley Montagu; London: Collier-Macmillan, 1967) contains articles by forty-nine different contributors.

Yet communications has proven as elusive as the unicorn. Each of the forty-nine contributors to *The Human Dialogue* has a theory of communications which is incompatible with all the others. The noise level has gone up so fast that no one can really listen any more to all that babble about communications. But there is clearly less and less communicating. The communications gap within institutions and between groups in society has been widening steadily—to the point where it threatens to become an unbridgeable gulf of total misunderstanding.

In the meantime, there is an information explosion. Every professional and every executive—in fact, everyone except the deaf-mute—suddenly has access to data in inexhaustible abundance. All of us feel—and overeat—very much like the little boy who has been left alone in the candy store. But what has to be done to make this cornucopia of data redound to information, let alone to knowledge?

We get a great many answers. But the one thing clear so far is that no one really has an answer. Despite "information theory" and "data processing", no one yet has actually seen, let alone used, an "information system", or a "data base". The one thing we do know, though, is that the abundance of information changes the communications problem and makes it both more urgent and even less tractable.

There is a tendency today to give up on communications. In psychology, for instance, the fashion today is the T-group with its "sensitivity training". The avowed aim is not communications, but self-awareness. T-groups focus on the "I" and not on the "thou". Ten or twenty years ago the rhetoric stressed "empathy"; now it stresses "doing one's thing". However needed self-knowledge may be, communication is needed at least as much (if, indeed, self-knowledge is possible without action on others, that is, without communications). Whether the T-groups are sound psychology and effective psychotherapy is well beyond my competence and the scope of this paper. But their popularity attests to the failure of our attempts at communications.

Despite the sorry state of communications in theory and practice, we have, however, learned a good deal about information and communications. Most of it, though, has not come out of the work on communications to which we have devoted so much time and energy. It has been the byproduct of work in a large number of seemingly unrelated fields, from learning theory to genetics and electronic engineering. We equally have a lot of experience—though mostly of failure—in a good many practical situations in all kinds of institutions. Communications we may, indeed, never understand. But communications in organizations—call it *managerial communications*—we do know something about by now. It is a much narrower topic than communications *per se*—but it is the topic to which this paper shall address itself.

We are, to be sure, still far away from mastery of communications, even in organizations. What knowledge we have about communications is scattered and, as a rule, not accessible, let alone in applicable form. But at least we increasingly know what does not work and, sometimes, why it does not work. Indeed, we can say bluntly that most of today's brave attempts at communication in organization—whether business, labor unions, government agencies, or universities—are based on assumptions that have been proven to be invalid—and that, therefore, these efforts cannot have results. And perhaps we can even anticipate what might work.

What We Have Learned

We have learned, mostly through doing the wrong things, the following four fundamentals of communications:

1. Communication is perception,
2. Communication is expectations,
3. Communication is involvement,
4. Communication and information are totally different. But information presupposes functioning communications.

Communication is Perception

An old riddle asked by the mystics of many religions—the Zen Buddhists, the Sufis of Islam, or the rabbis of the Talmud—asks: "Is there a sound in the forest if a tree crashes down and no one is around to hear it?" We now know that the right answer to this is "no". There are sound waves. But there is no sound unless someone perceives it. Sound is created by perception. Sound is communication.

This may seem trite; after all, the mystics of old already knew this, for they, too, always answered that there is no sound unless someone can hear it. Yet the implications of this rather trite statement are great indeed.

a. First, it means that it is the recipient who communicates. The so-called communicator, that is, the person who emits the communication, does not communicate. He utters. Unless there is someone who hears, there is no communication. There is only noise. The communicator speaks or writes or sings—but he does not communicate. Indeed he cannot communicate. He can only make it possible, or impossible, for a recipient—or rather percipient—to perceive.

b. Perception, we know, is not logic. It is experience. This means, in the first place, that one always perceives a configuration. One cannot perceive single specifics. They are always part of a total picture. *The Silent Language* (as Edward T. Hall called it in the title of his pioneering work ten years ago)—that is, the gestures, the tone of voice, the environment all together, not to mention the cultural and social referents—cannot be dissociated from the spoken language. In fact, without them the spoken word has no meaning and cannot communicate. It is not only that the same words, for example, "I enjoyed meeting you", will be heard as having a wide variety of meanings. Whether they are heard as warm or as icy cold, as endearment or as rejection, depends on their setting in the silent language, such as the tone of voice or the occasion. More important is that by themselves, that is, without being part of the total configuration of occasion, silent language, and so on, the phrase has no meaning at all. By itself it cannot make possible communication. It cannot be understood. Indeed, it cannot be heard. To paraphrase an old proverb of the Human Relations school: "One cannot communicate a word; the whole man always comes with it."

c. But we know about perception also that one can only perceive what one is capable of perceiving. Just as the human ear does not hear sounds above a certain pitch, so does human perception all together not perceive what is beyond its range of perception. It may, of course, hear physically, or see visually, but it cannot accept. The stimulus cannot become communication.

This is a very fancy way of stating something the teachers of rhetoric have known for a very long time—though the practitioners of communications tend to forget it again and again. In Plato's *Phaedrus*, which, among other things, is also the earliest extant treatise on rhetoric, Socrates points out that one has to talk to people in terms of their own experience, that is, that one has to use a carpenter's metaphors when talking to carpenters, and so on. One can only communicate in the recipient's language or altogether in his terms. And the terms have to be experience-based. It, therefore, does very little good to try to explain terms to people. They will not be able to receive them if the terms are not of their own experience. They simply exceed their perception capacity.

The connection between experience, perception, and concept formation, that is, cognition, is, we now know, infinitely subtler and richer than any earlier philosopher imagined. But one fact is proven and comes out strongly in the most disparate work, for example, that of Piaget in Switzerland, that of B. F. Skinner of Harvard, or that of Jerome Bruner (also of Harvard). Percept and concept in the learner, whether child or adult, are not separate. We cannot perceive unless we also conceive. But we also cannot form concepts unless we can perceive. To communicate a concept is impossible unless the recipient can perceive it, that is, unless it is within his perception.

There is a very old saying among writers: "Difficulties with a sentence always mean confused thinking. It is not the sentence that needs straightening out, it is the thought behind it." In writing we attempt, of course, to communicate with ourselves. An unclear sentence is one that exceeds our own capacity for perception. Working on the sentence, that is, working on what is normally called communications, cannot solve the problem. We have to work on our own concepts first to be able to understand what we are trying to say—and only then can we write the sentence.

In communicating, whatever the medium, the first question has to be, "Is this communication within the recipient's range of perception? Can he receive it?"

The "range of perception" is, of course, physiological and largely (though not entirely) set by physical limitations of man's animal body. When we speak of communications, however, the most important limitations on perception are usually cultural and emotional rather than physical. That fanatics are not being convinced by rational arguments, we have known for thousands of years. Now we are beginning to understand that it is not "argument" that is lacking. Fanatics do not have the ability to perceive a communication which goes beyond their range of emotions. Before this is possible, their emotions would have to be altered. In other words, no one is really "in touch with reality", if by that we mean complete openness to evidence. The distinction between sanity and paranoia is not in the ability to perceive, but in the ability to learn, that is, in the ability to change one's emotions on the basis of experience.

That perception is conditioned by what we are capable of perceiving was realized forty years ago by the most quoted but probably least heeded of all students of organization, Mary Parker Follett, especially in her collected essays, *Dynamic Administration* (London: Management Publications Trust, 1949). Follett taught that a disagreement or a conflict is likely not to be about the answers, or, indeed, about anything ostensible. It is, in most cases, the result of incongruity in perceptions. What *A* sees so vividly, *B* does not see at all. And, therefore, what *A* argues has no pertinence to *B*'s concerns, and vice versa. Both, Follett argued, are likely to see reality. But each is likely to see a different aspect thereof. The world, and not only the material world, is multidimensional. Yet one can only see one dimension at a time. One rarely realizes that there could be other dimensions, and that something that is so obvious to us and so clearly validated by our emotional experience has other dimensions, a back and sides, which are entirely different and which, therefore, lead to entirely different perception. The old story about the blind men and the elephant in which every one of them, upon encountering this strange beast, feels one of the elephant's parts, his leg, his trunk, his hide, and reports an entirely different conclusion, each held tenaciously, is simply a story of the human condition. And there is no possibility of communication until this is understood and until he who has felt the hide of the elephant goes over to him who has felt the leg and feels the leg himself. There is no possibility of communications, in other words, unless we first know what the recipient, the true communicator, can see and why.

Communication is Expectations

We perceive, as a rule, what we expect to perceive. We see largely what we expect to see, and we hear largely what we expect to hear. That the unexpected may be resented is not the important thing—though most of the writers on communications in business or government think it is. What is truly important is that the unexpected is usually not received at all. It is either not seen or heard but ignored. Or it is misunderstood, that is, mis-seen as the expected or misheard as the expected.

On this we now have a century or more of experimentation. The results are quite unambiguous. The human mind attempts to fit impressions and stimuli into a frame of expectations. It resists vigorously any attempts to make it "change its mind", that is, to perceive what it does not expect to perceive or not to perceive what it expects to perceive. It is, of course, possible to alert the human mind to the fact that what it perceives is contrary to its expectations. But this first requires that we understand what it expects to perceive. It then requires that there be an unmistakable signal—"this is different", that is, a shock which breaks continuity. A "gradual" change in which the mind is supposedly led by small, incremental steps to realize that what is perceived is not what it expects to perceive will not work. It will rather reinforce the expectations and will make it even more certain that what will be perceived is what the recipient expects to perceive.

Before we can communicate, we must, therefore, know what the recipient expects to see and to hear. Only then can we know whether communication can utilize his expectations—and what they are—or whether there is need for the "shock of alienation", for an "awakening" that breaks through the recipient's expectations and forces him to realize that the unexpected is happening.

Communication is Involvement

Many years ago psychologists stumbled on a strange phenomenon in their studies of memory, a phenomenon that, at first, upset all their hypotheses. In order to test memory, the psychologists compiled a list of words to be shown to their experimental subjects for varying times as a test of their retention capacity. As control, a list of nonsense words, mere jumbles of letters, were devised to find out to what extent understanding influenced memory. Much to the surprise of these early experimenters almost a century ago or so, their subjects (mostly students, of course) showed totally uneven memory retention of individual words. More surprising, they showed amazingly high retention of the nonsense words. The explanation of the first phenomenon is fairly obvious. Words are not mere information. They do carry emotional charges. And, therefore, words with unpleasant or threatening associations tend to be suppressed, word with pleasant associations retained. In fact, this selective retention by emotional association has since been used to construct tests for emotional disorders and for personality profiles.

The relatively high retention rate of nonsense words was a greater problem. It was expected, after all, that no one would really remember words that had no meaning at all. But it has become clear over the years that the memory for these words, though limited, exists precisely because these words have no meaning. For this reason, they also make no demand. They are truly neuter. In respect to them, memory could be said to be truly mechanical, showing neither emotional preference nor emotional rejection.

A similar phenomenon, known to every newspaper editor, is the amazingly high readership and retention of the fillers, the little three- or five-line bits of irrelevant incidental information that are being used to balance a page. Why should anybody want to read, let alone remember, that it first became fashionable to wear different-colored hose on each leg at the court of some long-forgotten duke? Why should anybody want to read, let alone remember, when and where baking powder was first used? Yet there is no doubt that these little titbits of irrelevancy are read and, above all, that they are remembered far better than almost anything in the daily paper except the great screaming headlines of the catastrophes. The answer is that these fillers make no demands. It is precisely their total irrelevancy that accounts for their being remembered.

Communications are always propaganda. The emitter always wants "to get something across." Propaganda, we now know, is both a great deal more powerful than the rationalists with their belief in "open discussion" believe, and a great deal less powerful than the myth-makers of propaganda, for example, a Dr. Goebbels in the Nazi régime, believed and wanted us to believe. Indeed, the danger of total propaganda is not that the propaganda will be believed. The danger is that nothing will be believed and that every communication becomes suspect. In the end, no communication is being received any more. Everything anyone says is considered a demand and is resisted, re-sented, and in effect not heard at all. The end results of total propagan-da are not fanatics, by cynics—but this, of course, may be even greater and more dangerous corruption.

Communication, in other words, always makes demands. It always demands that the recipient become somebody, do something, believe something. It always appeals to motivation If, in other words, commu-nication fits in with the aspirations, the values, the purposes of the recipient, it is powerful. If it goes against his aspirations, his values, his motivations, it is likely not to be received at all, or, at best, to be resisted. Of course, at its most powerful, communication brings about conversion, that is, a change of personality, of values, beliefs, aspira-tions. But this is the rare, existential event, and one against which the

basic psychological forces of every human being are strongly organized. Even the Lord, the Bible reports, first had to strike Saul blind
before he could raise him as Paul. Communications aiming at conversion demand surrender. By and large, therefore, there is no communication unless the message can key in to the recipient's own values, at
least to some degree.

Communication and Information are Different
and Largely Opposite—Yet Interdependent

a. Where communication is perception, information is logic. As
such, information is purely formal and has no meaning. It is impersonal rather than interpersonal. The more it can be freed of the human
component, that is, of such things as emotions and values, expectations and perceptions, the more valid and reliable does it become.
Indeed, it becomes increasingly informative.

All through history, the problem has been how to glean a little
information out of communications, that is, out of relationships between people, based on perception. All through history, the problem
has been to isolate the information content from an abundance of
perception. Now, all of a sudden, we have the capacity to provide
information—both because of the conceptual work of the logicians,
especially the symbolic logic of Russell and Whitehead, and because
of the technical work on data processing and data storage, that is, of
course, especially because of the computer and its tremendous capacity
to store, manipulate, and transmit data. Now, in other words, we have
the opposite problem from the one mankind has always been struggling
with. Now we have the problem of handling information *per se*,
devoid of any communication content.

b. The requirements for effective information are the opposite of
those for effective communication. Information is, for instance, always specific. We perceive a configuration in communications; but we
convey specific individual data in the information process. Indeed,
information is, above all, a principle of economy. The fewer data
needed, the better the information. And an overload of information,
that is, anything much beyond what is truly needed, leads to a complete information blackout. It does not enrich, but impoverishes.

c. At the same time, information presupposes communication. Information is always encoded. To be received, let alone to be used, the code must be known and understood by the recipient. This requires prior agreement, that is, some communication. At the very least, the recipient has to know what the data pertain to. Are the figures on a piece of computer tape the height of mountain tops or the cash balances of Federal Reserve member banks? In either case, the recipient would have to know what mountains are or what banks are to get any information out of the data.

The prototype information system may well have been the peculiar language known as *Armee Deutsch* (Army German), which served as language of command in the Imperial Austrian Army prior to 1918. A polyglot army in which officers, noncommissioned officers, and men often had no language in common, it functioned remarkably well with fewer than two hundred specific words, "fire", for instance, or "at ease", each of which had only one totally unambiguous meaning. The meaning was always an action. And the words were learned in and through actions, that is, in what behaviorists now call operant conditioning. The tensions in the Austrian Army after many decades of nationalist turmoil were very great indeed. Social intercourse between members of different nationalities serving in the same unit became increasingly difficult, if not impossible. But to the very end, the information system functioned. It was completely formal, completely rigid, completely logical in that each word had only one possible meaning; and it rested on completely preestablished communication regarding the specific response to a certain set of sound waves. This example, however, shows also that the effectiveness of an information system depends on the willingness and ability to think through carefully what information is needed by whom for what purposes, and then on the systematic creation of communication between the various parties to the system as to the meaning of each specific input and output. The effectiveness, in other words, depends on the preestablishment of communication.

d. Communication communicates better the more levels of meaning it has and the less possible it is, therefore, to quantify it.

Medieval aesthetics held that a work of art communicates on a number of levels, at least three if not four: the literal, the metaphorical, the allegorical, and the symbolic. The work of art that most consciously converted this theory into artistic practice was, of course, Dante's *Divina Commedia*. If, by information we mean something that can be quantified, then the *Divina Commedia* is without any information content whatever. But it is precisely the ambiguity, the multiplicity of levels on which this book can be read, from being a fairy tale to being a grand synthesis of metaphysics, that makes it the overpowering work of art it is, and the immediate communication which it has been to generations of readers.

Communications, in other words, may not be dependent on information. Indeed, the most perfect communications may be purely shared experiences, without any logic whatever. Perception has primacy rather than information.

I fully realize that this summary of what we have learned is gross oversimplification. I fully realize that I have glossed over some of the most hotly contested issues in psychology and perception. Indeed, I may well be accused of brushing aside most of the issues which the students of learning and of perception would themselves consider central and important.

But my aim has, of course, not been to survey these big areas. My concern is not with learning or with perception. It is with communications, and, in particular, with communications in the large organization, be it business enterprise, government agency, university, or armed service.

This summary might also be criticized for being trite, if not obvious. No one, it might be said, could possibly be surprised at its statements. They say what everybody knows. But whether this be so or not, it is not what everybody does. On the contrary, the logical implications of these apparently simple and obvious statements for communications in organizations are at odds with current practice and, indeed, deny validity to the honest and serious efforts to communicate which we have been making for many decades now.

What, then, can our knowledge and our experience teach us about communications in organizations, about the reasons for our failures, and about the prerequisites for success in the future?

1. For centuries we have attempted communication downwards. This, however, cannot work, no matter how hard and how intelligently we try. It cannot work, first, because it focuses on what we want to say. It assumes, in other words, that the utterer communicates. But we know that all he does is utter. Communication is the act of the recipient. What we have been trying to do is to work on the emitter, specifically on the manager, the administrator, the commander, to make him capable of being a better communicator. But all one can communicate downwards are commands, that is, prearranged signals. One cannot communicate downwards anything connected with understanding, let alone with motivation. This requires communication upwards; from those who perceive to those who want to reach their perception.

This does not mean that managers should stop working on clarity in what they say or write. Far from it. But it does mean that how we say something comes only after we have learned what to say. And this cannot be found out by "talking to", no matter how well it is being done. "Letters to the Employees", no matter how well done, will be a waste unless the writer knows what employees can perceive, expect to perceive, and want to do. They are a waste unless they are based on the recipient's rather than the emitter's perception.

2. But "listening" does not work either. The Human Relations school of Elton Mayo, forty years ago, recognized the failure of the traditional approach to communications. Its answer—especially as developed in Mayo's two famous books, *The Human Problems of an Industrial Civilization* (2nd ed, Boston: Harvard University, 1946) and *The Social Problems of an Industrial Civilization* (Boston: Harvard University, 1945)—was to enjoin listening. Instead of starting out with what I, that is, the executive, want to get across, the executive should start out by finding out what subordinates want to know, are interested in, are, in other words, receptive to. To this day, the human relations prescription, though rarely practiced, remains the classic formula.

Of course, listening is a prerequisite to communication. But it is not adequate, and it cannot, by itself, work. Perhaps the reason why it is not being used widely, despite the popularity of the slogan, is precisely that, where tried, it has failed to work. Listening first assumes that the superior will understand what he is being told. It assumes, in other words, that the subordinates can communicate. It is hard to see, however, why the subordinate should be able to do what his superior cannot do. In fact, there is no reason for assuming he can. There is no reason, in other words, to believe that listening results any less in misunderstanding and miscommunications than does talking. In addition, the theory of listening does not take into account that communications is involvement. It does not bring out the subordinate's preferences and desires, his values and aspirations. It may explain the reasons for misunderstanding. But it does not lay down a basis for understanding.

This is not to say that listening is wrong, any more than the futility of downward communications furnishes any argument against attempts to write well, to say things clearly and simply, and to speak the language of those whom one addresses rather than one's own jargon. Indeed, the realization that communications have to be upward—or rather that they have to start with the recipient, rather than the emitter, which underlies the concept of listening—is absolutely sound and vital. But listening is only the starting point.

3. More and better information does not solve the communications problem, does not bridge the communications gap. On the contrary, the more information the greater is the need for functioning and effective communication. The more information, in other words, the greater is the communications gap likely to be.

The more impersonal and formal the information process in the first place, the more will it depend on prior agreement on meaning and application, that is, on communications. In the second place, the more effective the information process, the more impersonal and formal will it become, the more will it separate human beings and thereby require separate, but also much greater, efforts, to reestablish the human relationship, the relationship of communication. It may be said that the effectiveness of the information process will depend increasingly on our ability to communicate, and that, in the absence of effective communication—that is, in the present situation—the information revolution cannot really produce information. All it can produce is data.

It can also be said—and this may well be more important—that the test of an information system will increasingly be the degree to which it frees human beings from concern with information and allows them to work on communications. The test, in particular, of the computer will be how much time it gives executives and professionals on all levels for direct, personal, face-to-face relationships with other people.

It is fashionable today to measure the utilization of a computer by the number of hours it runs during one day. But this is not even a measurement of the computer's efficiency. It is purely a measurement of input. The only measurement of output is the degree to which availability of information enables human beings not to control, that is, not to spend time trying to get a little information on what happened yesterday. And the only measurement of this, in turn, is the amount of time that becomes available for the job only human beings can do, the job of communication. By this test, of course, almost no computer today is being used properly. Most of them are being misused, that is, are being used to justify spending even more time on control rather than to relieve human beings from controlling by giving them information. The reason for this is quite clearly the lack of prior communication, that is, of agreement and decision on what information is needed, by whom and for what purposes, and what it means operationally. The reason for the misuse of the computer is, so to speak, the lack of anything comparable to the *Armee Deutsch* of yesterday's much-ridiculed Imperial Austrian Army with its two hundred words of command which even the dumbest recruit could learn in two weeks' time.

The Information Explosion, in other words, is the most impelling reason to go to work on communications. Indeed, the frightening communications gap all around us—between management and workers; between business and government; between faculty and students, and between both of them and university administration; between producers and consumers; and so on—may well reflect in some measure the tremendous increase in information without a commensurate increase in communications.

Can we then say anything constructive about communication? Can we do anything? We can say that communication has to start from the intended recipient of communications rather than from the emitter. In terms of traditional organization we have to start upward. Downward communications cannot work and do not work. They come *after* upward communications have successfully been established. They are reaction rather than action, response rather than initiative.

But we can also say that it is not enough to listen. The upward communication must first be focused on something that both recipient and emitter can perceive, focused on something that is common to both of them. And second, it must be focused on the motivation of the intended recipient. It must, from the beginning, be informed by his values, beliefs, and aspirations.

One example—but only an example: there have been promising results with organizational communication that started out with the demand by the superior that the subordinate think through and present to the superior his own conclusions as to what major contribution to the organization—or to the unit within the organization—the subordinate should be expected to perform and should be held accountable for. What the subordinate then comes up with is rarely what the superior expects. Indeed, the first aim of the exercise is precisely to bring out the divergence in perception between superior and subordinate. But the perception is focused, and focused on something that is real to both parties. To realize that they see the same reality differently is in itself already communication.

Second, in this approach, the intended recipient of communication—in this case the subordinate—is given access to experience that enables him to understand. He is given access to the reality of decision making, the problems of priorities, the choice between what one likes to do and what the situation demands and, above all, the responsibility for a decision. He may not see the situation the same way the superior does—in fact, he rarely will or even should. But he may gain an understanding of the complexity of the superior's situation, and above all of the fact that the complexity is not of the superior's making, but is inherent in the situation itself.

Finally, the communication, even if it consists of a "no" to the subordinate's conclusions, is firmly focused on the aspirations, values, and motivation of the intended recipient. In fact, it starts out with the question, "What would you *want* to do?" It may then end up with the command, "This is what I tell you to do." But at least it forces the superior to realize that he is overriding the desires of the subordinate. It forces him to explain, if not to try to persuade. At least he knows that he has a problem—and so does the subordinate.

A similar approach has worked in another organizational situation in which communication has been traditionally absent: the performance appraisal, and especially the appraisal interview. Performance appraisal is today standard in large organizations (except in Japan, where promotion and pay go by seniority so that performance appraisal would serve little purpose). We know that most people want to know where they stand. One of the most common complaints of employees in organizations is, indeed, that they are not being appraised and are not being told whether they do well or poorly.

The appraisal forms may be filled out. But the appraisal interview in which the appraiser is expected to discuss his performance with a man is almost never conducted. The exceptions are a few organizations in which performance appraisals are considered a communications tool rather than a rating device. This means specifically that the performance appraisal starts out with the question, "What has this man done well?" It then asks, "And what, therefore, should he be able to do well?" And then it asks, "And what would he have to learn or to acquire to be able to get the most from his capacities and achievements?" This, first, focuses on specific achievement. It focuses on things the employee himself is likely to perceive clearly and, in fact, gladly. It also focuses on his own aspirations, values, and desires. Weaknesses are then seen as limitations to what the employee himself can do well and wants to do, rather than as defects. Indeed, the proper conclusion from this approach to appraisal is not the question, "What should the employee do?" but "What should the organization and I, his boss, do?" A proper conclusion is not "What does this communicate to the employee?" It is "What does this communicate to both of us, subordinate *and* superior?"

These are only examples, and rather insignificant ones at that. But perhaps they illustrate conclusions to which our experience with communications—largely an experience of failure—and the work in learning, memory, perception, and motivation point.

The start of communications in organization must be to get the intended recipient himself to try to communicate. This requires a focus on (*a*) the impersonal but common task, and (*b*) on the intended recipient's values, achievements, and aspirations. It also requires the experience of responsibility.

Perception is limited by what can be perceived and geared to what one expects to perceive. Perception, in other words, presupposes experience. Communication within organization, therefore, presupposes that the members of the organization have the foundation of experience to receive and perceive. The artist can convey this experience in symbolical form: he can communicate what his readers or viewers have never experienced. But ordinary managers, administrators, and professors are not likely to be artists. The recipients must, therefore, have actual experience themselves and directly rather than through the vicarious symbols.

Communications in organization demands that the masses, whether they be employees or students, share in the responsibility of decisions to the fullest possible extent. They must understand because they have been through it, rather than accept because it is being explained to them.

I shall never forget the German trade-union leader, a faithful social- ist, who was shattered by his first exposure to the deliberations of the Board of Overseers of a large company to which he had been elected as an employee member. That the amount of money available was limited and that, indeed, there was very little money available for all the de- mands that had to be met, was one surprise. But the pain and complex- ity of the decisions between various investments, for example, be- tween modernizing the plant to safeguard workers' jobs and building workers' houses to safeguard their health and family life, was a much bigger and totally unexpected experience. But, as he told me with a half-sheepish, half-rueful grin, the greatest shocker was the realization that all the things he considered important turned out to be irrelevant to the decisions in which he found himself taking an active and responsi- ble part. Yet this man was neither stupid nor dogmatic. He was simply inexperienced—and, therefore, inaccessible to communication.

The traditional defence of paternalism has always been ''It's a complex world; it needs the expert, the man who knows best.'' But paternalism, as our work in perception, learning, and motivation is beginning to bring out, really can work only in a simple world. When people can understand what Papa does because they share his experi- ences and his perception, then Papa can actually make the decisions for them. In a complex world there is need for a shared experience in the decisions, or there is no common perception, no communications, and, therefore, neither acceptance of the decisions, nor ability to carry them out. The ability to understand presupposes prior communication. It presupposes agreement on meaning.

There will be no communication, in sum, if it is conceived as going from the "I" to the "thou." Communication only works from one member of 'us' to another. Communications in organization—and this may be the true lesson of our communications failure and the true measure of our communications need—are not a *means* of organization. They are a *mode* of organization.

23

Information and the Future of the City

In twenty years Japanese office workers may still commute, packed shoulder to shoulder, to downtown towers. But no one else in the developed world will. Office work, rather than office workers, will do the traveling. Tomorrow's big city is no longer going to be the office center.

The exodus is already under way. Citibank handles credit cards in North Dakota, clears checks in upstate New York and Delaware, and is moving data processing across the Hudson to suburban New Jersey. A large Boston-based mutual-fund group, Colonial Management Associates, has moved nationwide customer service and customer accounting to a Denver suburb. Insurance companies are rapidly shifting their labor-intensive work—claims handling, customer correspondence, record keeping—to the outskirts of metropolitan areas. And office parks especially built for back-office operations are now springing up in the suburbs as fast as shopping malls sprang up there in the 1960s and 1970s.

Acquiring Wheels

The modern big city is the creation of the nineteenth century's ability to move people. Everyone in Dickens's London walked to work except the owners, who lived over their shops or their countinghouses. But then, beginning in mid-century, people began to acquire wheels—

First published in the *Wall Street Journal*, 1989.

the railroad first, then the omnibus and the streetcar (horse-drawn, of course, for many decades), the subway and the elevated train, the automobile, the bicycle. Suddenly large masses of people could move over great distances to where the work was. And the elevator added vertical mobility. It was this ability to move people that, more than anything else, made possible large organizations, businesses, hospitals, government agencies, and universities.

By 1914, every single one of the means to move people into an office-centered large city—and to enable the office workers to live outside it—had been developed. But they did not have their full impact until after World War II. Until then only two cities had skyscrapers—New York and Chicago. Now every mid-sized city worldwide boasts a "skyline." And even in mid-sized cities people commute.

This trend has clearly reached its end, has indeed widely overshot the mark. Tokyo's office workers have to live more than two hours away just to get a seat on the morning train. In Los Angeles, traffic at six every weekday morning is bumper to bumper in all directions—people trying to get to their desks by 8:30 or 9. Things are not much better in Boston or New York or Philadelphia. London's Piccadilly Circus is chaos twenty-four hours a day, and so are those marvels of nineteenth-century city planning, the Grands Boulevards of Paris. Rome and Madrid are worse still.

Office workers in the world's big cities do not have eight-hour days; they have twelve-hour days. And all attempts in the past thirty years to relieve the traffic jams and their frustrations through new public transit have been total failures despite countless billions spent on them.

Yet none of this is necessary any more; indeed, commuting to office work is obsolete. It is now infinitely easier, cheaper and faster to do what the nineteenth century could not do: move information, and with it office work, to where the people are. The tools to do so are already here: the telephone, two-way video, electronic mail, the fax machine, the personal computer, the modem, and so on. And so is the receptivity: witness, for example, the boom in fax machines in the past eighteen months.

We already know how office work will be done in the future. Contrary to what futurists predicted twenty-five years ago, the trend is not toward individuals working in their homes. People greatly prefer to work where other people are. But even in Japan—where the need for community and companionship at work is greater than in the West—the exodus of such clerical work as data processing from downtown has begun.

But, equally important, clerical work increasingly will become "uncoupled," the way much physical office work—cleaning, maintaining equipment, running the cafeteria—already has been. Rather than being employees of the institution whose office work they do, more and more clerical workers will be employed by specialized and independent contractors. A growing amount of such work already is being done by people hired, trained, placed, and paid by temporary-help firms—with more and more of the "temporaries" actually holding down full-time, permanent assignments in the client companies. A good many of the new type of office parks provide a trained clerical force and the supervision for it. They provide office work rather than office space. And, according to some reports, that is where the demand is.

Office workers doing clerical and maintenance work are the largest single work group in the developed world's big cities—accounting for as much as half the working population. What, then, will the city of tomorrow look like when it is no longer an "office city"? It will, one can safely say, be a "headquarters city."

Twenty-five years ago a number of large U.S. companies—General Foods, IBM, General Electric—moved out of Manhattan and into suburbia, lock, stock, and barrel. At that time we did not know that we could move information. Thus, to free office workers from the need to commute, companies isolated top management people and professionals and imposed on them constant traveling into the city for business meetings.

Big companies tomorrow are almost certain to keep their management people—at least their senior ones—where other senior management people are: in the city. And so will government agencies and other large organizations. But this means that the big city will also house the purveyors of specialized skills and knowledge—the lawyer, the accounting firm, the architect, the consultant, the advertising agency, the investment banker, the financial analyst, and so on. But even these people will have their office work done outside the city.

A very big law firm is completing plans to have only one law library, in one suburban location. It will serve all 10 of its offices in the U.S. and overseas through a computer network supplemented by fax machines and two-way video. Within two or three years the firm expects to vacate all the space now occupied by the current 10 libraries—two floors each in every location.

We may be at the very end of the tremendous boom in office construction and office rents that was triggered when Napoleon III created the modern city's prototype in 1860 Paris and which has reached such a frenzy in all major Free World cities these past twenty years. (I, for one, am perfectly reconciled to the Japanese buying up more and more of the large office buildings in downtown America.) The big city of tomorrow is far more likely to resemble the preindustrial city than the nineteenth-century city that still shapes today's New York or Paris.

But will even those who work in the urban headquarters live in the cities? Where will the largest number of them—managers and professionals in particular—make their homes?

In continental Europe, where middle managers and professionals still tend to live in the city, the shift from office city to headquarters city may well prevent their moving out. But it is doubtful that the exodus to the suburbs will be reversed or even greatly slowed in the United States, in Britain, or in Japan, where middle-class people with children have already moved out of the core cities. And surely the headquarters city will have even less work for the poor and unskilled than the office city has. This will be a particular problem in the United States, where welfare payments have drawn so many of the least-skilled and least-schooled into the inner-city jungle.

What will the tax base of the headquarters city be, and can it remain a commercial center? Luxury shops do not depend on the office worker. But most other shops in the city do, especially department stores, and only in Japan do people not working in the city regularly come in to shop there. What about restaurants and hotels? Will theaters and opera houses become transmitters of shows through videotape and cable TV rather than places people go to? Will the big-city hospital become a center of teaching, of information, and of diagnosis for the suburban and exurban hospitals where the patients will be?

A Lecture to 10,000

And what about the large university? The costs of higher education in all developed countries are nearly as much out of control as the costs of health care. The only way, perhaps, to cap them may be to convert the university into a place form which learning flows to where the *students* are—something already done by the successful Open University in Britain.

Several times a year I lecture to 10,000 or more students, yet fewer than 100 are in the room with me. The rest see me via satellite in more than 100 "downlinks" and discuss their questions with me via telephone.

There is a great deal said and written these days about the technological impacts of information. But perhaps its social impacts are greater still, and more important.

24

The Information-Based Organization

The typical large business twenty years hence will have fewer than half the levels of management of its counterpart today, and no more than a third the managers. In its structure, and in its management problems and concerns, it will bear little resemblance to the typical manufacturing company, circa 1950, which our textbooks still consider the norm. Instead it is far more likely to resemble organizations that neither the practicing manager nor the management scholar pays much attention to today: the hospital, the university, the symphony orchestra. For like them, the typical business will be knowledge-based, an organization composed largely of specialists who direct and discipline their own performance through organized feedback from colleagues, customers, and headquarters. For this reason, it will be what I call an information-based organization.

Businesses, especially large ones, have little choice but to become information-based. Demographics, for one, demands the shift. The center of gravity in employment is moving fast from manual and clerical workers to knowledge workers who resist the command-and-control model that business took from the military 100 years ago. Economics also dictates change, especially the need for large businesses to innovate and to be entrepreneurs. But above all, information technology demands the shift.

Advanced data-processing technology isn't necessary to create an information-based organization, of course. As we shall see, the British

First Published in *Harvard Business Review*, 1988.

built just such an organization in India when "information technolo-gy" meant the quill pen, and barefoot runners were the "telecom-munications" systems. But as advanced technology becomes more and more prevalent, we have to engage in analysis and diagnosis—that is, in "information"—even more intensively or risk being swamped by the data we generate.

So far most computer users still use the new technology only to do faster what they have always done before, crunch conventional num-bers. But as soon as a company takes the first tentative steps from data to information, its decision processes, management structure, and even the way its work gets done begin to be transformed. In fact, this is already happening, quite fast, in a number of companies throughout the world.

We can readily see the first step in this transformation process when we consider the impact of computer technology on capital-investment decisions. We have known for a long time that there is no one right way to analyze a proposed capital investment. To understand it we need at least six analyses: the expected rate of return; the payout period and the investment's expected productive life; the discounted present value of all returns through the productive lifetime of the investment; the risk in not making the investments or deferring it; the cost and risk in case of failure; and finally, the opportunity cost. Every accounting student is taught these concepts. But before the advent of data-process-ing capacity, the actual analyses would have taken man-years of clerical toil to complete. Now anyone with a spreadsheet should be able to do them in a few hours.

The availability of this information transforms the capital-invest-ment analysis from opinion into diagnosis, that is, into the rational weighing of alternative assumptions. Then the information transforms the capital-investment decision from an opportunistic, financial deci-sion governed by the numbers into a business decision based on the probability of alternative strategic assumptions. So the decision both presupposes a business strategy and challenges that strategy and its assumptions. What was once a budget exercise becomes an analysis of policy.

The second area that is affected when a company focuses its data-processing capacity on producing information is its organization structure. Almost immediately, it becomes clear that both the number of management levels and the number of managers can be sharply cut. The reason is straightforward: it turns out that whole layers of management neither make decisions nor lead. Instead, their main, if not their only, function is to serve as "relays"—human boosters for the faint, unfocused signals that pass for communication in the traditional preinformation organization.

One of America's largest defense contractors made this discovery when it asked what information its top corporate and operating managers needed to do their jobs. Where did it come from? What form was it in? How did it flow? The search for answers soon revealed that whole layers of management—perhaps as many as six out of a total of fourteen—existed only because these questions had not been asked before. The company had had data galore. But it had always used its copious data for control rather than for information.

Information is data endowed with relevance and purpose. Converting data into information thus requires knowledge. And knowledge, by definition, is specialized. (In fact, truly knowledgeable people tend toward overspecialization, whatever their field, precisely because there is always so much more to know.)

The information-based organization requires far more specialists overall than the command-and-control companies we are accustomed to. Moreover, the specialists are found in operations, not at corporate headquarters. Indeed, the operating organization tends to become an organization of specialists of all kinds.

Information-based organizations need central operating work such as legal counsel, public relations, and labor relations as much as ever. But the need for service staffs—that is, for people without operating responsibilities who only advise, counsel, or coordinate—shrinks drastically. In its *central* management, the information-based organization needs few, if any, specialists.

Because of its flatter structure, the large, information-based organization will more closely resemble the businesses of a century ago than today's big companies. Back then, however, all the knowledge, such as it was, lay with the very top people. The rest were helpers or hands, who mostly did the same work and did as they were told. In the information-based organization, the knowledge will be primarily at the bottom, in the minds of the specialists who do different work and direct themselves. So today's typical organization in which knowledge tends to be concentrated in service staffs, perched rather insecurely between top management and the operating people, will likely be labeled a phase, an attempt to infuse knowledge from the top rather than obtain information from below.

Finally, a good deal of work will be done differently in the information-based organization. Traditional departments will serve as guardians of standards, as centers for training and the assignment of specialists; they won't be where the work gets done. That will happen largely in task-focused teams.

This change is already under way in what used to be the most clearly defined of all departments—research. In pharmaceuticals, in telecommunications, in papermaking, the traditional *sequence* of research, development, manufacturing, and marketing is being replaced by *synchrony:* specialists from all these functions work together as a team, from the inception of research to a product's establishment in the market.

How task forces will develop to tackle other business opportunities and problems remains to be seen. I suspect, however, that the need for a task force, its assignment, its composition, and its leadership will have to be decided on case by case. So the organization that will be developed will go beyond the matrix and may indeed be quite different from it. One thing is clear, though: it will require greater self-discipline and even greater emphasis on individual responsibility for relationships and for communications.

To say that information technology is transforming business enterprises is simple. What this transformation will require of companies and top managements is much harder to decipher. That is why I find it helpful to look for clues in other kinds of information-based organizations, such as the hospital, the symphony orchestra, and the British administration in India.

A fair-sized hospital of about 400 beds will have a staff of several hundred physicians and 1,200 to 1,500 paramedics divided among some 60 medical and paramedical specialities. Each specialty has its own knowledge, its own training, its own language. In each specialty, especially the paramedical ones like the clinical lab and physical therapy, there is a head person who is a working specialist rather than a full-time manager. The head of each specialty reports directly to the top, and there is little middle management. A good deal of the work is done in ad hoc teams as required by an individual patient's diagnosis and condition.

A large symphony orchestra is even more instructive, since for some works there may be a few hundred musicians on stage playing together. According to organization theory then, there should be several group vice president conductors and perhaps a half-dozen division VP conductors. But that's not how it works. There is only the conductor-CEO—and every one of the musicians plays directly to that person without an intermediary. And each is a high-grade specialist, indeed an artist.

But the best example of a large and successful information-based organization, and one without any middle management at all, is the British civil administration in India.[1]

The British ran the Indian subcontinent for 200 years, from the middle of the eighteenth century through World War II, without making any fundamental changes in organization structure or administrative policy. The Indian civil service never had more than 1,000 members to administer the vast and densely populated subcontinent—a tiny fraction (at most 1 percent) of the legions of Confucian mandarins and palace eunuchs employed next door to administer a not-much-more populous China. Most of the Britishers were quite young; a thirty-year-old was a survivor, especially in the early years. Most lived alone in isolated outposts with the nearest countryman a day or two of travel away, and for the first hundred years there was no telegraph or railroad.

The organization structure was totally flat. Each distinct officer reported directly to the "COO," the provincial political secretary. And since there were nine provinces, each political secretary had at least 100 people reporting directly to him, many times what the doctrine of the span of control would allow. Nevertheless, the system worked remarkably well, in large part because it was designed to ensure that each of its members had the information he needed to do his job.

Each month the district officer spent a whole day writing a full report to the political secretary in the provincial capital. He discussed each of his principal tasks—there were only four, each clearly delineated. He put down in detail what he had expected would happen with respect to each of them, what actually did happen, and why, if there was a discrepancy, the two differed. Then he wrote down what he expected would happen in the ensuing month with respect to each key task and what he was going to do about it, asked questions about policy, and commented on long-term opportunities, threats, and needs. In turn, the political secretary "minuted" every one of those reports—that is, he wrote back a full comment.

On the basis of these examples, what can we say about the requirements of the information-based organization? And what are its management problems likely to be? Let's look first at the requirements. Several hundred musicians and their CEO, the conductor, can play together because they all have the same score. It tells both flutist and timpanist what to play and when. And it tells the conductor what to expect from each and when. Similarly, all the specialists in the hospital share a common mission: the care and cure of the sick. The diagnosis is their "score"; it dictates specific action for the X-ray lab, the dietitian, the physical therapist, and the rest of the medical team.

Information-based organizations, in other words, require clear, simple, common objectives that translate into particular actions. At the same time, however, as these examples indicate, information-based organizations also need concentration on one objective or, at most, on a few.

Because the "players" in an information-based organization are specialists, they cannot be told how to do their work. There are probably few orchestra conductors who could coax even one note out of a French horn, let alone show the horn player how to do it. But the conductor can focus the horn player's skill and knowledge on the musician's joint performance. And this focus is what the leaders of an information-based business must be able to achieve.

Yet a business has no "score" to play by except the score it writes as it plays. And whereas neither a first-rate performance of a symphony nor a miserable one will change what the composer wrote, the performance of a business continually creates new and different scores against which its performance is assessed. So an information-based business must be structured around goals that clearly state management's performance expectations for the enterprise and for each part and specialist and around organized feedback that compares results with these performance expectations so that every member can exercise self-control.

The other requirement of an information-based organization is that everyone take information responsibility. The bassoonist in the orchestra does so every time she plays a note. Doctors and paramedics work with an elaborate system of reports and an information center, the nurse's station on the patient's floor. The district officer in India acted on this responsibility every time he filed a report.

The key to such a system is that everyone asks: Who in this organization depends on me for what information? And on whom, in turn, do I depend? Each person's list will always include superiors and subordinates. But the most important names on it will be those of colleagues, people with whom one's primary relationship is coordination. The relationship of the internist, the surgeon, and the anesthesiologist is one example. But the relationship of a biochemist, a pharmacologist, the medical director in charge of clinical testing, and a marketing specialist in a pharmaceutical company is no different. It, too, requires each party to take the fullest information responsibility.

Information responsibility to others is increasingly understood, especially in middle-sized companies. But information responsibility to oneself is still largely neglected. That is, everyone in an organization should constantly be thinking through what information he or she needs to do the job and to make a contribution.

This may well be the most radical break with the way even the most highly computerized businesses are still being run today. There, people either assume the more data, the more information—which was a perfectly valid assumption yesterday when data were scarce, but leads to data overload and information blackout now that they are plentiful. Or they believe that information specialists know what data executives and professionals need in order to have information. But information specialists are tool makers. They can tell us what tool to use to hammer upholstery nails into a chair. We need to decide whether we should be upholstering a chair at all.

Executives and professional specialists need to think through what information is for them, what data they need: first, to know what they are doing; then, to be able to decide what they should be doing; and finally, to appraise how well they are doing. Until this happens MIS departments are likely to remain cost centers rather than become the result centers they could be.

Most large businesses have little in common with the examples we have been looking at. Yet to remain competitive—maybe even to survive—they will have to convert themselves into information-based organizations, and fairly quickly. They will have to change old habits and acquire new ones. And the more successful a company has been, the more difficult and painful this process is apt to be. It will threaten the jobs, status, and opportunities of a good many people in the organization, especially the long-serving, middle-aged people in middle management who tend to be the least mobile and to feel most secure in their work, their positions, their relationships, and their behavior.

The information-based organization will also pose its own special management problems. I see as particularly critical:

1. Developing rewards, recognition, and career opportunities for specialists.
2. Creating unified vision in an organization of specialists.
3. Devising the management structure for an organization of task forces.
4. Ensuring the supply, preparation, and testing of top management people.

Bassoonists presumably neither want nor expect to be anything but bassoonists. Their career opportunities consist of moving from second bassoon to first bassoon and perhaps of moving from a second-rank orchestra to a better, more prestigious one. Similarly, many medical technologists neither expect nor want to be anything but medical technologists. Their career opportunities consist of a fairly good chance of moving up to senior technician, and a very slim chance of becoming lab director. For those who make it to lab director, about one out of every twenty-five or thirty technicians, there is also the opportunity to move to a bigger, richer hospital. The district officer in India had practically no chance for professional growth except possibly to be relocated, after a three-year stint, to a bigger district.

Opportunities for specialists in an information-based business organization should be more plentiful than they are in an orchestra or hospital, let alone in the Indian civil service. But as in these organizations, they will primarily be opportunities for advancement within the specialty, and for limited advancement at that. Advancement into "management" will be the exception, for the simple reason that there will be far fewer middle-management positions to move into. This contrasts sharply with the traditional organization where, except in the research lab, the main line of advancement in rank is out of the specialty and into general management.

More than thirty years ago General Electric tackled this problem by creating "parallel opportunities" for "individual professional contributors." Many companies have followed this example. But professional specialists themselves have largely rejected it as a solution. To them— and to their management colleagues—the only meaningful opportunities are promotions into management. And the prevailing compensation structure in practically all businesses reinforces this attitude because it is heavily biased towards managerial positions and titles.

There are no easy answers to this problem. Some help may come from looking at large law and consulting firms, where even the most senior partners tend to be specialists, and associates who will not make partner are outplaced fairly early on. But whatever scheme is eventually developed will work only if the values and compensation structure of business are drastically changed.

The second challenge that management faces is giving its organization of specialists a common vision, a view of the whole.

In the Indian civil service, the district officer was expected to see the "whole" of his district. But to enable him to concentrate on it, the government services that arose one after the other in the nineteenth century (forestry, irrigation, the archaeological survey, public health and sanitation, roads) were organized outside the administrative structure, and had virtually no contact with the district officer. This meant that the district officer became increasingly isolated from the activities that often had the greatest impact on—and the greatest importance for—his district. In the end, only the provincial government or the central government in Delhi had a view of the "whole," and it was an increasingly abstract one at that.

A business simply cannot function this way. It needs a view of the whole and a focus on the whole to be shared among a great many of its professional specialists, certainly among the senior ones. And yet it will have to accept, indeed will have to foster, the pride and professionalism of its specialists—if only because, in the absence of opportunities to move into middle management, their motivation must come from that pride and professionalism.

One way to foster professionalism, of course, is through assignments to task forces. And the information-based business will use more and more smaller self-governing units, assigning them tasks tidy enough for "a good man to get his arms around," as the old phrase has it. But to what extent should information-based businesses rotate performing specialists out of their specialties and into new ones? And to what extent will top management have to accept as its top priority making and maintaining a common vision across professional specialties?

Heavy reliance on task-force teams assuages one problem. But it aggravates another: the management structure of the information-based organization. Who will the business's managers be? Will they be task-force leaders? Or will there be a two-headed monster—a specialist structure, comparable, perhaps, to the way attending physicians function in a hospital, and an administrative structure of task-force leaders?

The decisions we face on the role and function of the task-force leaders are risky and controversial. Is theirs a permanent assignment, analogous to the job of the supervisory nurse in the hospital? Or is it a function of the task that changes as the task does? Is it an assignment or a position? Does it carry any rank at all? And if it does, will the task-force leaders become in time what the product managers have been at Procter & Gamble: the basic units of management and the company's field officers? Might the task-force leaders eventually replace department heads and vice presidents?

Signs of every one of these developments exist, but there is neither a clear trend nor much understanding as to what each entails. Yet each would give rise to a different organizational structure from any we are familiar with.

Finally, the toughest problem will probably be to ensure the supply, preparation, and testing of top management people. This is, of course, an old and central dilemma as well as a major reason for the general acceptance of decentralization in large businesses in the last forty years. But the existing business organization has a great many middle-management positions that are supposed to prepare and test a person. As a result, there are usually a good many people to choose from when filling a senior management slot. With the number of middle-management positions sharply cut, where will the information-based organization's top executives come from? What will be their preparation? How will they have been tested?

Decentralization into autonomous units will surely be even more critical than it is now. Perhaps we will even copy the German *Gruppe* in which the decentralized units are set up as separate companies with their own top managements. The Germans use this model precisely because of their tradition of promoting people in their specialties, especially in research and engineering; if they did not have available commands in near-independent subsidiaries to put people in, they would have little opportunity to train and test their most promising professionals. These subsidiaries are thus somewhat like the farm teams of a major-league baseball club.

We may also find that more and more top management jobs in big companies are filled by hiring people away from smaller companies. This is the way that major orchestras get their conductors—a young conductor earns his or her spurs in a small orchestra or opera house, only to be hired away by a larger one. And the heads of a good many large hospitals have had similar careers.

Can business follow the example of the orchestra and hospital where top management has become a separate career? Conductors and hospital administrators come out of courses in conducting or schools of hospital administration respectively. We see something of this sort in France, where large companies are often run by men who have spent their entire previous careers in government service. But in most countries this would be unacceptable to the organization (only France has the *mystique* of the *grandes écoles*). And even in France, businesses, especially large ones, are becoming too demanding to be run by people without firsthand experience and a proven success record.

Thus the entire top management process—preparation, testing, succession—will become even more problematic than it already is. There will be a growing need for experienced businesspeople to go back to school. And business schools will surely need to work out what successful professional specialists must know to prepare themselves for high-level positions as *business* executives and *business* leaders.

Since modern business enterprise first arose, after the Civil War in the United States and the Franco-Prussian War in Europe, there have been two major evolutions in the concept and structure of organizations. The first took place in the ten years between 1895 and 1905. It distinguished management from ownership and established management as work and task in its own right. This happened first in Germany, when Georg Siemens, the founder and head of Germany's premier bank, *Deutsche Bank,* saved the electrical apparatus company his cousin Werner Siemens had founded after Werner's sons and heirs had mismanaged it into near collapse. By threatening to cut off the bank's loans, he forced his cousins to turn the company's management over to professionals. A little later, J. P. Morgan, Andrew Carnegie, and John D. Rockefeller, Sr. followed suit in their massive restructurings of U.S. railroads and industries.

The second evolutionary change took place twenty years later. The development of what we still see as the modern corporation began with Pierre S. du Pont's restructuring of his family company in the early twenties and continued with Alfred P. Sloan's redesign of General Motors a few years later. This introduced the command-and-control organization of today, with its emphasis on decentralization, central service staffs, personnel management, the whole apparatus of budgets and controls, and the important distinction between policy and operations. This stage culminated in the massive reorganization of General Electric in the early 1950s, an action that perfected the model most big businesses around the world (including Japanese organizations) still follow.[2]

Now we are entering a third period of change: the shift from the command-and-control organization, the organization of departments and divisions, to the information-based organization, the organization of knowledge specialists. We can perceive, though perhaps only dimly, what this organization will look like. We can identify some of its main characteristics and requirements. We can point to central problems of values, structure, and behavior. But the job of actually building the information-based organization is still ahead of us—it is the managerial challenge of the future.

Notes

1. The standard account is Philip Woodruff, *The Men Who Ruled India,* especially the first volume, *The Founders of Modern India* (New York: St. Martin's 1954). How

the system worked day by day is charmingly told in *Sowing* (New York: Harcourt Brace Jovanovich, 1962), volume two of the autobiography of Leonard Woolf (Virginia Woolf's husband).

2. Alfred D. Chandler, Jr. has masterfully chronicled the process in his two books *Strategy and Structure* (Cambridge: MIT Press, 1962) and *The Visible Hand* (Cambridge: Harvard University Press, 1977)—surely the best studies of the administrative history of any major institution. The process itself and its results were presented and analyzed in two of my books: *The Concept of the Corporation* (New York: John Day, 1946) and *The Practice of Management* (New York: Harper Brothers, 1954).

Part Seven

Japan as Society
and Civilization

Introduction to Part Seven

In the United States I am usually viewed as an apologist for Japan. In Japan I am more often seen as a "Japan-Basher." To be sure, the Japanese consider me one of three Americans who gave them the concepts, provided the tools, and did the training that led to their economic rebirth after World War II (the other two are Edwards Deming and Joseph Juran). But ever since I coined the term "adversarial trade" (in the 1978 article reprinted as chapter 26 in this book) to describe Japan's economic policy, I am seen in Japan as a critic, and by no means a friendly one.

Actually, I was the first observer—in 1961—to see and to report the emergence of Japan as a major economic power and, above all, her emergence as a major power in the world markets. I was also the first—in 1971—to write on Japanese management in an article entitled "What We Can Learn From the Japanese"—the first Western report on such now thoroughly familiar practices as consensus decisions, lifetime employment, long-range strategy, quality control (it appeared in the *Harvard Business Review*). But I was also the first Westerner to warn that the West's—and especially the United States'—traditional responses and policies were not appropriate to deal with Japan and that the Japanese have a very different view of economics and economic policy and play the economic game by different rules. Thus both those who consider me an admirer of Japan and those who consider me a detractor are quite correct. But my point all along has not been that Japan is "good" or that Japan is "bad," but that Japan is different. And the differences are not economic; they are social. In fact my approach to Japan has not been through economics and business. It has been through Japanese art and through Japanese history—the result of my falling in love with Japanese painting when still a very young man and an economist working for a London merchant bank. I did not actually travel to Japan until the mid-fifties. But I had by then lived with Japanese culture for twenty years. And thus what I see in Japan— or at least what I watch for in Japan—is not the new. It is the continuity in Japanese society, in Japanese culture, in Japanese community, the tension between it and "Modern Japan," and its resolution—often in compromise, more often in a new synthesis.

It is thus fitting, I believe, that these chapters on Japan begin with an essay on Japanese art. Japan has always imported the Conceptual, first from China, then from the West. There are no great Japanese theologians, logicians, philosophers, mathematicians. Japan's native culture, unlike that of any other country, is totally perceptual. It is built around painting and calligraphy. Even the theater in Japan—the one Japanese art form totally uninfluenced by Chinese or Buddhist models—is more pictorial than dramatic. Kabuki, for instance, the theater of post-feudal Japan (it began around 1600 and reached its peak in the eighteenth century) is cinema without camera and without film. And thus, the essays in this part approach modern Japan, her up-to-date economy, her advanced technology and her business culture—in no country is the successful business executive venerated more these days—from a vantage point in Japanese society, Japanese aesthetics, Japanese history.

25

A View of Japan through Japanese Art

Japan, as everybody knows, is a country of rigid rules and of individual subordination to a collective will. It is the country where the young college student goes hiking in the mountains, but turns boot and pack over to a younger brother or sister upon graduation. It is the country where the student is a radical in college but becomes a faithful conservative upon being hired by Mitsubishi Bank or Ministry of Finance. It is the country where a young woman wears one kind of kimono until the day of her wedding, and then puts on the married woman's kimono for the rest of her life.

Japan is a country where junior high school graduates become manual workers, high school graduates become clerks, and college graduates become managers and professionals—all three thus slotted for the rest of their lives by the school-leaving diploma. It is a country where there is lifetime commitment to one employer. Japan is also, as everyone knows, the country of mutual obligations, in which speech is minutely regulated by social relationship and status. It is the country of "Japan Inc.," where conflicting interests pull together for the greater glory of the common economy. The best-known—and best—book on Japanese social organization and institutions, *Japanese Society* by Chie Nakane, depicts the *ie,* the community of the clan or tribe, as the organizing reality within which the individual exists as a member

First published in *Song of the Brush, Japanese Paintings from the Sansō Collection,* ed. John M. Rosenfield and Henry Trubner (Seattle, Wash.: Seattle Art Museum, 1979).

rather than as a person. Whenever Japanese and Western (especially American) scholars meet, in any discipline and on any subject, the Japanese at once contrast Japanese cooperation with the excessive competition and rampant diversity of the West.

Yet the most pervasive trait of all Japanese art is its individualism. In every major period of artistic activity in the West there has been one universal style; we speak of the Hellenistic, of the Romanesque and the Gothic, the Renaissance and the Baroque. But every period of great artistic activity in Japan has been characterized by diversity. Indeed, in the arts, and especially in painting, the contrast is properly between Western conformity and the "excessive diversity" of Japan. During the Edo period (1603–1867), the Japanese tendency to diversity reached its apogee. In painting alone, over a dozen major schools flourished, along with countless subschools. There is nothing comparable in other cultures to the flamboyant diversity of the last great artistic era of premodern Japan.

The Japanese scholars and experts who castigate American excessive competition, and who contrast it to its disadvantage with Japanese cooperation, think of competition among businesses in the marketplace or of competition for promotion within the management group in a company. They never, it seems, think of the Japanese school system. Yet every American recoils in horror when told that a Japanese schoolboy, ten years old, will applaud with joy upon hearing that his best friend is ill and will have to miss a week or two of school. The friend will thus fall behind in the competition for the examination that will decide on the few who will make it into the prestigious junior high school.

And as to "Japan Inc.," there is no commercial rivalry and competition in the West that compares with the fierce ruthlessness with which the major Japanese industrial groups, the *zaibatsu*, fight one another. If Mitsubishi goes into a new field, be it synthetic fibers or electronics or shipbuilding. Mitsui and Sumitomo have to go into it too—never mind that overcapacity already exists in that industry worldwide. And Japanese political parties are not disciplined monoliths; they are not an *ie*. They are loose congeries of fiercely competing factions.

The Japanese are probably the world's best animal painters. In the West, the few animal painters are specialists, a Rosa Bonheur or George Stubbs, for instance. In Japan almost every painter painted animals. The Japanese took some traditions of animal painting from the Chinese: the *kachōga* (flower-and-bird) painting, for instance. But most animals, and especially the birds, in Japanese painting express purely native Japanese values, traditions, and perceptions.

Nothing I know expresses one basic trait of the Japanese as well as these bird paintings do: the capacity for pure enjoyment. It is the same capacity one gets at a Japanese picnic or at a simple folk dance in an empty lot on a summer's night. It is the capacity for pure enjoyment that makes pompous company presidents and grave scholars play at the silliest children's games at a party, without embarrassment or reticence. It is the capacity for pure enjoyment that can be seen in parks on Sundays where young Japanese fathers romp with their children. It is a quality of immediacy that is present in the most sophisticated Japanese artwork or novel, and that is the essence of the *haiku*. The traditional Japanese animal or bird painting always looks ludicrously simple—just a few strokes of the brush. Yet it is done with complete control of brush, ink, and composition. It also expresses the artists' intuitive, immediate projection of their own selves into the spirit of the birds or the frog. These Japanese paintings are a hymn to diversity and spontaneity in keeping with the first of the modern English poets, the late Victorian Gerard Manley Hopkins, who sang, "Glory be to God for dappled things."

And yet the cooperation, the mutual obligations, the lifetime commitment to one employer, the *ie*, and even "Japan Inc.," are not myths. Central to Japan is constant and continuing polarity between tight, enveloping community—supportive but demanding subordination to its rules—and competitive individualism demanding spontaneity.

Japanese artists of the eighteenth century were highly individualistic, yet most considered themselves as belonging to a school—Nanga or Rimpa or Shijō, for instance. The few who did not—Shōhaku, Rosetsu, Jakuchū—are called "eccentrics" in Japan. And if an artist starts in a school and then outgrows it and develops his own style, Japanese propriety demands that there be a violent break, like the confrontations of a Kabuki drama. Nagasawa Rosetsu (1755–99), for instance, is reported to have broken violently with Maruyama Okyo (1733–95), whose student he originally was, though the record shows unambiguously that the two actually kept on working together and that

Okyo entrusted Rosetsu with important and confidential commissions. Similarly, a century earlier, Kusumi Morikage (d. before 1700) was reported to have been excommunicated by, and exiled from, the atelier of Kanō Tanyū (1602–74) when he went his own way, though the record shows a close and continuing family relationship between the two artists.

Even today, in a modern, Westernized Japan, it is not considered proper for a young man to be on his own and not to belong to an organization, an *ie*. My interpreter on my first lecture tour in Japan, more than twenty years ago, was a young Japanese who had gone to graduate school in the United States and who had then established his own marketing consulting practice in Tokyo. He was, I found out, not welcome in his father-in-law's house. When I met the father-in-law, who was dean at a university where I lectured, I asked him what he had against his son-in-law. "He is barely thirty," he replied, "and on his own; that's quite improper. He has no organization to back him up, no boss to bail him out when he gets into trouble. What's worse, he is successful and thus sets a dangerous precedent." The point of the story is that the father-in-law was known all over Japan as the Red Dean, who delivered himself every Saturday evening on national radio of a violent philippic against the remnants of feudalism in Japanese family life and against the evils of the organization man.

Art history (or art anecdote) may provide the answer to the paradox, and a key to understanding the relationship between the rigid community of the *ie* and the spontaneity and individualism that characterize so much of Japanese art as well as of Japanese life and society. Sakai Hōitsu (1761–1828), the last of the great Rimpa masters, started out studying under a Kanō painter around 1790. He then became the pupil of a distinguished Nanga artist, Kushiro Unsen (1759–1811). Next he apparently went for career advice to Tani Bunchō (1764–1840), the recognized Nanga master in the city of Edo, Japan's political capital. Bunchō did not tell the young Hōitsu to stick to Nanga, but counseled him to study the works of Ogata Kōrin (1663–1743) and to become a Rimpa painter. A great teacher in the West might have said to such a highly gifted young man: "Find the style that fits you." Bunchō said, in effect: "Find the school that fits you."

The tension between the pressure to belong and to conform and the stress on spontaneity, independence, and individuality is one, but only one, of the polarities that characterize Japanese art and Japanese culture. A well-known Japanese art collection contains for instance three works by famous seventeenth-century masters; *Two Wagtails,* attributed to Kanō Sanraku (1559–1635); *Child Holding a Spray of Flowers,* attributed to Tarawaya Sōtatsu (early seventeenth-century); and a circular fan, *Autumnal Ivy Leaves with Bamboo,* by Ogata Kōrin. Each epitomizes the Japanese talent for simplicity refined to the point of austerity. Yet Sanraku's best-known paintings, such as his screens of birds and trees and flowers, are ornate and sumptuous, with gold and silver and ostentatious colors. Sōtatsu founded the decorative Rimpa School, with its strong lyricism and colorful elegance, and Kōrin perfected it with his rich designs. Thus the three paintings in this one collection may be called atypical of their painter—and yet each is also completely typical of him.

In the same collection there is a landscape by the early sixteenth-century painter Kantei that simplifies and makes more austere the already simplified and austere style of the fifteenth-century Japanese landscape painter. But there is also a pair of flower-and-bird paintings by the same master that is ornate, decorative, almost sumptuous. Almost two hundred years later, in the early nineteenth century, Watanabe Kazan, the most austere of Neo-Confucians, painted a lush, sensuous, colorful picture of *Lotus Flowers and Swimming Fish.*

To a Westerner these seem to be contradictions; to a Japanese they are polarities. A Westerner may feel that an artist should be attracted either to the austerity and empty space of a fifteenth-century landscape or to the colorful, decorative design of a Kantei flower-and-bird painting or a Sanraku bird screen, but not to both. To the Japanese, however, these are necessary tensions, poles of expression within the same person.

Any visitor to Kyoto sees examples of these tensions within a few miles of each other: Nijō Castle—ornate, sensuous, boastful, the official Kyoto residence of the military dictator, the Tokugawa shōgun; and Katsura Villa—simple to the point of being austere, exquisite, without ornaments, and totally disciplined, the summer villa of an imperial prince. Both were built within the same generation and by the same ruling class. And there is Nikkō, north of Tokyo, the great seventeenth-century mausoleum of the first Tokugawa shōgun Ieyasu—the "shōgun" of the movie of that name—with its extreme ornateness, almost too much, even for Baroque tastes. But the same shōgun, in his castle, lived in restrained austerity. To the Japanese, the two belong together. The tension is not one of opposites, but one of poles; and where there is a north pole there has to be a south pole.

This tension, this polarity, extends through all of Japanese culture. It is found in the tension between the official Confucian male supremacy, which dictates that in public the woman is invisible and subservient, and the reality of family life, where the woman holds the power and the purse strings, and where a recent prime minister could say in parliament: "I have no position on this measure yet; my mother-in-law has been sick and I could not get her guidance," to which the opposition spokesman nodded, replying: "Please convey my wishes for a speedy recovery to your respected mother-in-law."

A similar polarity is found in the upbringing of children. Until they reach school age, children are indulged in a way that goes beyond any American permissiveness. Then they go to school, and on the first day there is discipline and the children are expected to behave—and they do. There is a remarkable tension between the genius of the Japanese language, in which everything focuses on human relationship, and the nature of Chinese ideographs, which are built up of representations of objects. The Japanese very early invented syllabaries in which the sounds of Japanese can easily be written. Every Japanese learns the two national syllabaries in the elementary grades. But then the syllabaries are used mainly as auxiliaries to Chinese ideographs. To the Japanese, the tension between the Japanese language and the Chinese ideograph is essential, no matter how heavy a burden it puts on learning and literacy.

There are strict rules for proper behavior that tell every Japanese what form of address to use when talking to his aunt and to his uncle's boss and to his cousin's mistress. But there is also the encouragement of the eccentric, who is given almost infinite leeway. Sengai, the last and perhaps greatest of the "Zenga Expressionists"—he died in 1838, aged almost ninety—for instance, was the most respected cleric, abbot of an ancient and most sacred temple; but at the age of eighty-five he was also a free spirit, who traveled around the country, often in low company, and liked to paint satirical frogs to look like the Buddha, circus riders, and balloon vendors at county fairs.

This polarity can be found today in Japanese industry and its human relations. To a Westerner, an organization can be either autocratic or democratic, but the Japanese organization is both. Surely no more perfect example of the autocratic personality exists than the head of a big Japanese organization, whether government agency or business. Yet decision making is by consensus and participation, and starts at the bottom rather than at the top. In every Japanese organization from ancient times to the present the word of the chief has been absolute law; the chief could order a retainer to commit suicide or to divorce his wife. And yet no chief could make one step without the consent of his retainers, and indeed without active participation of the clan elders in the decision. Similarly, today the top people in a company or a government agency are obeyed without argument or reservation—and yet every decision comes up from below and is an expression of a general will. Every Japanese organization is in Western terms both an extreme of autocracy and an extreme of democratic participation.

The tension is not dialectic, and resolved in a higher synthesis, nor does one principle overcome the other. It is not the dualism of the Chinese yin and yang. The Japanese do not mix their principles any more than one mixes north pole and south pole. For the Japanese tension is not contradiction or contrast or conflict—the tensions of the analytical mind. It is polarity—the tension of perception, of configuration, of existence. To understand Japanese art and Japanese life, one has to accept the polarity between the ornate and the austere; between male supremacy and female power; between spoiled, indulged brats and disciplined scholars; between the Japanese language with its inflected verbs and syllabary script and the complexities of the Chinese ideograph. Such polarities are essential to Japan and, to my knowledge, to Japan alone.

It is this tension, this polarity, that has made Japan throughout its history a country of contrasts, of sharp and sudden swings: from wide-open receptivity to foreign cultures and foreign commerce to self-imposed isolation, for instance, in the seventeenth century. But it is also this polarity that gives Japanese art, Japanese literature, and Japanese industry their dynamics and creativity.

A Westerner who has business in Japan—the professor who goes there to lecture or the businessman who negotiates a contract—soon becomes familiar with the phrase *"Wareware Nihon-jin,"* which means "We Japanese." But whenever it is used (and it is used all the time), it conveys: "We Japanese are so different that you will never understand us." To understand what a Japanese friend or business partner, or the student in the audience who gets up and asks a question, means when he starts out with *Wareware Nihon-jin,* one needs to look at Japanese landscape paintings. But where in the landscapes are the people—the *Nihon-jin?* Yet it is precisely their absence, or their subordination to the land, that is the point. For *Nihon-jin* does not just mean Japanese. It means: "We who belong to the land of Japan." The landscape painting is the soul of Japanese art because the Japanese landscape has formed the soul of Japan.

Some of the features of these landscape paintings the Japanese took from the Chinese—the bizarre rock formation of the eroded Chinese karst limestone that can be seen in so many Japanese landscape paintings is an example. But most of the features of these landscapes can be found in Japan; indeed, a Japanese friend of mine claims that he knows the valley, someplace near Gifu, that Gyokudō, the great nineteenth-century landscape painter, so often painted in his small lyrical landscapes. The Japanese landscape looks like the landscape of the Japanese painters, as everyone knows who has traveled in the Japanese countryside. And yet the Japanese landscape does not look a bit like the landscape of the Japanese landscape painter, nor does any landscape on earth. The landscape of the Japanese painter is a spiritual landscape, a landscape of the soul.

The Japanese feeling for this landscape is part of "Shinto." What Shinto really means, probably no Westerner has ever been fully able to understand. It surely does not mean a religion in the Western sense; it became a religion only after 1867, when the Meiji Restoration created a monstrosity known as State Shinto because it felt it had to emulate the way religions are set up in the West. Far more ancient and pervasive are the many Shinto shrines and rituals, but there is, above all, a Shinto feeling—the feeling of the uniqueness of Japan as an environment. I did not write human environment; it is far more than that. It is an environment fully as much for the supernatural, for the forces that control the universe, as it is an environment for man and beast, plant and rock. It is unique and it is complete. And it is different, which is the point of *Wareware Nihon-jin.* Underlying the phrase is the feeling that Japan is unique; that Japan is by itself. What this means is expressed in the landscape paintings. Their hills and trees are the visible surface, the skin, of a spiritual landscape that is invisible and unique. There may be landscapes elsewhere that look like it. Taiwan has similar hills, and so does Korea. But there is no landscape, to a Japanese, that means the same thing. A painting of the Japanese landscape can be a realistic image that serves as a valid legal document to determine the boundary lines of a Shinto shrine, as some of the earliest Japanese landscape paintings were intended to do; but even then it also means an inner space, a landscape of the soul that is the center of gravity of Japanese existence. This landscape is, so to speak, Japan *an sich*.

I am not saying that the Japanese are, in fact, unique; I am saying that the Japanese feel that they are. It is not that they feel superior; nationalism has been a Japanese vice only in rare, short moments of aberration. They feel different because they feel at home only in this landscape of their soul. This may explain why, of all the foreign students in the United States and Europe, the Japanese are the only ones, who, with few exceptions, cannot wait to go back home.

I now come to what I would call "Japanese aesthetics," or the "topological approach," or "What makes the Chinese so uneasy when they look at a Japanese painting?"

Almost any Japanese landscape painting could be used to demonstrate Japanese aesthetics. The fifteenth-century paintings deliberately set out to follow the Chinese; and so did the Nanga painters of the eighteenth century. And yet, put a Chinese connoisseur or a Chinese art historian in front of these painters' works and he will be uneasy. "Yes, these hills look like so-and-so in China. And yes, this rock looks like some other painter in China. And the brushwork is this or that school. And the brush technique, of course, follows another Chinese example. And yet, and yet, and yet . . ." What he is saying, if he is candid, is: "And yet these are definitely not Chinese paintings. They make me very uneasy and I do not understand why. But I do not want to have them around."

One only has to put these works next to Chinese paintings to understand his feelings. I am not saying that one cannot mistake a Chinese painting for a Japanese one, and vice versa. The technique is the same, the brushstrokes are the same, the ink values are the same—and the painting is different. What makes it different is the Japanese sense of beauty. The Japanese paintings are dominated by empty space. It is not only that so much of the canvas is empty. The empty space organizes the painting. This is the opposite to what most Chinese would do, but it is basic to Japanese aesthetics. The same aesthetics are found in Japanese paintings of all schools, and in the works of painters who followed the Chinese as well as in the works of painters who rejected the Chinese.

If I were to define these aesthetics in contrast to those in Western and Chinese painting, I would say that Western painting is basically geometric. It is no coincidence that modern Western painting begins with the rediscovery of linear perspective, around 1425, that is, with the subordination of space to geometry. Chinese painting, on the other hand, is algebraic. In Chinese painting, proportion governs, as it does in Chinese ethics. Japanese painting is by contrast topological—that branch of mathematics that began around 1700 and that deals with the properties of surface and space in which shapes and lines are defined by space, so that there is no difference between a straight line and a curve, such as a hyperbola. Topology deals with angles and vortices and boundary lines. It deals with what space imposes rather than with what is imposed on space. The Japanese painter is topological in his aesthetics. He sees space and then he sees lines. He does not start out with the lines.

It has been commonplace for Western art critics and art historians for almost a hundred years to say that painters do not see objects but configurations. But the *Gestalt* that the Japanese painter sees is what we today would call a design rather than a structure. This is what the topologist means when he says that, topologically speaking, it is space that determines the line rather than the line that determines space. In discussions of Japanese painting, one usually finds a reference to the Japanese tendency to become "decorative." The Nanga painters of the eighteenth century abhorred the decorative as totally incompatible with the values and aesthetics of their Chinese literati models; and yet, as we are told by all authorities, they always became decorative. Like so many words in art criticism, "decorative" is misleading; the right word might be "designed." And this irrepressible tendency toward design—the tendency that explains why ceramics, lacquer, and painting in Japanese art tend to be closely allied, while the Chinese kept them strictly separate artistically and socially—is based on a Japanese vision that is neither perspectival (i.e., geometric) nor proportional (i.e., algebraic), but topological in design.

Both the fifteenth-century black-ink painters and the eighteenth-century Nanga painters looked to the Chinese as models and masters. Both learned techniques from the Chinese, but also motives and style and form. But both transmuted Chinese algebra into Japanese topology. This ability to receive a foreign culture and then to "Japanize" it is a continuing thread in Japanese history and experience.

Around A.D. 500, Buddhism, and with it the highly advanced and most refined civilization of China, swept into Japan. At first the impact seemed to inundate Japan completely. Everything was brought from China or Korea, including monks and architects and artists and artisans and scribes and poetry and artworks and textiles. After only two centuries, by the Nara period, Japan was producing religious sculpture that was completely Buddhist and yet deeply Japanese, even though the techniques were still those of the Chinese and Korean sculptors. But Japan equally transmuted China's governmental and social structures. It made both Buddhism and Confucianism serve a tribal, and soon thereafter, a warrior society. It made Chinese concepts of land tenure, grounded in family ownership of soil, serve a system in which there was no ownership of land at all, except by temples and the throne. There were only graduated rights to the product of the land— that is, graduated rights to tax and tribute rather than ownership rights in land as such. The same thing happened in ceramics, in poetry, and in architecture.

It is happening again today; only now it is the West rather than China that is the foreign culture that is being Japanized. Forms, techniques, and concepts are used very skillfully. As the fifteenth-century and the eighteenth-century painters did, the Japanese rapidly improve on the techniques. There are few Chinese painters whose control and command of the brush equal the artistry of the great fifteenth-century Japanese landscape painter Sesshū. There are few Western companies that have the control and command of the corporate form and of managerial techniques that the large Japanese trading companies possess. But the essence is Japanese. The Japanese are not unaffected by the foreign influence; it becomes part of their own experience. Yet they distill out of the foreign influence whatever serves to maintain and strengthen Japanese values, beliefs, traditions, purposes, relationships. The result is not a hybrid. It is, as the fifteenth-century paintings or the eighteenth-century paintings show, all of one piece. This is a truly unique Japanese characteristic.

Again and again, Japanese society has lived through periods when it was wide open to foreign influences. But then it closes in again, to digest, transmute, and transform almost alchemically. What is considered base metal in the foreign culture sometimes becomes gold in Japan, as with the Chinese thirteenth-century painters Mu-ch'i or Yin T'o-lo. Both were rejected by the Chinese as "coarse" or "vulgar." They then became the models and masters for the most austerely refined Japanese painters of the fourteenth and fifteenth centuries. But sometimes the metal in a foreign culture may become dross in Japan, as happened in this century to the idea of the national state, imported from the West and transmuted into a poisonous parody of an old and peculiarly Japanese political form, the shōgunate, or military government. Always before the shōgun had served to eliminate fighting, to make war both unnecessary and impossible, and, above all, to prevent foreign adventure.

The Japanese aesthetics are a way to understand, or at least to perceive, a fundamental and central element: the very special (I would say unique) relationship between Japan and the outside world. It is a relationship based on receptivity, on an ability to learn quickly and to improve on what is being taught, while at the same time accepting, or at least retaining, only what makes Japan more Japanese: what fits topology rather than geometry or algebra; what fits Japanese human relations; what fits the inner experience of the uniqueness of Japan, and what might be called, with a Western term, Japanese spirituality. We are talking here of an existential phenomenon; and by the way, the best translation of the peculiar word "*Shinto*" is probably spirituality.

Whether it can maintain these abilities is the great question ahead of Japan, I think. Japan is now becoming integrated into the outside world, and not just economically (perhaps least of all economically) in a way which neither the Japan of the sixth century, that of the Buddhist and Chinese tidal wave, nor the Japan of Sesshū's time around 1500, nor perhaps even the Japan of a hundred years ago, could have imagined. Is it still possible for Japan to encapsulate and transmute into Japanese the foreign, the non-Japanese, culture, behavior, ethics, and even aesthetics?

I dare not even speculate—but there are a few straws in the wind. If one looks at the visual arts that today prosper in Japan—the modern Japanese woodblock print; the Japanese movie; modern Japanese ceramics; and perhaps one could add Japanese architecture—one would say that there is a possibility, even a probability, that the Japanese are again Japanizing the imported culture. The Japanese woodblock print is modern and Japanese very much in the way in which Nara sculpture was Buddhist and Japanese. So too, to a large extent, are the ceramics of Japan today. I can hope only that the Japanese will do again what they have done before so many times. The world needs a culture that is both modern and distinctly, uniquely, non-Western. It needs a Japanese Japan rather than a Japanese version of New York or Los Angeles or Frankfurt.

"Ten minutes and eighty years," Hakuin Ekaku, the great eighteenth-century Zen master, is said to have answered when asked how long it took him to paint one of his paintings of Daruma, the founder of the Zen sect. Of course, Rembrandt might have given the same answer when asked how long it took him to paint the self-portraits of his old age, Claude Monet when asked how long it took him to paint one of the hymns to light in his versions of Rouen Cathedral, or Pablo Casals when asked how long it took him to play one of Bach's "Unaccompanied Suites for Cello." But Hakuin's answer has two levels of meaning beyond that of the Western artist: it expresses a Japanese view of the nature of man, and a Japanese view of the nature of learning.

These may be seen in the one dimension in Japanese figure and portrait painting for which the West has no real parallel, nor China either—the spiritual self-portrait. If the Westerner says that it takes eighty years to be able to do what Rembrandt's last self-portraits represent, or Monet's pure light and Casals' Bach, he talks of the decades of practice needed to attain the skill. But the "eighty years" of the Japanese saying refer, above all, to the spiritual self-realization needed to become the person who can paint Daruma. An old Zen saying has it that "Every painting of Daruma is a [spiritual] self-portrait." The Zen painter who has not worked for decades on control of the self will not be the person to paint Daruma. Daruma is not a god, he is not a saint. He is man, but one who has realized man's full spiritual potential, who has attained man's full spiritual power, and who has transformed himself into a spiritual being. And only the painter who has himself become the spiritual man which Daruma represents can then paint a portrait on which he can inscribe, as Hakuin did on one of his *Darumas,* "This Is IT!" The spiritual power, the spiritual qualities of Daruma cannot be faked. No matter how great a painter's skill, if he lacks these qualities, his *Daruma* will lack them too.

Kanō Tanyū, in the mid-seventeenth century, and Rosetsu, a century later, very great masters and without peers in painting skill, both painted Daruma. Tanyū's *Daruma* looks like an elderly bureaucrat or successful banker; Rosetsu's like the urbane and witting chairman of a university graduate department. Both are excellent paintings—but neither has spirituality, power, total compelling control. But the *Daruma* by a painter who himself has the spirituality will have that power even if, as in the case of a *Daruma* painted by Hakuin in his extreme old age, the body is weighed down by the physical infirmity of advanced old age, the eyes are almost blind despite heavy glasses, the legs have given out, and death is very near.

Daruma is mortal. He is a sentient being. But unlike the saints of Christianity or of Buddhism, he is not dependent on divine grace, on a Supreme Being, or on redemption. He had attained spiritual perfection through his own efforts and by fulfilling the divinity within him. This is not a "humanist" view of man but a spiritual and existential one. It is a view that focuses on wisdom rather than on knowledge; on self-control rather than on power; on excellence rather than on success.

The Zen saying of "ten minutes and eighty years" also expresses a uniquely Japanese concept of continuous learning. In the West and in China, one learns to prepare oneself for the next job, for a promotion, for a new challenge. The most extreme example was the Confucian examination system of Imperial China, where one actually had to unlearn what the first examination tested to get ready for the next, the second one. The Western modern medical school is not too different either, nor are most Advanced Management courses. But in Zen, one learns so as to do better what one already does well. One keeps on painting Daruma until the control becomes completely spontaneous. One draws, as did the early seventeenth-century calligrapher Konoe Nobutada, a picture of Tenjin, the patron of learning, every morning—the same picture, but with ever-increasing mastery. Or, like Nakabayashi Chikutō, around 1800, one paints the same landscapes over and over again. Of course, in the West the artist does that too—Casals practiced the Bach cello suites until his death, well past ninety years of age. But in the West—and in China—only the artist does this; the rest of us are like Confucian scholars who pass one examination to be qualified to sit for the next, and for whom one promotion is the stepping stone to the next.

In Japan there is to this day the specialist in the trading company, the specialist on cotton, for instance, or on woodworking machinery, who gets more money and a bigger title but stays the specialist on cotton or on woodworking machinery all his working life, becoming more accomplished with every year. There is the continuous learning process in the Japanese factory, where employees get more money with seniority but keep on doing the same job, and meet every week to discuss how they can do their present jobs better. And there is the uniquely Japanese concept of the Living National Treasure, the great craftsman or artist who has excelled through doing the same work. The Western theory of the learning curve is not accepted in Japan—the theory that people reach a plateau of accomplishment after a certain time and then stay on it. The Japanese learning curve has them break out of that plateau again by continuing to practice—until they reach a new plateau, when they again, after a time, start learning and growing, and so on, always approaching perfection. The Japanese learning curve, like the Zen master's ten minutes and eighty years, sees learning as an act of spiritual perfection and personal self-development as much as an acquisition of skills. It is a way of changing the person, and not just a way of acquiring performance capacity.

Again this is but one strand. Japanese history and Japanese society are as full of climbers, of ambitious schemers, of people on the make and people on the take, as any other history or society. But there is also the counterpoint—the ten minutes and eighty years of continuous learning to do better what one already does very well.

The Japanese or Zen concept of learning is not without dangers. It can degenerate into imitation and repetition. This is what happened to the Kanō School of painting. It was Japan's "official" art for almost three hundred years, from the mid-sixteenth to the mid-nineteenth century. It maintained itself in this position by insistence on continuous learning, on meticulous technique, on close adherence to the models. Thus it retained its technical competence. But it also, after 1650 or so, rapidly degenerated into mechanical copying. And it was still at mechanical copying when the Meiji Restoration opened Japan to the West more than two centuries later. But though capable of degenerating into mechanical copying and mindless repetition, the Zen tradition is closer to being a genuine theory of learning than the Western and Chinese concept of learning for the sake of advancement, promotion, or moving on. With its focus on developing the strengths of a person, it anticipated by hundreds of years modern theories of the person and of self-realization. There is indeed profound wisdom in the insight that work is an extension of personality and personality a distillation of work, so that one cannot paint Daruma's spiritual qualities without having them oneself, but so that one also becomes Daruma by painting him every day for decades.

The insight and wisdom that lie in the Zen conception of the person and of learning are endangered in today's Japan. The Japanese educational system has opted for an extreme of the Western and Confucian position, which sees the purpose of learning as getting ready for the next examination, the next promotion, the next external reward. Infants are drilled to pass the entrance examination to the right nursery school, so as to be admitted to the entrance examination to the right kindergarten, which in turn leads to the entrance examination to the right elementary school, and on to high school, the university, and the corporation. Is there still room for the emphasis on learning to become, on learning to be, on learning to say, "This is IT!" when painting *Daruma* as a spiritual portrait?

I have so far used Japanese painting to look at Japan. Now I shall use—or abuse—Japanese painting to look at the West and at Western modern art. Rosetsu painted *The Temple Bell at Dōjō-ji* in the 1780s. The title refers to a well-known Kabuki play. But the painting itself is virtually abstract and non-objective. Yet it was painted a century and a half before there were abstract painters in the West. It is by no means Japan's oldest abstract painting; in fact, such paintings can be traced back to the Heian period of the tenth century. Tani Bunchō, the great master-painter in Edo—today's Tokyo—painted a *Flowering Plum Tree in the Moonlight* shortly after 1800. It anticipates what Turner or

Monet tried to do half a century later in the West: to make light the subject of a painting. A Hakuin *Daruma* is an expressionist painting, like those of Klimt, Schiele, and Kubin, and Picasso in his expressionist years, and Matisse; but with a power very few of them had. Modernism in Western art is thus anticipated by Japanese tradition. There is a story—perhaps apocryphal—that Picasso was taken in 1953 through an exhibition in Paris of Japanese paintings that featured the works of the Zen priest and painter Gibon Sengai, who had died in 1838, and that he stormed out of the exhibition exclaiming furiously that it was a hoax since no one could possibly have painted like this without first having seen his, Picasso's work. Modernism in Western art is in fact anticipated, if not prefigured, by the Japanese tradition.

Yet, of course, Westerners had never laid eyes on the Japanese originals or even heard of them. Other than *ukiyo-e*, the woodblock prints, Japanese art was virtually unknown in the West until fairly recent years. The West, in other words, has developed within the last century elements of a modern vision and sensibility that were ancient in Japan. The West has learned to see in somewhat the same way that the Japanese have seen all along. The West has shifted from description and analysis to design and configuration.

Marshall McLuhan has announced that the electronic media have changed our ways of seeing and interpreting the world, and are making us perceive rather than conceive. But a view of Western perception as informed by an understanding of Japanese art would lead to the conclusion that this shift began much earlier and owed nothing to electronic technology. On the contrary, it would appear more probable that the West became ready for the electronic technology and receptive to it because its perception had shifted from traditional description and analysis to the perception of design and configuration that Japan had known all along.

A distinguished historian of modern Western painting, Robert Rosenblum, in his *Modern Painting and the Northern Romantic Tradition: Friedrich to Rothko*, asserts that modern Western painting has its roots in the northern, mostly North German, painters of the nineteenth century—Caspar David Friedrich and Otto Runge—who shifted from description to design. But this, it could be argued, is precisely what had occurred in Japan far earlier: perception as against conception, design as against description, topology as against geometry, and configuration as against analysis, have indeed been continuing characteristics of Japanese art from the tenth century on.

Edwin O. Reischauer, the former American ambassador to Japan and the foremost authority on Japanese history and society, wrote in his book, *The Japanese*, that Japan has never produced a great or original thinker of the first rank. This has been read as severe criticism, especially in Japan; but Reischauer's point was that Japan's genius is perceptual rather than conceptual.

The towering achievement of the high Middle Ages in the West was Thomas Aquinas' *Summa Theologica*, perhaps the boldest conceptual and analytical feat in human history. The proudest achievement of Japan's "Middle Ages," the eleventh century, is the world's first novel, Lady Murasaki Shikibu's *Tale of Genji*, filled with intimate descriptions of men and women in court life, of love and illness and death. Japan's greatest playwright, Chikamatsu Monzaemon (1643–1724), had neither camera nor screen, but his Kabuki and Bunraku (puppet) plays are highly cinematic. They are song, dance, costumes, and music, as well as the spoken word. The characters are defined not so much by what they say as by how they appear. People rarely quote a line that Chikamatsu wrote. Yet no one ever forgets a scene. Chikamatsu was not a dramatist but a scriptwriter of genius. And without benefit of cinematic tools, his Kabuki theater invented cinema techniques; the *mie* in which the actors freeze, is, for instance, the equivalent of the movie close-up.

The perceptual in Japanese tradition largely underlies Japan's rise as a modern society and economy. It enabled the Japanese to grasp the essence, the fundamental configuration of things foreign and Western, whether an institution or a product, and then to redesign. The most important thing that can be said about Japan as viewed through its art may well be that Japan is perceptual.

26

Japan: The Problems of Success

1

Everyone I met in Japan last fall, during my tenth long trip to that country in eighteen years, talked economics and only economics. Even the theoretical mathematician and the elderly abbot of the famous Zen temple were obsessed with the dollar/yen exchange rate, the export surplus, and the cost of petroleum. Japan is indeed undergoing traumatic economic changes. Yet the basic issues facing Japan are not economic. They are changes in social structure and social values. Social policies that have served Japan superbly well for a century are rapidly becoming untenable. These policies were designed to change Japan from a poor, and poorly educated, rural society with low life expectancies into a wealthy, highly educated, industrial society with high life expectancies. Their very success is rendering them obsolete and is turning them into dangers to Japan's social cohesion and her ability to compete economically.

This is true with respect to the Japanese seniority-wage system, under which incomes of all three kinds of employees—manual workers, clerks, and managers and professionals—are determined solely, or at least primarily, by length of service, with an employee at the entrance level getting about one-third of what the same employee, regardless of his job or title, will receive as income after twenty-five years of service, that is, in his forties. But equally obsolescent is Japan's traditional linkage of education to career opportunities, under

First published in *Foreign Affair*, 1978.

which people who finish their formal schooling with "middle school," that is, at age fifteen, are slotted for a lifetime as manual workers in manufacturing, farming, or service work; with high school graduates becoming clerks or technicians for their entire working life; and with university graduates becoming managers and professionals—and with practically no crossover from one group into the other. The employee's lifetime commitment to one employer and one place of employment—often, and misleadingly, called "lifetime employment"—may equally turn into a serious threat to social harmony, rather than remain its strongest pillar. And equally untenable by now is the policy, as old as "modern Japan," of preventing social dislocation by using economic measures to shield the social order of the "old Japan" while building a radically different "new Japan."

2

The villain is not the petroleum cartel, Japan's dependence on raw material imports, or the "world recession"—the monsters of today's popular Japanese demonology. If the world recession were really the serious threat to Japan that almost any Japanese maintains it to be, Japan could hardly export enormous quantities of advanced and technologically complex near-luxuries, such as motor cars, color television sets and calculators, to practically every market in the noncommunist world. And while Japan's petroleum bill has, of course, gone up sharply since 1973, her bill for *all* industrial raw materials, including petroleum and petroleum products, has actually gone down as a percentage both of industrial production and of gross national product. For the petroleum cartel has had the impact cartels always have had, and have been known to have for over seventy years, since a German economist, Arthur Spiethoff, first studied cartels around the turn of the century. Every cartel depresses the prices of other goods competing for a share of the same income category—for example, other raw materials in the case of petroleum—by as much as, if not by more than, the percentage increase in the price of the product the cartel controls. The drop in raw material prices is altogether a major contributory cause of Japan's record balance-of-trade surplus. Since Japan imports virtually all her raw materials, and since petroleum is a fairly small fraction of total material needs—less than ten percent—Japan, in her terms of trade and in her balance of payments, has actually benefited from the high petroleum price and from the resulting depression in the price of all other raw materials. (Conversely, the fact that the United States is most nearly self-sufficient of all major developed countries in raw materials means that it is the country hardest hit in its trade accounts by higher petroleum prices. We pay more for imported petroleum while getting less for the raw materials that we produce for export.)

Rather, the villain in Japan's story—if "villain" be the right word —is Japan's population dynamics, and especially Japan's great success in raising life spans faster than any country has ever done before. Seventy-five years ago, the average life expectancy was age forty-two. Fifty years ago, it was still only fifty-three. It was, therefore, completely realistic for the Japanese to set their "official" retirement age in government service at age fifty-five in the 1880s and to extend this retirement age to most private employees in the 1920s. Japanese life expectancies, by now, are seventy-three for men and seventy-eight for women—the same as in the West. Infant mortality rates in Japan fell sharply in the 1920s to the low levels of the developed West. But the birthrate stayed high until World War II. And after World War II, Japan, very much like the United States or Western Europe, experienced a short-lived, but sharp, "baby boom." A policy that based economic growth on a large supply of young workers—the policy that is expressed in the seniority-wage system—therefore made abundant sense as late as 1965. But the "baby boom" in Japan ended in a "baby bust" in the mid-1950s—some six years before a similar "baby bust" in the United States and ten years before the birthrate dropped sharply in Germany. And as in the West, the birthrate in Japan has stayed at, or below, the net reproduction level ever since.

As late as 1946, at the end of World War II, three-fifths of the Japanese population lived on the land and almost half the labor force worked in farming, and yet Japan had to import rice. Japan reacted by giving top priority to higher rice production regardless of costs. Japan has more than doubled her farm output in the last thirty years and has an unsalable rice surplus. Yet only one-eighth of the Japanese population now lives on the farm, and less than one-twelfth of the labor force is employed in farming.

The Japanese people have shifted drastically the educational structure of the population. As late as 1938—the last "normal" year before Japan began to mobilize for war—only 4 to 5 percent of the young men—one of every twenty—went to the university, and half of the young men went to work after graduating from middle school. Today, more than half of the young men go to the university. There are literally no middle-school leavers of either sex. Practically every Japanese, whether male or female, goes on to high school. Yet Japan's social structures and policies still push heavily the expansion of college education.

In 1970, less than ten years ago, Japan was still the youngest of all developed countries. People over sixty-five amounted to less than 7 percent of the population, which meant that only one out of every fifteen Japanese was sixty-five years or older. And for every Japanese over sixty-five, there were about seven to eight adult Japanese employed and working. The comparable U.S. figures for 1970 were 10 percent of the population over sixty-five and five people in the labor force for every one past retirement age. In Sweden and France, the two oldest Western countries, the comparable figures were then 12 percent and one retired to three-and-a-half working people, respectively. By 1977, Japan had moved to the U. S. level of 1970. By 1990, just a little more than ten years hence, Japan will be beyond the Swedish level of 1970 and will have close to 15 percent of her population in the over-sixty-five age bracket, with a ratio of only three people in the labor force to every one over sixty-five and past traditional Western retirement age. Japan will, by then, be one of the very oldest of the developed countries—somewhat holder even than the United States is likely to be.

The overwhelming majority of Japanese who reach retirement age now are middle-school graduates who therefore have spent their entire working lives as manual workers. At least half the males entering the labor force in Japan are now university graduates and too highly schooled to be employed in any but managerial or professional jobs.[1] At the same time, the reservoir of surplus labor on the farm available for manufacturing and service jobs in the cities has by now completely dried up.

3

Similar developments have taken place in every single developed country. However, they have taken place much faster in Japan than in the West—much faster than in any country on record. The transformation of the Japanese age structure, for instance, required twenty-five years. The same transformation took seventy years in the United States, 100 years in Germany, and 200 years in France. The educational shift that Japan performed in thirty years took well over sixty years in the United States and a century in Western Europe. And the shift that Japan performed in thirty years, from a near-majority of the labor force employed in farming to agriculture being the employer of the smallest group in the working population, took almost a century in the United States and is not yet complete in some West European countries.

Above all, the impact of these shifts is infinitely greater in Japan than in any country in the West. For the basic policies on which Japan's society is based presume yesterday's age structure, yesterday's educational structure, yesterday's rural society, and yesterday's consumption pattern.

In particular, Japan's economic growth and competitive strength are based on the availability of large numbers of manual workers at the entrance age—that is, on the availability of young middle-school graduates. Under the seniority-wage system, productivity automatically increases, the more young people are being employed. It makes little sense to talk of an "average Japanese wage" or "average Japanese labor costs." Two plants of the same company, using the same equipment and the same methods, turning out the same products and paying exactly the same wages and benefits to workers of the same age, may still have labor-cost differentials of 100 percent, depending on the age distribution of their work force. For every year by which the average age of the labor force in a plant goes up, the plant's labor costs rise by 5 to 7 percent.

This explains why Japanese businesses have put the emphasis on sales volume and have seemed to neglect profitability. If sales volume goes up sufficiently so that young people in large numbers can be hired, profitability takes care of itself. Conversely, if sales volume does not go up, or goes down, so that no young workers can be hired, profitability will inexorably go down—no matter how high the profit margin. And now Japan faces a period in which the labor force will inevitably become older. The more labor-intensive any particular Japanese industry is—and even the most modern steel mill is still highly labor-intensive—the more serious will be the drop in productivity from population dynamics.

No one, to my knowledge, has tried to separate productivity increases in Japanese industry that resulted from improved working methods and modernization from productivity increases that resulted from population dynamics. The best guesses I can find credit population dynamics with about half of the productivity increases of the last twenty years. For during these last twenty years, young people were available in very large numbers, as a result of the "baby boom" and of the large-scale migration of young people from the farms into the cities. On that basis, real productivity growth in Japanese industry—as against the impact of population dynamics—would work out at no more than 5 to 7 percent a year for the postwar period—still respectable by any Western standard, but barely enough to offset the downturn in productivity caused by the imminent aging of the Japanese labor force. And for Japan to maintain her competitive position against such newcomers as South Korea or Brazil, with their lower labor costs, would obviously require an even greater growth rate in real productivity.

This leads to what would appear to be an insoluble growth dilemma for Japan. A growth rate of 6 percent is the minimum for Japan to sustain her competitive position in the world, given the seniority-wage system and its impact on productivities and competitive position. But the Japanese population, and especially the population available for manual work—that is the middle-school leavers—is not going to grow at anywhere near that rate. A 6 percent growth rate in the traditional industries is simply not sustainable on the basis of available manpower and existing retirement policies.

But the population dynamics also threaten Japan's traditional capital supply. Japan has, for 100 years, had the highest rate of savings and capital formation of any industrial country in the world, with personal savings of 35 percent of gross personal income being the norm rather than the exception. This, too, very largely rested on the simple fact that until recently few Japanese survived to retirement age. We have learned, in the West, that as life expectancies go up, capital formation inexorably goes down. A larger and larger share of personal savings becomes "transfer payments" to the older retired people and to "survivors," that is widows of retirees who have died. These "transfer payments," through social security taxes and through payments into employers' pensions funds, look like savings. But older people—that is the recipients of these "savings"—do not save; they consume.

Japanese government economists calculate that the Japanese personal savings rate would be lower by at least one-third, if not a full half— that is, it would run at 17-24 percent of gross personal income—if Japan today had the same age structure the United States has, where there are 32 million people who are on social security, including eight million "survivors," as against a working population of 92 million. This would still be one of the highest savings rates in the world, and on a par with those of West Germany and Switzerland. But it would be a low savings rate for Japan, and perhaps too low a savings rate in the light of Japan's future needs. For Japan, to offset the impact on productivity of an aging labor force under the seniority-wage system, will have to step up capital investment tremendously to obtain the real productivity increases she needs to remain competitive. Yet capital formation will almost certainly drop sharply, just when Japan needs capital formation the most.

4

There is one category of young people in which Japan will experience no shortage: university-educated people, prepared to be managers or professionals, either in the private or the public sector. In fact, there is already a surplus. Despite the sharp drop in the birthrate, the supply of university-educated young people in Japan during the next ten years will be many times what is was only twenty years ago. It is still likely to increase, for a university degree is the only way to have career opportunities in Japan.

Knowledge workers will thus be Japan's major, and most important, resource. And Japan's economic strategy will increasingly have to be to export knowledge work and its fruits, rather than manual work and its fruits. Increasingly, Japan will have to go in for the form of economic integration that, for want of a better term, I have called "production-sharing"—that is, for economic integration under which the developed country performs those stages in the productive process that are management-intensive, capital-intensive, and technology-intensive, while the abundant labor resources of young manual workers in the developing country perform the stages that are labor-intensive, with the final product sold and consumed in the large markets, that is in the developed countries. Japan, alone of all the developed countries, has seen the need for systematic work on "production-sharing." The Ministry of International Trade and Industry (MITI), several years ago, switched to a policy of encouraging the export of entire manufacturing plants rather than of the products of such plants. And while MITI so far still expects these plants, for example, a petrochemical plant designed and built by Japanese in Algeria, to sell its products in other markets, MITI is increasingly willing to accept payment for such a plant in the products of the plant—that is, payment for the Japanese-designed and Japanese-built petrochemical plant in Algeria in the form of petrochemicals to be sold in Japan.

Increasingly, this will mean that Japanese managers and professionals will design, build and perhaps manage consumer goods plants in developing countries—plants for shoes, textiles, automobiles and electronic products—that are going to be paid for by their own output, to be sold to the Japanese consumer in Japan. It is hard to see any other way of finding employment for the one resource in which Japan will have a continuing surplus—that is, highly schooled people, qualified under the Japanese system only for managerial and professional work.

Such a policy is obviously exceedingly difficult in any country, and likely to run into determined opposition on the part of unions, old industries and politicians alike. It is doubly difficult in Japan because of the traditional policy under which the employee is committed to one employer and cannot easily find other employment, once he is past the entrance age. Under the Japanese seniority-wage system, an employee, and especially a low-paid and low-skilled employee, such as a manual worker, cannot normally find a job past age thirty or thirty-five. At that age, he would have to be paid twice the entrance wage. But since wage is based on seniority, he can only be paid the entrance wage. He is, therefore, basically unemployable. If he loses his job, he becomes a "problem worker," very much the way in which the *samurai* of old became a *ronin* and reduced to banditry if his lord dismissed him or if the lord lost his fief. The worker of thirty-five who loses his job cannot, normally, expect to become a "permanent employee" anyplace. He becomes a "casual employee," who will never again have employment security, never again have seniority, never again have a good job, and never again, in fact, work for a major company.

"Lifetime employment," that is the right of the worker to his job, developed as an answer to the burden that lifetime commitment to the job imposed on the Japanese worker. At that, only about one-quarter of the Japanese labor force has the security of lifetime employment. Women, by definition, are "temporary employees," who are expected to leave the labor force when they marry—and are expected to marry by their mid-twenties. And employees in the "old industries"—that is in the service industries, the small shops, the small artisans' factories and on the farm—have no lifetime employment either. It is reserved for the male employees in government and large businesses—and is not even universal there. But precisely because of the Japanese work-er's constant and nagging fear of becoming a *ronin*, lifetime employ-ment has been the core of Japanese labor relations and the pillar of social order and harmony in Japan.

Now it is rapidly becoming a threat, both to the Japanese worker and to Japanese society. But any effort to change the system would not only cause the labor unions directly involved to fight very hard indeed in opposition, but would engage a host of other groups that are in one way or another taken care of by the economic enterprises or institutions in which they work, albeit on a basis that falls short of strict lifetime employment. In essence, the practice reflects a whole concept of responsibility in the society, as well as a symbol of fairness. Any attempt to modify or to eliminate it would strike at deep emotional roots throughout Japanese society.

Today, to deploy her one abundant resource—highly educated peo-ple—Japan needs to be able to liquidate fast the most labor-intensive and least productive stages of production, that is, old low-skill indus-tries. But these industries—whether textile industries or shoe facto-ries—are, of course, precisely the industries that employ the largest number of low-skilled and older workers. Even without the barrier to reemployment that lifetime employment and the seniority-wage system together create, these workers, while relatively small in numbers, would present a serious economic and social problem. These workers will, of necessity, cling the more tenaciously to lifetime employment, the less productive they and their industries become. Yet, unless they can move out of their old employments and into new ones, Japan will find it increasingly difficult to find work for the tremendous potential of highly schooled knowledge workers who constitute Japan's major and potentially most productive capital investment.

The alternative—and one that is hardly realistic anymore—is dumping high-cost, labor-intensive products, at a loss, in foreign markets that are bound to have surplus capacity and redundant over-aged labor in precisely the same industries as Japan.

5

No other country has had as purposeful and as successful an educational policy as has Japan. That career opportunities are dependent almost entirely on educational attainment was deliberate and planned. It was meant to make economic progress the engine of educational advancement. A famous story is told of Shibusawa Eiichi (1840–1931), the Meiji statesman who left a high government post to become Japan's leading industrial entrepreneur and her first banker. When approached for a loan by a successful businessman, Shibusawa turned the application down because the man was not a university graduate. This was not snobbery. And Shibusawa knew perfectly well that success in business does not depend on a sheepskin. But he had decided earlier to make managerial positions the monopoly of university graduates so as to give young Japanese an incentive to stay in school and to acquire higher education.

As a result, Japan has become the only genuine "meritocracy" around, in which birth and wealth count for almost nothing and educational attainment for almost everything. By now this has outlived its usefulness. School is turning into a nightmare. There is increasingly an oversupply of university graduates. But this has not, at least so far, resulted in a lessening of the pressure on young people to get into the right educational track and to acquire the university degree that is the passport to all opportunity and preferment.

On the contrary, during the last few years the pressure has reached a fever pitch. It now begins with the child's application to nursery school, that is before the age of three. There is an entrance examination—and little tots, barely able to walk, are being pushed into taking ballet lessons, into learning arithmetic, and into learning a few words of English. And the pressure keeps on building up. More than half of all middle-school students in Japan attend *juku*—afternoon and evening cram schools that drill for the exams. According to a study that the Ministry of Education released last fall, middle-school students aged twelve to fifteen spend almost nine hours each day *after* school doing homework and going to the *juku*. It is normal for a middle-school child to work until two o'clock in the morning each day, and to take a sample test of the high school examinations every Sunday for years on end. Senior high school students spend "only" eight hours a day in after-school work and *juku*. The pressure is becoming more and more intense with the suicide rate among teenagers, and even among preteens, reaching alarming proportions.

But worse, school instead of being accessible to everyone on the basis of merit is becoming increasingly expensive. To be sure, school —from kindergarten on—is nominally free. But to get into the "right" school requires larger and larger "voluntary contributions." The most expensive school to get into is medical school. A candidate, even if his father is a physician, will have to pay about $100,000 in "voluntary gifts" and "voluntary presents" to the faculty members on the admissions' committee to have his application even considered. A friend of mine, the widow of a Japanese clergyman, who is working as an administrative assistant in a small government agency, will have to find $4,000 by next April to have her son admitted to the entrance examination of a supposedly free public high school. If he fails, she will not get the money back—and only one out of every ten candidates passes. Then the boy will have to go for a year to a *juku*—at a cost of about $7,500—pay another $4,000 in "voluntary contributions" and try again. And the Ministry of Education study referred to above makes clear that these are not unusual but, on the contrary, fairly standard fees and financial contributions. For not to get into the "right" high school means not to get into the "right" university four years later; this would debar the boy forever from access to better jobs and career opportunities. Not to go to high school at all is a life sentence to manual labor, from which there is no appeal.

At the same time, the rewards are beginning to lose their luster. With the shortage of middle-school leavers, and therefore of manual workers, and the abundance of university graduates, the differential in the pay of the two groups is bound to shrink. In the 1950s and 1960s, when the demand for university graduates was high while the supply was still low, the differential became greater than it had been traditionally. It is already back to where it was before World War II. It is bound to go down, and fairly rapidly. University graduates are already beginning to have difficulty finding "good" jobs, and in many cases jobs of any kind—especially if they have no technical qualifications. And while the salaries of university graduates, once they have found employment, still go up with seniority, the promotional opportunities are rapidly shrinking as most organizations find themselves well supplied with highly schooled people in their early and mid-thirties.

Again the lifetime employment system aggravates the problem. The university graduate can, as a rule, not leave the employer with whom he starts. And more and more of the graduates even of the "prestige" universities have to start in dead-end jobs with employers who have no future themselves—without hope of being able to switch jobs, no matter how well they perform. For the time being, these conditions are likely to increase the premium on going to the "right" schools, and with it both the emotional and the financial pressures. But there is a distinct possibility of a collapse of the system, or of an explosion—and surely a system that puts such extreme pressure on children and literally forbids them to have any leisure time, any interests of their own, any learning of their own, can hardly produce educated adults. It is likely to produce people who can take examinations, but who do not know how to learn.[2]

6

Japan is considered the most "protectionist" country by foreigners—and of all the complaints foreigners have about Japan, this is the one the Japanese themselves understand the least. For, what to the foreigner seems to be economic protection, appears to the Japanese to be self-defense of Japanese society against the Westerner's cultural imperialism.

From the early days of "modern Japan,"after the Meiji Restoration of 1867, Japan has always defined her policy as using Western instrumentalities to maintain the essence of Japan. And economic policy is the foremost Western instrumentality thus used. "Old Japan" means the distribution system in which large numbers of people are employed at rather low wages by small wholesalers, in "Papa, Mama and Rosie" retail shops, and in small workshops making traditional goods and acting as suppliers to modern industry. It also means a system of small family farms growing rice. But while there was once a cultural rationale for the farm policy, it has long become economic insanity. The underlying economic premises of Japan's farm policy are the fear of a rice shortage; the belief that a country as dependent as Japan on imported raw materials has to keep as large as possible a proportion of the population on the farm and out of the consumer society; the conviction that as poor a country as Japan has to discourage its people from eating anything but rice and fish; and finally the notion that the fish catch is capable of being expanded indefinitely while the rice crop is finite and will always be inadequate. All these premises have been proven fallacies during the last ten years.

It has been Japanese farm policy to encourage rice growing and to discourage growing anything else. A Japanese farmer gets a large subsidy, the less suitable his land is for growing rice. The origin of this policy was the fear that Japan could not produce enough rice for her own domestic demands—which was still a rational fear twenty-five years ago. By now the policy has produced an unmanageable rice surplus, while at the same time the cost of the rice subsidy has become so high that the rice subsidy fund has become bankrupt. And since the rice that the subsidy produces is grown on land not suitable for growing rice, the surplus rice is of poor quality and could not possibly be sold on the world market. In respect to proteins, Japan has similarly pursued a policy that, however rational its origins, has become bankrupt. It has been national policy to encourage fishing and to discourage and to penalize every other production of proteins, whether vegetable, such as oil-bearing crops, or animal, such as livestock or poultry. But the fishing yield has long been stagnant and is now going down fast, with the growing resistance of other nations against Japanese fishing practices that threaten to overfish and to destroy the fish population of major fishing zones. The extension of sovereignty, and with it the control of fishing 200 miles from the shore, which by now has become general, was directed in large measure against the Japanese threat to marine ecology, whether real or perceived.

These policies were designed to ensure low-cost food for Japan—they have had the opposite effect for at least ten years now. Japan has, without doubt, the highest food costs of any major country—twice those of the United States and perhaps 50 percent higher than those of the most expensive and most protectionist countries of Western Europe. Meat costs about four times what it costs in the United States or in Western Europe—and even twice what it costs in the Soviet bloc countries. Fish, at the same time, has become almost as expensive as meat, and food costs are still rising fast. The Japanese inflation rate in the last few years has rarely fallen below ten percent a year. Food prices have been rising at a rate of 30 percent a year. At the same time, food production—other than rice production—has remained stagnant, and in some cases has actually gone down.

The high cost of animal proteins has acted like a suction pump on the cost of all other food products. 1977 was a year of bumper harvests of such things as Japanese oranges and vegetables—with crops so large as to be almost unsalable. Yet the costs of these products have kept going up. There is now virtually no dish—even *soba* (buckwheat noodles), which used to be what peanut butter sandwiches are to the American diet—that can be bought for less than $1.50 a portion.

Food and education are basic economic costs. They are the costs of forming the human resource. Both, in Japan, are out of control. And as they keep rising, the real income of the Japanese family is going down. Even well-to-do people, during my last trip to Japan, had begun to ration food. Even well-to-do people are hard pressed to find the increasing sums to finance their children's access to education. That the basic costs of the Japanese economy are going up so fast will endanger the long-range cohesion of Japanese society. It is the greatest threat, both to Japan's economic position in a competitive world and to her social order.

7

Some of the policies that Japan will have to develop—including some that only a few years ago were unthinkable—are emerging with reasonable clarity, and are beginning to be discussed in Japan. The government is beginning to press for delayed retirement—at age sixty instead of fifty-five. There is talk of retraining redundant workers and of placing them in new employments where there is demand for labor, for example, shipyard workers in automobile plants. There are the beginnings of mobility for highly qualified young professionals. And there is some slight movement toward tempering the seniority-wage system with wage measures based on productivity.

But the policies that are most needed—and needed most urgently— are still beyond Japanese political will and social imagination. What is needed most is a sharp cut in prices for the domestic consumer. Japan suffers from excess inventory, both of raw materials and of finished goods. Her policy, these last few years, has been to maintain domestic prices—if only to protect the domestic distribution system—by conducting a gigantic clearance sale abroad. That policy has to come to an end—it can only destroy Japan's capacity to export altogether. Now is the time to bring prices down in the domestic market, both to answer the charge of "dumping" and to boost the domestic economy.

There is need for a shift in agricultural policy from one that encourages the production of unsalable high-cost rice to one that, like the British farm policy, maintains farm incomes while giving the consumer low food prices. Especially there is need for a shift to a farm program that encourages farmers to take poor rice land out of rice production and to shift it into producing feed crops, oil-bearing crops, and animal proteins. Japan could probably supply far more—one estimate is at least two-fifths—of her growing demand for animal and vegetable protein, other than fish (which is likely to keep on declining), by domestic production on land that today produces rice at high cost.

There is need to encourage "production-sharing"—that is, the offshore performance of labor-intensive work, with the product sold on the Japanese market, in exchange against export of knowledge and technology.

Seventy years ago, the general manager of the Mitsui Group, then as now one of Japan's leading industrial concerns, worked out the solution to the seniority-wage conundrum. When some of the Mitsui companies had to lay off employees in the recession that followed the Russo-Japanese War, he organized their systematic placement with other Mitsui companies, in which the new employer would pay the wage befitting the employee's seniority, that is, the entrance wage, while the Mitsui Group itself would make up the difference between that wage and the wage befitting the employee's age. The sum that the Mitsui Group had to pay was ludicrously small, yet Mitsui could maintain employment security without employment rigidity. Some such policy will be needed on a national scale in Japan, together with a policy of systematically retraining and replacing employees of industries that cannot be maintained under Japanese labor costs, population structures and education structures.

What needs to be done is not so very difficult to discern, albeit difficult to do. The one area, however, in which no one in Japan can see a solution is that of education. At most, there is talk—and only talk so far—of loosening the linkage between the "right" university and the "prestige employments," under which careers in major companies, major universities, and practically all government agencies are reserved to the graduates of a few schools. But the "tracking" of people that reserves career opportunities to the graduate from the "prestige" university and that, in turn, creates what are increasingly unbearable emotional pressures and financial costs, is still beyond challenge, with no substitute for it in sight or even imaginable.

For the first time in twenty years, I left Japan not wholly confident about the country's future. So far, it seemed to me, few people in Japan have faced up to the fundamental issues. But Japan has shown unparalleled capacity throughout her history for the 180-degree turn and for forging, almost overnight, a national consensus to impose on herself radical change. Japan has also shown unparalleled capacity, throughout her history, for social innovation. This capacity, going back almost 1,500 years, may indeed be Japan's greatest strength. Japan may well suddenly face up to her problems of success and to the need to maintain the essence of Japan in new forms and through new policies. But also, Japan faces a turbulent, a difficult and a dangerous period—a period in which there will be surprises for Japan, as well as for the Western world, with which Japan is now so strongly linked, and on which she is so heavily dependent. The economic problems and needs—the resistance to Japan's trade offensive, the currency chaos, the petroleum cartel—may be the catalysts. But the real decisions will have to be on social structure, social policies and social values.

Notes

1. The best available Japanese study concludes that university graduates make up 58 percent of the males now entering the labor force in Japan. This is the highest percentage in the world. The comparable U. S. figure is 48 percent.
2. So far as I can tell, there are as yet few signs of the kinds of explosion of resentment and rejection that I am suggesting is possible. During this visit to Japan, I heard for the first time of apparently qualified students dropping out of the university—but these appeared to be only among members of families with such strong connections that the children would still have substantial advantages. My hunch is that the system may not show many signs of erosion up to a certain point, but that it may then change very rapidly once, so to speak, it boils over.

27

Behind Japan's Success

"I am more afraid of the Japanese than I am of the Russians," a young lawyer, partner in a leading law firm, said recently to me. "To be sure, the Russians are out to conquer the world. But their unity is imposed from the top and is unlikely to survive a challenge. The Japanese too are out to conquer us, and their unity comes from within. They act as one superconglomerate." But this is myth rather than reality. The Japanese indeed have learned how to act in the world economy effectively and with national consensus behind their policies. But their unity is not the result of a "Japan Inc.," of a monolith of thought and action. It is the result of something far more interesting and perhaps far more important: of policies aimed at using conflict, diversity, dissent to produce effective policy and effective action.

To any Japanese, "Japan Inc." is a joke, and not a very funny one. He sees only cracks and not, as the foreigner does, a "monolith." What he experiences in his daily life and work are tensions, pressures, and conflicts rather than "harmony." There is, for instance, the intense, if not cutthroat competition between the major banks and between the major industrial groups. And almost every Japanese is himself involved personally every day in the bitter factional infighting which—rather than unity and cooperation—characterizes Japanese institutions: the unremitting guerrilla warfare which each ministry wages against all other ministries; the factional sniping and bickering within the political parties and within the Cabinet, but also within each

First published in *Harvard Business Review*, 1981.

business and each university. Where the foreigner sees close cooperation between government and business, the Japanese businessman sees
government attempts to meddle and to dictate, and a constant tug-of-
war. "To be sure," the chief executive officer of a big company
remarks, "we pull at the same rope, but we pull in opposite directions." Nor is government always successful in making industry work
together and subordinate itself to what government sees as the national
interest. Despite twenty years of continuous pressure, the supposedly
all-powerful Ministry of International Trade and Industry (MITI) has,
for instance, not gotten the major Japanese computer manufacturers to
pool their efforts—something that Germany, France, and Britain have
all accomplished.

One foreigner after another extols Japan's uniquely harmonious
industrial relations. But the Japanese public curses at the all-too-
common wildcat strikes on the government-owned National Railways.
It is only where the labor unions are exceedingly weak, that is, in the
private sector, that labor relations are harmonious. There is no sign of
"harmony" in the public sector, where (a legacy of the U.S. occupation) unions are strong. Indeed, Japanese labor leaders are inclined to
point out, somewhat acidly, that Western firms without unions—IBM,
for instance—tend to have exactly the same labor policies and the same
"harmony" as "Japan Inc.," so that the Japanese situation signifies
management's hostility to unions rather than the fabled "harmony."

And yet—while "Japan Inc." may be more myth than reality, Japan
has developed habits of political behavior which make it singularly
effective as a nation in economic policy and in international economic
competition. One of these habits is thorough consideration of a proposed policy's impact on the productivity of Japanese industry, on
Japan's competitive strength in the world markets, and on Japan's
balances of payment and of trade. This has become almost second
nature with Japanese policymakers, whether in the ministries, in the
Diet, or in business, and equally with analysts and critics in the
popular newspapers or the university economics departments. The
Japanese are far too conscious of their dependence on imports for the
bulk of their energy and of their raw materials and for two fifths of
their food to shrug off the external world or to push it out of their field
of vision altogether, as American lawmakers, American government

departments, and so many American economists are wont to do.

The Japanese do not go in for formal "productivity impact statements." And its impact on competitive position and productivity is by no means the sole criterion in adopting or rejecting a proposed policy. Even if the most powerful government agency opposes a policy because of its deleterious impacts on Japan's position in the world economy, the Japanese public or Japanese industry may yet embrace it—as they did in respect to the expansion of the Japanese automobile industry.

MITI, the powerful Ministry of International Trade and Industry has, since 1960 and 1961, steadily opposed expansion of the automobile industry—in large part because it views the private automobile as "self-indulgence" and as the opening wedge of the "consumer society" which a puritanical MITI abhors. There was also, at least in the early years, considerable skepticism about the ability of untried Japanese automobile manufacturers to compete against the likes of GM, Ford, Fiat, and Volkswagen. And there was, and is, great fear that a large automobile market in Japan will provoke irresistible demands to open Japan to foreign imports—the one thing MITI is determined to prevent. But MITI also held—and quite sincerely—that expansion of the automobile industry would have an adverse, indeed a deleterious, effect on Japan's balance of trade, on its ability to earn its way in the world economy, and on its productivity altogether. The more successful the Japanese automobile industry, MITI economists argued, the worse the impacts on Japan. The automobile, they pointed out, requires the two raw materials that are in shortest supply in Japan: petroleum and iron ore. It also requires diversion of scarce resources, both of food-growing land and of capital, to highways and highway construction. What MITI wanted was massive investment to upgrade the railroads' freight-handling capacity.

There are plenty of diehards around—and not only at MITI—who still maintain that to let the Japanese automobile industry expand was a serious mistake. The industry's export earnings, the diehards will argue, are only a fraction of what the automobile costs Japan in foreign exchange for petroleum and iron ore imports, even with record automobile sales to North America and Western Europe. A small part of the sums spent on highways would have given the Japanese railroads the freight-carrying capacity which the country needs and still lacks.

Yet, though enormous amounts have been spent on roads, it has not been nearly enough to build an adequate highway system—thus resulting in trucks clogging the inadequate roads, in high transportation costs for Japanese industry, in unhealthy concentration of people and factories around a few already overcrowded port cities such as Tokyo, Yokohama, Nagoya, Osaka, and Fukuoka, and in growing air pollution.

MITI lost its fight against the automobile, despite its reputation as a kind of economic superman. It was defeated in part by the automobile industry, which forged ahead despite MITI's disapproval. In large part MITI was defeated by the infatuation of "Nabe-san," the Japanese "man in the street" (and of his wife) with the motor car, despite its high costs, despite the lack of places to park, despite the traffic jams which make commuting a nightmare in every Japanese city, and despite air pollution, about which no one complains louder than "Nabe-san," sitting in the driver's seat.

But at least—and this is the point—the automobile's impact on Japan's productivity, competitive position, and balance of trade was seriously considered. And even the automobile company executives who fought MITI the hardest admit that it was the ministry's duty to make sure that these impacts were taken seriously, no matter how popular "wheels" are with the Japanese consumer and voter.

The impact on Japan's competitive position in the world economy is only one of the considerations Japanese leaders are expected to think through and weigh carefully before espousing a policy or taking a course of action. They are expected altogether to start out with the question: "What is good for the country?" rather than the question: "What is good for us, our institution, our members and constituents?"

In no other country are interest groups so well organized as in Japan, with its endless array of economic federations, industry associations, professional societies, trade groups, special interest "clubs," guilds, and what-have-you. Each of these groups lobbies brazenly and openly uses its voting power and its money to advance its own selfish ends in ways that would have made a Tammany boss blush. Yet if it wants to be listened to and to have influence on the policymaking process, every group must start out in its thinking and in its deliberations with the national interest rather than with its own concerns. It is not expected to be "unselfish" and to advocate policies that might cost it money, power, or votes—Japan's Confucian tradition rather distrusts self-sacrifice as unnatural. But the group is expected to fit what serves self-interest into a framework of national needs, national goals, national aspirations and values. Sometimes this is blatant hypocrisy, as when the Japanese physicians claim that the only thought behind their successful demand for near-total exemption from taxes is concern for the nation's health. Still, the physicians pay lip service to the rule which demands that the question, "What is the national interest?" be asked first. That it fails to do even that and instead is forced by the very logic of trade unionism to assert that "What is good for labor is ipso facto good for the country," is probably largely responsible for the Japanese union's lack of political influence and public acceptance, despite the union's impressive numbers. And that, conversely, business management in Japan—or at least a substantial minority among business leaders—has for a hundred years subscribed to the rule that the national interest comes first, that indeed the rule was first formulated by one of the earliest of modern Japanese business leaders, the nineteenth-century entrepreneur, banker, and business philosopher Eiichi Shibuzawa (1840-1931), may also explain why business management is respectfully listened to whenever it discusses economic and social policies, even by the two fifths of the Japanese population who faithfully vote every time for avowedly Marxist and stridently anti-business parties and candidates.

The demand that they take responsibility for thinking through the policies which the national interest requires forces the leadership groups—and especially the business leaders—to lead. It demands that they take the initiative and formulate, propound, and advocate national policies *before* they become issues. Indeed, it forces the leadership groups to define what the proper issues are and should be.

In the West, and especially in the United States, the "interests"—such as the conventional "interests" in the economic sphere: business, labor, and the farmer—are expected to start out with their own concerns and their own needs and wants. This then means as a rule that they can rarely act at all in any matter that is general rather than sectional. They can only react. They cannot lead; they can only oppose what someone else proposes. For whenever a matter of general concern comes up, someone within the group is bound to fear being harmed, someone else will be opposed to doing anything at all, and a third will drag his feet. In Japan too, of course, any proposal is likely to run into opposition within any group. But the special interests and concerns of the members of the group which form the starting point for policy deliberations in the West are pushed aside in Japan until the national interest has been thought through. In the West, the individual, sectional, specific interests and concerns are the focus; in Japan, they are the qualifiers. The Western approach tends to lead to inaction—or to "another study"—until someone from the outside proposes a law or a regulation that can then be fought as "unacceptable." But this is only rearguard action to prevent defeat or to contain damage and, even worse, the other side then determines what the issues are or should be. Yet, as the Japanese see clearly, to define the issue is the first duty of a leader.

But the Japanese approach also means that business—and the other leadership groups in society—are rarely "surprised." It is their job, after all, to anticipate and to define the issues. This does not always work, of course. Both the bureaucracy and the business leaders of Japan were totally unprepared for the explosive emergence of the environmental issue ten years ago—even though by that time it had already erupted in the United States, so that there was plenty of warning. Today the leadership groups in Japan—the bureaucracy, business, labor, and academia—prefer to ignore the challenge of women moving into professional and managerial jobs; yet the movement is gathering momentum and is grounded in irreversible demographics. But whereas in the United States business, labor, government, and academia talked of lowering the mandatory retirement age at the very time when the growing power of the older people made first California and then the U.S. Congress enact laws postponing retirement or prohibiting mandatory retirement altogether, big business in Japan anticipated the issue. And although the costs are very high, Japan's largest companies on their own and without any pressure from government, from labor, or from public opinion have raised the mandatory retirement age. "It's what the country needs," was the explanation.

The Western approach worked as long as national policy could effectively be formed through adversary proceedings and by balancing the conflicting reactions of large, well-established "blocs" or "interests"—the traditional economic policy triad of business, labor, and farmer. But with the fragmentation of politics in all Western countries where now small, single-cause zealots hold the swing vote and the balance of power, the traditional approach is clearly not adequate any longer. Thus the Japanese rule under which leadership groups, and especially those of the "interests," derive their legitimacy and authority from their taking responsibility for the national interest, and from anticipation, definition, resolution of issues ahead of time, might better serve in a pluralist society.

The third of the Japanese habits of effective behavior also originated with the banker-entrepreneur-business philosopher Eiichi Shibuzawa, in the closing years of the nineteenth century: Leaders of major groups, including business, have a duty, so Shibuzawa taught, to understand the views, behavior, assumptions, expectations, and values of all other major groups, and an equal duty to make their own views, behavior, assumptions, expectations, and values in turn known and understood. This is not "public relations" in the Western sense. It is, rather, very "private" relations—relations between individuals; relations made not by speeches, pronouncements, press releases, but by the continuous interaction of responsible men in policymaking positions.

Irving Shapiro, the chairman and CEO of DuPont de Nemours, the world's largest chemical company, was recently quoted in the American press for pointing out in a public speech that he was now being forced to devote four fifths of his time to "relations," especially with individual policymakers in the Congress and in the Washington bureaucracy, and could only spend one fifth of his time on managing his company. The only thing that would have surprised a Japanese CEO in a business of comparable importance was the one fifth Mr. Shapiro has available to run the company which he heads; very few CEOs of large Japanese companies have *any* time available for managing their companies. All their time is spent on "relations." And what time they have for the company is spent on "relations" too, rather than on "managing." They keep control through thorough and careful attention to personal decisions in the upper ranks and through meticulous financial and planning reports. But they do not "manage"—that is left to lower levels. The top people spend their time sitting, sipping a cup of green tea, listening, asking a few more questions. They sit with the people from their own industry. They sit with suppliers, with the trading company people, with the managements of subsidiaries. They sit with top people from other companies in their group—as, for instance, in the famous four-hour "luncheons" in which the presidents of all companies in the Mitsubishi group come together once a week. They sit with the people from the banks. They sit with the senior bureaucrats of the various ministries, and on half a dozen committees in each of half a dozen economic and industry federations. They sit with the people of their own company in an after-hours party in a Ginza bar. They sit and sit and sit.

In these sittings they do not necessarily discuss business—and surely not their own business. Indeed, to a Westerner the conversation at times appears quite pointless. It ranges far afield—or so it seems—from issues of economic policy to personal concerns, to the other fellow's questions and his problems, to the topics of the day, to expectations for the future, and to reappraisal of lessons of the past. The aim is not to "solve" anything, but to establish mutual understanding. Then one knows where to go when there is a "problem"—and there is, of course, always one, sooner or later. Then one knows what the other person and his institution expect, can do and will do—but also what they cannot or will not do. And then, when either crisis or opportunity arrives, these immobile "sitters" suddenly act with amazing speed and decisiveness, and indeed ruthlessly. But also when the crisis comes, the others are ready to support, or equally, if they see the need, to oppose. For the purpose of all this sitting is not to like one another; not to agree with one another; not even primarily to trust one another: it is to know and understand one another and, above all, to know and understand where—and why—one does not like the other, does not agree, does not trust.

And finally, the effectiveness of the Japanese is based on their having learned that living together cannot be based on adversary relations, but must have a foundation in common interest and mutual trust.

Adversary relations in Japan have historically been fiercer, fought more violently and with less forgiveness or compassion than in the West. The popular movie *Shōgun* does not exaggerate the violence of Japanese history, however much it may romanticize other aspects. Neither "Love thine enemy" nor "Turn the other cheek" is to be found in any of Japan's creeds. Even nature is violent in Japan, a country of typhoons, volcanoes, and earthquakes. Indeed, Japanese convention dictates that relations be adversary—or at least be made to appear so—where the Westerner sees no need for feuding and re- crimination when, for instance, a painter or another artist of old parted company with his teacher and established his own style or school. This tradition extends today to divorce, which has reached epidemic propor- tions in Japan and is approaching the California rate, especially among young, educated couples. An "amicable" divorce is apparently not considered proper; it must be made to look "adversary," even if the couple parts by mutual consent and on reasonably good terms.

But all these are situations in which the relationship is dissolved for good. Where people or parties live together, let alone where they have to work together, the Japanese make sure that the relationship has at its core a mutuality of interest and a common concern. Then there can be conflict, disagreement, even combat; for then conflict, disagreement, and combat can still be confined and subsumed in a positive bond.

One of the main—though rarely voiced—reasons why the Japanese automobile companies are reluctant to build plants in the United States is their bafflement at management-union relations in the American automobile industry. They simply cannot understand them. "Our unions," said a young Toyota engineer, an avowed "leftist" and "socialist" with strong pro-union leanings, "fight management. But yours fight the company. How can they not *know* that for anything to be good for the company's employees it has to be good for the company? Where this is not taken for granted—and it's completely obvious to every one of us—no Japanese could be a manager, but no Japanese could be an employee and subordinate either."

One does not have to live and work with a competitor; hence competition tends to be ruthless between different companies in the same field and between different groups, for example, between Sony and Panasonic or between Mitsui Bank and Fuji Bank. But whenever there has to be a continuing relationship with the opponent, common ground must be found. Then the question that always comes up first— the question indeed to which all these endless sittings and meetings between the leaders of different groups are largely devoted is: "What interests do we share?" or: "On what issues are we in agreement?" or: "What can we do together that will help both of us attain our respec- tive goals?" And great care is then taken to avoid destruction or damage to the unity and common purpose.

Great care is also taken to see that there be no "final victory" over some group or interest with whom one has to live and work. For then to win the war means to lose the peace. Thus whenever groups or interests in Japan have to live together, both will be more concerned with making their conflict mutually productive than with winning—even though the same people in the same group will go all out for total victory and for unconditional surrender against an opponent with whom their group does not have to live and who therefore can (indeed, should) be destroyed.

These are rules and, like all rules of this kind, are ideals and normative rather than descriptive of what everybody does all the time. Every Japanese can point to dozens of cases where the rules were broken or disregarded, and with impunity. The rules are also not necessarily accepted by everybody as being right. Some of Japan's most success-ful entrepreneurs and business builders—Honda, for instance, or Mat-sushita at Panasonic, or Sony—have shown scant respect for some of them. These successful leaders do not, for example, give a great deal of time and top management attention to outside relationships, and do not care much whether they are accepted in the "club" or not. They do not necessarily agree that putting the national interest first in one's thinking and policies is the responsibility of the business leader; and they may even on occasion have been quite willing to inflict crushing defeats on opponents with whom they still have to live and work.

There is also a good deal of criticism within Japan—especially within business—of some of the rules, and grave doubt whether they are still fully appropriate to Japan's needs. Can top management, for instance, devote practically all its time to outside relations or will it lose touch with the reality of its business at a time of rapid change in economics, markets, and technologies? And there is a good deal of grumbling to the effect that concern for finding common ground with other groups—especially for business finding common ground with government—has led to spineless appeasement and has only encour-aged bureaucratic arrogance.

The rules, in other words, are similar to all other such rules in that they have weaknesses, limitations, shortcomings—and in that they do not apply universally and without exception. But they surely also have unique strengths and have been uniquely effective. What then is their essence, what is the "secret" of their success?

The most common answer, in Japan as well as in the West, is that these rules represent uniquely Japanese traditions and values. But it is surely not the whole answer, and is indeed largely the wrong answer. Of course, rules of social and political behavior are part of a culture and have to fit it, or at least be acceptable to it. And *how* the Japanese handle their policies, their rules, and their relations is very Japanese indeed. But the rules themselves represent *a* rather than *the* Japanese tradition. They represent a choice between widely different, but equally traditional, alternatives rather than historical continuity. Some of the rules, moreover, have no foundation whatever in Japanese tradition. The industrial harmony of Japan is usually attributed to history and traditional values. But the only historical tradition of relations between superior and subordinate in Japanese history is violence and open warfare. As late as the 1920s, that is, through the formative stage of modern Japanese industry, Japan had the world's worst, most disruptive and most violent labor relations of any industrial country. And for the hundred and fifty years before modern Japan was born in the Meiji Restoration of 1867, relations between "bosses" and "workers"— between the lords and their military retainers, the Samurai, who were the "bosses," and the peasants, who were the "workers"—meant at least one bloody peasant rebellion per year, more than two hundred during the period, which was then suppressed just as bloodily. "Government by assassination" rather than the building of relationships or the finding of common ground was still the rule for relationships between different groups in the 1930s. And it is not entirely coincidence that both student violence and terrorism began in Japan in the 1960s and took their most extreme form there; both are surely as much a "Japanese tradition" as the attempt to find common ground between opponents, and maybe more so.

Also these rules did not just evolve. They were strongly opposed when first propounded and considered quite unrealistic for a long time. The greatest figure in Japanese business history is not Eiichi Shibuzawa, who formulated the most important rules of behavior for today's Japanese society. It is Yataro Iwasaki (1834–85), the founder and builder of Mitsubishi, who was to nineteenth-century Japan what J. P. Morgan, Andrew Carnegie, and John D. Rockefeller, Sr., combined were to the United States. And Iwasaki rejected out of hand Shibuzawa and his rules—whether the demand that business leadership take responsibility and initiative in respect to the national interest, or the demand that it build and nurture its relationships, or in particular the idea of finding common ground with opponents and of embedding conflict in a bond of common interest and unity. Shibuzawa was greatly respected. But his teachings had little influence with "practical men," who were far more impressed by Iwasaki's business success.

Whatever their roots in Japanese traditions, these rules became accepted and the approved behavior only after World War II. Then, when a defeated, humiliated, and almost destroyed Japan began painfully to rebuild, the question was asked: "What are the right rules for a complex modern society, and one that is embedded in a competitive world economy and dependent on it?" Only then did the answers which Shibuzawa had given sixty years earlier come to be seen as right and relevant.

Why and how this happened goes well beyond the scope of this essay, and the author is hardly qualified to answer the question. There was no one single leader, no great figure, to put Japan on a new path. Indeed, the historians will be as busy trying to explain what happened in the 1950s in Japan as they have been to explain what happened at the time of Meiji, eighty years earlier, when an equally humiliated and shocked Japan organized itself to become a modern nation and yet to remain profoundly Japanese in its culture. One might perhaps speculate that the shock of total defeat and the humiliation of being occupied by foreign troops—no foreign soldier ever before had landed on Japanese soil—created a willingness to try things that had never been tried before, even though powerful forces in Japan's history had urged and advocated them. In respect to industrial relations, for instance, we know that there was no one single leader. Yet the strong need of Japanese workers, many of them homeless, many of them discharged veterans from a defeated army, many of them without employment of any kind, to find a "home" and a "community" was surely an important factor, as was the strong pressure by workers on management to protect them from the pressures of the American occupation and its "liberal" labor experts to join left-wing unions and to become a "revolutionary" force. The conservatism of the Japanese worker in the late forties and early fifties, but also the need of the Japanese worker to have a little security when his emotional, his economic, and in many cases his family ties had been severed, undoubtedly played a large part in the course Japan then took. But why Japanese management found itself able to respond to these needs and in an effective form, no one yet knows.

Indeed, the Japanese "rules" could just as well be explained with purely "Western" teachings and traditions. That business leadership, especially in big business, needs to take active responsibility and initiative for the national interest and must start out with what is good for the nation rather than what is good for business, was for instance preached in the West around 1900 by such totally un-Japanese leaders as Walter Rathenau in Germany, and Mark Hanna in the United States. That an enemy that cannot be destroyed must be made into a friend and must never be "defeated" and humiliated was first taught around 1530 by the first modern political thinker, Nicolò Machiavelli. And Japan's embedding of conflict in a core of unity is also Machiavelli—the Machiavelli of *The Discourses* rather than that of *The Prince*. Four hundred years later, in the 1920s, Mary Parker Follett, most proper of proper Bostonians, concluded again that conflicts must be made constructive by being embedded in a core of common purpose and common vision. All these Westerners—Rathenau and Hanna, Machiavelli and Follett—asked the same questions: How can a complex modern society, a pluralist society of interdependence, a society in rapid change, be effectively governed? How can it make productive its tensions and conflicts? How can it evolve unity of action out of diversity of interests, values, and institutions? And how, as Machiavelli asked, can it derive strength and cohesion from being surrounded by, and dependent on, a multitude of competing powers?

And why then did the West, and especially the United States, reject this tradition while Japan accepted it? Again the scope of this question greatly exceeds this essay and the expertise of its author. But one might speculate that the Great Depression and its trauma had something to do with it. For before it, there was indeed leadership that subscribed to these values. Both Herbert Hoover in the United States and Heinrich Bruening, the last Chancellor of a democratic Germany, represented a tradition that saw in the common interest of all groups the catalyst of national and social unity. It was their defeat by the Great Depression—for instance, in Franklin D. Roosevelt's New Deal—

which ushered in a belief in "countervailing power," in adversary relations, as leading to a compromise solution acceptable to all because it does not offend any one group too much, and therefore unites them on the least common denominator. And surely the victory of Keynesian economics in the West, and especially in the United States, with their apotheosis of national government as omnipotent, omniscient, and able to control the national economy almost irrespective of what happens outside, had something to do with our forgetting the old adage of American politics that "Politics (and economic disputes) stop at the water's edge." But this is speculation.

What is fact is that the "secret" behind Japan's achievement is not a mysterious "Japan Inc.," which belongs in a Hollywood Grade B movie anyhow. It is perhaps not even the particular values of behavior that Japan has been practicing. It may be that Japan, so far alone among major industrial countries, has asked the right questions: What are the rules for a complex modern society, a society of pluralism and large organizations which have to coexist in competition and antagonism, a society that is embedded in a competitive and rapidly changing world and increasingly dependent on it?

28

Misinterpreting Japan and the Japanese

For forty years I've been trying to explain to my American friends that they misinterpret Japan. I haven't made much progress. The first challenge is explaining what any Japanese civil servant understands perfectly well: despite the country's economic miracle since World War II, Japan has not had an economic policy; it has had a *social* policy.

When I first started to work with the Japanese government and Japanese businesses in the early 1950s, Japan was not only a war-ravaged country, but an incredibly fragile society. Half the people lived on the land, and there was an exceedingly high number of small shopkeepers and small factories with a few dozen employees and pre-World War I machinery.

The policy Japan adopted at that time was to avoid taking any social risks—to protect domestic society, especially domestic employment, and at the same time to push a few carefully groomed industries into export opportunities. Nobody then believed that Japanese exports would ever earn enough foreign currency to pay for the imported food and raw materials on which the island country depends.

Japan today remains the world's largest importer of food and commodities. But the basic social, demographic, and economic foundations of the past forty years are quickly shifting. Today, of the tens of thousands of small shops in Japan, many are franchises of large chains

First published in the *Wall Street Journal*, 1990.

such as 7-Eleven and Kentucky Fried Chicken. Only about 5 percent of the population now makes its living as farmers.

The new largest single group, though not the majority yet, is people who did not even exist in 1954: educated, middle-class, salaried employees. They are up for grabs politically. In the last election they still voted for the Liberal Democrats, not because they have any use for them, but because the opposition is even less credible. For the next ten to twenty years, the big political and economic challenge in Japan will be to find the new consensus. The need for policies to protect Japanese society is gone, and in fact those policies are becoming increasingly unpopular among consumers, as the new dominant groups do not feel any need to be protected. What is now needed is a truly *economic* policy.

Without understanding the fundamental changes that are reshaping Japan, Americans cannot hope to make sense of the United States' much lamented trade deficit with that country—or know whether the Japanese indeed pose the economic threat to the United States that some people in Washington would have us believe.

For all the headlines about Japan's trade surplus, I do not know a single Japanese in a responsible government or academic position who believes in it. The export surplus is a mirage. It exists on paper, but nowhere else.

To be sure, a good many of our industries are lagging behind the Japanese. But the United States' trade deficit with Japan has almost nothing to do with manufacturing, great though Japan's own industrial prowess may be. The Japanese trade surplus is largely the result of low prices in the raw materials and food that Japan imports.

There has been a glut of petroleum for nearly a decade, to the delight of Japan, which imports all of its oil. In the developed nations, farm productivity has increased dramatically in the last forty years. At the same time, food consumption is not going up. As people become affluent, they do not eat more; in fact, they buy magazines that tell them how to eat less.

The Japanese have also benefited because food and commodities, with their low prices, are traded in dollars. And since 1985 the dollar has been devalued—at times by as much as half—against the yen. As a result of these various factors, the Japanese pay little more than one-third of what might be considered the "normal" value of the food and raw materials they import.

Yet from the Japanese point of view, if prices ever go up, Japan will have difficulty earning enough through foreign exchange to pay for these imports. And there is a very good chance that in the next few years—not forever, but for a while—world food prices will skyrocket. Where is the food going to come from to feed Eastern Europe, where the food shortage is already great and growing day by day?

As we feed the declining Soviet empire, the food surpluses of the free world will diminish over the next five years. And the Japanese, quite rightly, see the need for a policy that is based on maintaining their present export surplus. Maybe food prices will stay low. Maybe petroleum will stay affordable. But what if not? That is Japan's nagging fear.

At the risk of sounding totally absurd, I've always considered Japanese-American trade negotiations Punch-and-Judy shows. Anyone who thinks the Japanese do not buy foreign goods need only go into any Tokyo shop. You will see nothing but foreign brand names—they just happen to be made in Japan. Does anyone seriously think that IBM is going to ship computers to Japan when IBM Japan controls 40 percent of the Japanese market and is one of the most profitable parts of IBM? Or that Coca-Cola is going to ship anything to Japan?

And so I've never seen that mythical great, untapped Japanese market. When I sit down with my sharp pencil and try to figure out how much more Japan is likely to buy—with its American trade surplus of $50 billion a year—I come up with a maximum of $5 billion.

We are so hypnotized by the trade surplus that we do not understand how dependent upon the United States Japan has become. In economic history, the point at which a nation's dependence on one market becomes economically and politically dangerous is somewhere around 25 percent. Japan has surpassed that point with the United States, which buys more than 40 percent of Japanese exports.

Now, if you have a $50 billion trade surplus, you may be able to take a small portion home in goods. But what do you do with the rest of the money—dump it in the ocean? You must invest.

Since the Japanese know that there is tremendous political risk in buying American real assets—movie companies, Manhattan land-marks and so on—United States Treasury bills are almost the only thing they can buy. And here, we have them over a barrel. Were they to pull out their money, as so many people seem to fear, there would be a run on the market and the dollar might go down to 100 to the yen, which penalizes only the Japanese.

We might not feel the effects at all, but the Japanese would lose a lot of dough. As the old saw goes, if you owe your bank $10,000, the bank owns you. If you owe the bank $1 million, you own the bank. In this sense, the United States owns Japan; or, at least, the Japanese are much more dependent upon this country than they would like to be.

Last year Sony's chairman, Akio Morita, and a former cabinet minister and nationalist politician, Shintaro Ishihara, irritated many Americans with a book published in Japan, *The Japan That Can Say No*. The premise was that Japan's supposed technological superiority has made the United States dependent on the Japanese.

One of the most publicized examples was that if Japan stopped shipping one particular microchip to this country, we would have to stop making a major missile. In fact, most of the microchips for this missile are made here, and it would take us about six weeks to get the rest, if the Japanese cut us off. We know how to make chips. It is a matter of price, not of technical capability.

The same politician, Shintaro Ishihara, recognized the danger of single-market dependence during the 1960s and said, in effect, that Japan needed an outrigger on the left side, to balance its canoe on the right. He is the man who began to beat the drum for Japanese invest-ment in China, and it has been a total disaster—not for political, but for economic reasons.

Some of us tried to tell him that no underdeveloped country has ever been a satisfactory market for the goods of a developed one. The first country to learn this was Great Britain in respect to India in the nineteenth century. Ishihara did not want to hear it, but it has turned out to be true. Toyota, for example, shipped 6,032 vehicles to China last year; the company shipped more than twice that number each *week* to the United States.

In looking to Europe as another possible outlet, the Japanese see major obstacles. The industries in which the Japanese are strongest are the ones in which Europe has the most overcapacity and incredible overemployment. Not counting Eastern Europe, there are still almost a million people too many in the European steel industry. The automobile and consumer electronics industries also have far more capacity than demand for products. And so, in a period when the European market is rapidly integrating, the Japanese are viewed as a tremendous external threat.

How are the Japanese to integrate themselves into the world economy in which—because of domestic political changes and external forces—they can no longer pursue the strategies of the past forty years? The need to protect society in Japan is gone, and is in fact meeting stiff resistance from Japanese importers. Pushing more exports to the West is no longer a viable option.

Is it necessary to try to form an East Asian bloc? And how does one do this with countries of such incredibly uneven social and economic development as, let's say, Japan and Thailand? And can it be done without the coastal cities of China? And is Southeast Asia likely to allow itself to become dependent on Japan, considering the chilling memories of the 1930s that are still very much alive in China, Korea, Indonesia and Thailand?

If China breaks up again into regions controlled by economic warlords, which cannot be ruled out, then we could see an East Asian bloc organized around Japan. It wouldn't be easy, in view of the tension between those two cultures and two peoples, but I think Ishihara's China card might nevertheless finally come into play. So long as there remains a unified Communist China, however, Japan must look westward.

A very drastic course would be for Japan to shift 180 degrees and become the leader of "freer" trade—do not call it free trade; no such thing exists except in textbooks—and enlist the United States in trying to prevent European-imposed protectionism. Japanese society could adjust to that.

It would basically require Japan to forget forty years in which the fundamental policy was to never take a social risk or expose protected industries to the risk of competition. If you are one of the major corporations such as Sony or Toshiba or Toyota that has proved its ability to compete in the world market, you may be willing to take such a risk; it may be the only way you can maintain access to the American market and gain access to Europe.

One thing you can count on with the Japanese: behind the policy of "take it easy, one step at a time," there is furious, hard thinking going on. Nobody in the world is as good at making decisions as the Japanese.

29

How Westernized Are the Japanese?

"I DETEST the nepotism that pervades Japanese academic live," says the distinguished historian. "Young scholars should make their careers on their merits and not on their family ties. That's why I married my four daughters to the ablest and brightest of my doctoral students. This way I could do what Japanese tradition expects and place my sons-in-law into the best of professorships—and I could do it with a good conscience for I know they deserve it."

"We are completely Western in this outfit," says the successful independent movie producer in his modernistic Tokyo studio. "We even have a woman vice president in charge of finance and administration. But, Professor Drucker, could you act as her go-between and find a husband for her in the United States? She is thirty now and should be married soon."

"It would be wonderful if you'd find an American husband for me and arrange the marriage," chimes in the attractive woman VP. "No Japanese will marry a women's libber like myself, who is a professional and executive." "Does it have to be an arranged marriage?" I ask. "Definitely," answers the women's libber, "the other way is much too risky."

"Young Ohira will be the chief executive of this company in ten to fifteen years," I had been told repeatedly by the chairman of one of the leading high-technology firms. But when I inquire about Ohira on my latest Japanese trip, there is embarrassed silence. "We had to let him

Published first in *The Wall Street Journal*, 1980.

go," the chairman says. "He is an oldest son and his father, who owns a small wholesale business in Kobe, demanded that he take over the family company. We tried to talk the old man out of it, but he is stubborn and so we had to let Ohira go."

"Did he want to leave?" I ask. "Of course not, but he had no choice. He could never have been promoted if he had stayed. Executives, after all, have to set an example—and in Japan an oldest son is still expected to follow his father in his business."

The young woman who interpreted for me at a press conference asks whether she and her biochemist husband might come and get my advice. Their problem? Interpreters are very well paid and so the young woman makes more than her husband, who, under Japanese seniority rules, won't be a full professor for six more years. Then their positions will be reversed. He'll make about three times what he makes now; and she plans to take time out to have children. But in the meantime both his and her family disapprove and nag.

"Does it bother you that the wife has the larger income?" I ask. "Not in the least," they both answer. "Then why do you have to tell your families?" I say. They beam, tell me I've saved their marriage, and thank me profusely. "Did you really need me to tell you that?" I ask. "Of course not," they say, "but this is Japan; to do anything unconventional you have to have a *sensei* [master] tell you so—and we were pretty sure what you'd tell us when we came for your advice."

There is thus a good deal to support the "old Japan hands" in their contention that the Japanese are Westernized "from nine to five" only. It is certainly a much safer assumption in dealing with Japanese than the advice I heard a Swiss banker give to his successor as the bank's representative in Tokyo: "Treat them as if they were American MBAs with German-sized attaché cases," let alone the wondrous description by a group of European Common Market economists that is quoted up and down Japan with mixed amusement and indignation: "The Japanese are economic animals who live in rabbit hutches."

"For twenty years," says an experienced executive recruiter, "I have placed Japanese executives with Western companies. The firms that have been successful in attracting and holding truly able Japanese are the ones who know that they'll behave like Japanese no matter how impeccable their English or how much they prefer whiskey to sake."

Indeed the Japanese—except "from nine to five"—may well have become more "Japanese" and less "Western" these last few years.

Ten or fifteen years ago, for instance, performances of the Noh—the traditional and stylized dance-opera—played to empty houses, with the few spectators mainly elderly men who came in, one suspected, because the place was air-conditioned. When I went to a Noh performance in June of 1980 it was sold out, with every seat taken, mostly by professionals or young executives by their looks.

But things are rarely that simple—and never in Japan. Consider, for instance:

The twenty-year-old daughter of old friends—we have known her since she was a toddler—tells us that she is majoring in philosophy. "Last term I took a nifty seminar reading Plato," she says. "Do you have good translations of Plato into Japanese?" I ask. "We don't read translations," she answers, quite indignant. "We read Plato in Greek. And this term we are reading Kant and Schopenhauer, in German. And I am also taking a very interesting course in Whitehead, Russell, Wittgenstein, and Symbolic Logic, in English, of course."

"And what are you doing for fun?" I ask. "But *this* is the fun," she answers. "Of course I also have to prepare myself for a job and for earning a living, and so I am doing judo. I got the Black Belt eighteen months ago and am now studying for the instructor's exam. I am already adviser to the judo club at my university and hope the university will hire me as a judo teacher when I graduate next year. Don't smirk," she cries. "I am in dead earnest. Japanese girls are now studying medicine and accounting and even engineering. But these are all imports from the West. To be accepted as equals our women will have to make it in something purely Japanese—and what could be more Japanese than judo?"

Miyeko, whom we first knew as a college sophomore who interpreted for us on a hiking and camping trip, visits us with her six-year-old daughter and her husband, a middle-level executive in a big trading company. She confides that they both very much want to have another child but have decided against it.

"It might be a boy and, of course, that's what we'd hope for," Miyeko says. "And then the firm would not send us overseas or would demand that I stay in Japan with the children and send him alone. You know that a Japanese boy has to grow up in Japan ever to be accepted as a Japanese. And my husband is in line, just now, for New York or

Los Angeles.'' ''Why do you want so badly to go abroad?'' I ask.
''Are the career opportunities better?''

''On the contrary,'' answers her husband. ''If I stay in the home
office I have a good chance to be in top management in ten years; if I
go overseas I'll be tagged as a foreign specialist and never make it. But
that's a cheap price to pay for the freedom one has outside of Japan. I
can't tell you how much Miyeko and I enjoyed the seven years I was
posted to Dusseldorf when we could go together in the evening to
concerts or to the theater and could go hiking and camping on weekends. Now Japanese convention demands that we live with Miyeko's
parents and they expect us to look after them on weekends. And I
never see my wife and daughter. I have to sit and drink in a Ginza bar
every night till eleven during the week, either with my bosses or with
my subordinates. You can't imagine how sick and tired Miyeko and I
are of all this never-ending Japanese togetherness. And all my colleagues who have served overseas feel the same way.''

There is a revival under way of the old and charming Japanese
custom of having one really good and expensive piece of art serve as
the sole decoration of the small and bare Japanese apartment. But what
the young couples now buy when they start out in their own home is
rarely Japanese art; they buy a Picasso etching or pre-Colombian
pottery from Mexico or Peru, or a miniature from Moghul India, or a
terra-cotta figurine supposedly found in an Etruscan tomb.

When, on a hot June Sunday, I made my way through a traffic jam
of strollers on the beaches south of Tokyo, there were, it seemed, the
same young families—father, mother, and two children—that had
been there twenty years earlier. Twenty years ago the father strode
ahead, carrying nothing; the mother followed, dragging one infant,
carrying the other, and weighted down with paraphernalia. Now it was
the young woman who walked ahead holding the older child by the
hand, with the husband following, carrying the baby, the portable TV,
the ice bucket, the sand pails and spades, the lunch boxes, the bal-
loons, and the blow-up animals. Up and down the beach road, cutting
in and out of the stalled traffic, roared young men on motorcycles
picking up single girls. But lo and behold, there suddenly appeared a
brigade of young women on motorbikes looking for young men to pick
up.

According to Japanese folklore one is reborn on one's sixtieth birthday and starts life all over again as a baby. What then did the Empress of Japan choose as most appropriate to this uniquely Japanese tradition when he picked a gift for her husband, the Emperor, on his eightieth (i.e., second twentieth) birthday last spring: an electric razor!

Part Eight

Why Society Is Not Enough

Introduction to Part Eight

The Lutheran Protestantism of my childhood was so "liberal" that it consisted of little more than a tree at Christmas and Bach cantatas at Easter. And the Pastor—attendance at whose religion class, two hours a week, was still mandatory in the Austrian *Gymnasium* of my childhood—hardly aimed much higher. Thus, I was totally unprepared when, barely nineteen—a bored trainee at an export house in Hamburg, employed mainly in copying invoices for shipments of padlocks to India or to East Africa—I encountered Kierkegaard's masterpiece *Fear and Trembling*—accidentally, but, of course, providentially. It was many years before I *understood* what had happened. But I *knew* immediately that something had happened. I knew immediately that I had found a new, a crucial, an existential dimension. I even then probably knew that my own work would be totally in society—though by then I also did suspect that it was not going to be in business and that I was, indeed, most unlikely to become a "commercial success." And even though I later did teach religion for a few years (though only on the side) my *work* has indeed been totally in society. But I knew at once, in those far-back days of 1928, that my *life* would not and could not be totally in society, that it would have to have an existential dimension which transcends society.

Still, my *work* has been totally in society—except for this essay on Kierkegaard. It was written out of despair. For the years after World War II were years of deep despair. I had predicted Hitler's "Final Solution," the killing of all Jews, in my 1939 book *The End of Economic Man* (to be reissued in the Transaction series of my books). Still, the reality of the Holocaust as it unfolded after the end of World War II was infinitely more horrible then the abstractions of my book. I

425

had in the same book portrayed Stalinism as obscene, nightmarish terror and tyranny—and had earned the everlasting hatred of all well-meaning "progressives" for doing so. Still, the reality of Stalinist terror as it unfolded after World War II—the ruthless persecution and mass execution of Russian soldiers, for instance, whose only "crime" was that, being wounded in battle, they had been captured by the Germans, or the murder on Stalin's orders of Czechoslovakia's last true Democrat, Jan Masaryk, were so much worse than anything I could possibly have imagined that I despaired of humanity. And where Communism before World War II was content to try to foment revolution everywhere, Stalin, after the "World Revolution" had proven to be a wet squib, was clearly determined on world conquest by military might. Having already conquered Eastern Europe, he was in those days attacking Greece and Iran and was threatening Norway. And then there was the atom bomb! And all this horror was aggravated by the infatuation of the intellectuals with Stalin and with communist tyranny, and by the prattle of the "Establishment" about our having attained "eternal peace" and "the new world order"—and I, by that time, had of course myself become a member (albeit a junior one) of this Establishment.

"The Unfashionable Kierkegaard" was thus written as an affirmation of the existential, the spiritual, the *individual* dimension of the Creature. It was written to assert that society is not enough—not even for society. It was written to affirm hope.

30

The Unfashionable Kierkegaard

1

The Kierkegaard boom of the last few years is showing the first signs of fatigue. For Kierkegaard's sake I hope it will burst soon. The Kierkegaard of the literary boom is a fellow wit and fellow modern, distinguished from the other members of the smart set mainly by his having lived a hundred years earlier. But this Kierkegaard of the psychologists, existentialists, and assorted ex-Marxists bears hardly any resemblance to the real Kierkegaard, who cared nothing for psychology or dialectics (save to show them to be inadequate and irrelevant), but concerned himself solely with religious experience. And it is this real Kierkegaard who is meaningful for the modern world in its agony. We have neither saint nor poet to make whole the shards of our experience; in Kierkegaard we have at least a prophet.

Like all religious thinkers, Kierkegaard places in the center the question, How is human existence possible?

All through the nineteenth century this question—which before had been the core of Western thought—was not only highly unfashionable; it seemed senseless and irrelevant. The era was dominated by a radically different question, How is society possible? Rousseau asked it; Hegel asked it; the classical economists asked it. Marx answered it one way; liberal Protestantism another way. But in whatever form it is asked, it must always lead to an answer which denies that human existence is possible except in society.

First published in *Sewanee Review*, 1949.

Rousseau formulated this answer for the whole era of progress: whatever human existence there is; whatever freedom, rights, and duties the individual has; whatever meaning there is in individual life—all is determined by society according to society's objective need of survival. The individual, in other words, is not autonomous. He is determined by society. He is free only in matters that do not matter. He has rights only because society concedes them. He has a will only if he wills what society needs. His life has meaning only insofar as it relates to the social meaning and as it fulfills itself in fulfilling the objective goal of society. There is, in short, no human existence; there is only social existence. There is no individual; there is only the citizen.

It is hardly possible to exaggerate the differences between Rousseau's "General Will," Hegel's concept of history as the unfolding of ideas, and the Marxian theory of the individual's determination through his objectively given class situation. But they all gave the same answer to the question of human existence: there is no such thing, there is no such question! Ideas and citizens exist, but no human beings. What is possible is merely the realization of ideas in and through society.

For if you start with the question, How is society possible?, without asking at the same time, How is human existence possible?, you arrive inevitably at a negative concept of individual existence and of freedom: individual freedom is then what does not disturb society. Thus freedom becomes something that has no function and no autonomous existence of its own. It becomes a convenience, a matter of political strategy, or a demagogue's catch phrase. It is nothing vital.

To define freedom as that which has no function is, however, to deny the existence of freedom. For nothing survives in society save it have a function. But the nineteenth century believed itself far too secure in the possession of freedom to realize this. Prevailing opinion failed to see that to deny the relevance of the question, How is human existence possible?, is to deny the relevance of human freedom. It actually saw in the question, How is society possible?, a key to the gospel of freedom—largely because it aimed at social equality. And the break of the old fetters of inequality appeared equivalent to the establishment of freedom.

We now have learned that the nineteenth century was mistaken. Nazism and Communism are an expensive education—a more expensive education, perhaps, than we can afford; but at least we are learning that we cannot obtain freedom if we confine ourselves to the question, How is society possible? It may be true that human existence in freedom is not possible; which is, indeed, asserted by Hitler and the Communists as well as, less openly, by all those well-meaning "social engineers" who believe in social psychology, propaganda, re-education, or administration as a means of molding and forming the individual. But at least the question, How is human existence possible?, can no longer be regarded as irrelevant. For those who profess to believe in freedom, there is no more relevant inquiry.

I am not trying to say that Kierkegaard was the only thinker during the nineteenth century who saw the direction in which Rousseau was leading the Western world. There were the Romanticists, some of whom, especially in France, sensed what was coming. There was the futile and suicidal revolt of Nietzsche—a Samson whose gigantic power pulled down nothing but himself. There was, above all, Balzac, who analyzed a society in which human existence was no longer possible and drew an Inferno more terrible than Dante's in that there is not even a Purgatory above it. But although they all asked, How is human existence possible?, none but Kierkegaard answered.

2

Kierkegaard's answer is simple: human existence is possible only in tension—in tension between man's simultaneous life as an individual in the spirit and as a citizen in society. Kierkegaard expressed the fundamental tension in a good many ways throughout his writings— most clearly and centrally when he described the tension as the consequence of man's simultaneous existence in eternity and in time. He took his formulation from St. Augustine; it is the intellectual climax of the *Confessions*. But Kierkegaard gave to the antithesis a meaning that goes far beyond St. Augustine's speculation in dialectical logic.

Existence in time is existence as a citizen in this world. In time we eat and drink and sleep, fight for conquest or for our lives, raise children and societies, succeed or fail. But in time we also die. And in time there is nothing left of us after our death. In time we do not, therefore, exist as individuals. We are only members of a species, links in a chain of generations. The species has an autonomous life in

time, specific characteristics, an autonomous goal; but the member has
no life, no characteristics, no aim outside the species. He exists only in
and through the species. The chain has a beginning and an end, but
each link serves only to tie the links of the past to the links of the
future; outside the chain it is scrap iron. The wheel of time keeps on
turning, but the cogs are replaceable and interchangeable. The individual's death does not end the species or society, but it ends his life in
time. Human existence is not possible in time; only society is possible
in time.

In eternity, however, in the realm of the spirit, "in the sight of
God," to use one of Kierkegaard's favorite terms, it is society which
does not exist, which is not possible. In eternity only the individual
does exist. In eternity each individual is unique, he alone, all alone,
without neighbors and friends, without wife and children, faces the
spirit in himself. In time, in the sphere of society, no man begins at the
beginning and ends at the end; each of us receives from those before us
the inheritance of the ages, carries it for a tiny instant, to hand it on to
those after him. But in the spirit, each man is beginning and end.
Nothing his fathers have experienced can be of any help to him. In
awful loneliness, in complete, unique singleness, he faces himself as if
there were nothing in the entire universe but him and the spirit in
himself. Human existence is thus existence on two levels—existence
in tension.

It is impossible even to approximate eternity by piling up time; mere
time, even infinitely more time, will still only be time. And it is also
impossible to reach time by subdividing eternity; eternity is inseparable and immeasurable. Yet it is only as simultaneous existence on
both planes, existence in the spirit and existence in society, that human
existence is possible. St. Augustine had said that time is within
eternity, created by eternity, suspended in it. But Kierkegaard knew
that the two are on different planes, antithetic and incompatible with
each other. And he knew it not only by logic and by introspection but
by looking at the realities of nineteenth-century life.

It is this answer that constitutes the essential paradox of religious
experience. To say that human existence is possible only in the tension
between existence in eternity and existence in time is to say that human
existence is only possible if it is impossible: what existence requires on
the one level is forbidden by existence on the other. For example,

existence in society requires that the society's objective need for survival determine the functions and the actions of the citizen. But existence in the spirit is possible only if there is no law and no rule except that of the person, alone with himself and with his God. Because man must exist in society, there can be no freedom except in matters that do not matter; but because man must exist in the spirit, there can be no social rule, no social constraint in matters that do matter. In society, man can exist only as a social being—as husband, father, child, neighbor, fellow citizen. In the spirit, man can exist only personally—alone, isolated, completely walled in by his own consciousness.

Existence in society requires that man accept as real the sphere of social values and beliefs, rewards and punishments. But existence in the spirit, "in the sight of God," requires that man regard all social values and beliefs as pure deception, as vanity, as untrue, invalid, and unreal. Kierkegaard quotes from Luke 14:26, "If any man come to me, and hate not his father, and mother, and wife, and children, and brethren, and sisters, yea, and his own life also, he cannot be my disciple." The Gospel of Love does not say: *love* these *less* than you love me; it says *hate*.

To say that human existence is possible only as simultaneous existence in time and in eternity is thus to say that it is possible only as one crushed between two irreconcilable ethical absolutes. And that means (if it be more than the mockery of cruel gods): human existence is possible only as existence in tragedy. It is existence in fear and trembling; in dread and anxiety; and, above all, in despair.

3

This seems a very gloomy and pessimistic view of human existence and one hardly worth having. To the nineteenth century it appeared as a pathological aberration. But let us see where the optimism of the nineteenth century leads to. For it is the analysis of this optimism and the prediction of its ultimate outcome that gave Kierkegaard's work its vision.

It was the very essence of all nineteenth-century creeds that eternity can and will be reached in time; that truth can be established in society and through majority decision; that permanence can be obtained through change. This is the belief in inevitable progress, representative of the nineteenth century and its very own contribution to human thought. You may take the creed of progress in its most naïve and therefore most engaging form—the confidence that man automatically and through his very sojourn in time becomes better, more nearly perfect, more closely approaches the divine. You may take the creed in its more sophisticated form—the dialectic schemes of Hegel and Marx in which truth unfolds itself in the synthesis between thesis and antithesis, each synthesis becoming in turn the thesis of a new dialectical integration on a higher and more nearly perfect level. Or you may take the creed in the pseudo-scientific garb of the theory of evolution through natural selection. In each form it has the same substance: a fervent belief that by piling up time we shall attain eternity; by piling up matter we shall become spirit; by piling up change we shall become permanent; by piling up trial and error we shall find truth. For Kierkegaard, the problem of the final value was one of uncompromising conflict between contradictory qualities. For the nineteenth century, the problem was one of quantity.

Where Kierkegaard conceives the human situation as essentially tragic, the nineteenth century overflowed with optimism. Not since the year 1000, when all Europe expected the Second Coming, has there been a generation which saw itself so close to the fulfillment of time as did the men of the nineteenth century. Certainly there were impurities in the existing fabric of society. But the liberal confidently expected them to be burnt away within a generation or, at the most, within a century by the daily strengthening light of reason. Progress was automatic. And though the forces of darkness and superstition might seem to gain at times, that was only a momentary illusion. "It is always darkest just before the dawn" is a truly liberal maxim (and one, incidentally, as false in its literal as in its metaphorical sense). The apogee of this naïve optimism was the book which the famous German biologist Ernst Haeckel wrote just before the turn of the century—the one which predicted that all the remaining questions would be finally and decisively answered within a generation by Darwinian biology and Newtonian physics. It is perhaps the best commentary on the fate of the nineteenth-century creed that Haeckel's *Weltraetsel* sold by the millions in the generation of our grandfathers (and still hides out on old bookshelves) at the very moment when the universe of Darwinian biology and Newtonian physics was completely disintegrating.

To those whom the optimism of liberalism or Darwinism failed to satisfy, Marx offered the more complicated but also infinitely more profound vision of a millennium that had to come precisely because the world was so corrupt and so imperfect. His was a truly apocalyptic message in which the impossible, the attainment of the permanent perfection of the classless society, is promised precisely because it is impossible. In Marx the nineteenth-century optimism admits defeat, only to use defeat as a proof of certain victory.

In this creed of imminent perfection, in which every progress in time meant progress toward eternity, permanence, and truth, there was no room for tragedy (the conflict of two absolute forces, of two absolute laws). There was not even room for catastrophe. Everywhere in the nineteenth-century tradition the tragic is exorcised, catastrophe suppressed. A good example is the attempt—quite popular these last few years—to explain so cataclysmic a phenomenon as Hitlerism in terms of "faulty psychological adjustment," that is, as something that has nothing to do with the spirit, but is exclusively a matter of techniques. Or, in a totally different sphere, compare Shakespeare's *Antony and Cleopatra* with Flaubert's *Madame Bovary* and see how the essentially tragic "eros" becomes pure "sex"—psychology, physiology, even passion, but no longer a tragic, that is, an insoluble, conflict. Or one might, as one of the triumphs of the attempt to suppress catastrophe, take the early Communist explanation of Nazism as "just a necessary stage in the inevitable victory of the proletariat." There you have in purest form the official creed that whatever happens in time must be good, however evil it is. Neither catastrophe nor tragedy can exist.

There has never been a century of Western history so far removed from an awareness of the tragic as the one that bequeathed to us two world wars. Just over two hundred years ago—in 1755, to be exact— the death of 15,000 people in the Lisbon earthquake was enough to bring down the tottering structure of traditional Christian belief in Europe. The contemporaries could not make sense of it; they could not reconcile this horror with the concept of an all-merciful God; they could not see any answer to a paradox of catastrophe of such magnitude. For years now we have learned daily of vastly greater destruction, of whole peoples being starved or exterminated. And it is far more difficult to comprehend these man-made catastrophes in terms of our modern rationality than it was for the eighteenth century to comprehend the earthquake of Lisbon in the terms of traditional Christianity. Yet our own catastrophes make no impression on the optimism of those thousands of committees that are dedicated to the belief that permanent peace and prosperity will "inevitably" issue from today's horrors. To be sure, they are aware of the facts and duly outraged by them. But they refuse to see them as catastrophes. They have been trained to deny the existence of tragedy.

4

Yet however successful the nineteenth century was in suppressing the tragic, there is one fact that could not be suppressed, one fact that remains outside of time: death. It is the one fact that cannot be made general, but remains unique; the one fact that cannot be socialized, but remains personal. The nineteenth century made every effort to strip death of its individual, unique, and qualitative aspect. It made death an incident in vital statistics, measurable quantitatively, predictable according to the actuarial laws of probability. It tried to get around death by organizing away its consequences. Life insurance is perhaps the most significant institution of nineteenth-century metaphysics; its proposition "to spread the risks" shows most clearly the nature of the attempt to consider death an incident in human life rather than its termination. And the nineteenth century invented spiritualism—an attempt to control life after death by mechanical means.

Yet death persists. Society might make death taboo, might lay down the rule that it is bad manners to speak of death, might substitute "hygienic" cremation for those horribly public funerals, and might call gravediggers morticians. The learned Professor Haeckel might hint broadly that Darwinian biology is just about to make us live permanently; but he did not make good his promise. And so long as death persists, the individual remains with one pole of his existence outside of society and outside of time.

So long as death persists, the optimistic concept of life, the belief that eternity can be reached through time and that the individual can fulfill himself in society, must have only one outcome—despair. Suddenly every man finds himself facing death; and at this point he is all alone, all individual. If his existence is purely in society, he is lost—for now this existence becomes meaningless. Kierkegaard diagnosed the phenomenon and called it the "despair at not willing to be an individual." Superficially, the individual can recover from this encounter with the problem of existence in eternity; he may even forget it for a while. But he can never regain his confidence in his existence in society. Basically he remains in despair.

Society must make it possible for man to die without despair if it wants him to be able to live exclusively in society. And it can do so in only one way: by making individual life meaningless. If you are nothing but a leaf on the tree of the race, a cell in the body of society, then your death is not really death; you had better call it a process of collective regeneration. But then, of course, your life is not a real life either; it is just a functional process within the life of the whole, devoid of any meaning except in terms of the whole. Thus, as Kierkegaard foresaw a hundred years ago, an optimism that proclaims human existence as existence in society leads straight to despair. And this despair can lead only to totalitarianism. For totalitarianism—and that is the trait that distinguishes it so sharply from the tyrannies of the past—is based on the affirmation of the meaninglessness of life and of the nonexistence of the person. Hence the emphasis in the totalitarian creed is not on how to live, but on how to die; to make death bearable, individual life had to be made worthless and meaningless. The optimistic creed, that started out by making life in this world mean everything, led straight to the Nazi glorification of self-immolation as the only act in which man can meaningfully exist. Despair becomes the essence of life itself.

5

The nineteenth century arrived at the very point the pagan world had reached in the late Roman Empire. And like antiquity, it tried to find a way out by escaping into the purely ethical—by basing virtue on man's reason. The great philosophical systems of German idealism—above all Kant's, but also Hegel's—dominated the age because they identified reason with virtue and the good life. Ethical culture and that brand of liberal Protestantism that sees in Jesus the "best man that ever lived," with its slogans of the Golden Rule, of the "categorical imperative," and of the satisfaction of serve—these and related ethical formulas became as familiar in the nineteenth century as most of them had been in antiquity. And they failed to provide a basis for human existence in modernity, just as they had failed two thousand years before.

In its best representatives, the ethical concept leads, indeed, to moral integrity and moral greatness. Nineteenth-century humanism, based half on Plutarch, half on Newton, could be a noble thing. (We have only to remember the great men of the last nineteenth-century generation, such as Woodrow Wilson, Masaryk, Jaurès, or Mommsen.) Kierkegaard himself was more attracted by it than he realized. Though fighting every inch of the way, he could never quite free himself from the influence of Hegel; and Socrates, symbol of the ethical life, remained to him the apogee of man's natural history.

But Kierkegaard also saw that the ethical concept, while it may give integrity, courage, and steadfastness, cannot give meaning—neither to life nor to death. All it can give is stoic resignation. Kierkegaard considered this position to be one of even greater despair than the optimistic one; he calls it "the despair at willing to be an individual." And only too often the ethical position does not lead to anything as noble and as consistent as the Stoic philosophy, but turns into sugar coating on the pill of totalitarianism. This is, I feel, the position of many an apologist for Soviet Russia; he hopes that man will find individual fulfillment in the ethical attempt at making his neighbor happy and that this will suffice to offset the reality of totalitarianism. Or the ethical position becomes pure sentimentalism—the position of those who believe that evil can be abolished and harmony established by good intentions.

And in all cases the ethical position is bound to degenerate into relativism. For if virtue is to be found in man, everything that is accepted by man must be virtue. Thus a position that starts out—as did Rousseau and Kant some two hundred years ago—to establish man-made ethical absolutes must end in the complete denial of absolutes and, with it, in the complete denial of the possibility of a truly ethical position. This way there is no escape from despair.

Is then the only conclusion that human existence can be only existence in tragedy and despair? Are the sages of the East right who see the only answer in the destruction of self, in the submersion of man into the Nirvana, the nothingness?

Kierkegaard has another answer: human existence is possible as existence not in despair, as existence not in tragedy; it is possible as existence in faith. The opposite of Sin (to use the traditional term for existence purely in society) is not Virtue; it is Faith.

Faith is the belief that in God the impossible is possible, that in Him time and eternity are one, that both life and death are meaningful. Faith is the knowledge that man is creature—not autonomous, not the master, not the end, not the center—and yet responsible and free. It is the acceptance of man's essential loneliness, to be overcome by the certainty that God is always with man, even "unto the hour of our death."

In my favorite among Kierkegaard's books, a little volume called *Fear and Trembling*, he raises the question: What distinguished Abraham's willingness to sacrifice his son, Isaac, from ordinary murder? If Abraham had never intended to go through with the sacrifice, but had intended all the time only to make a show of his obedience to God, then Abraham, indeed, would not have been a murderer, but he would have been something more despicable: a fraud and a cheat. If he had not loved Isaac but had been indifferent, he would have been willing to be a murderer. Yet Abraham was a holy man; God's command was for him an absolute command to be executed without reservation; and we are told that he loved Isaac more than himself. The answer is that Abraham had faith. He believed that in God the impossible would become possible—and he could carry out God's command and yet retain Isaac.

Abraham was the symbol for Kierkegaard himself, and the sacrifice of Isaac the symbol for his own innermost secret, his great and tragic love—a love he had slaughtered although he loved it more than he loved himself. But the autobiographical allusion is only incidental. The story of Abraham is a universal symbol of human existence which is possible only in faith. In faith the individual becomes the universal, ceases to be isolated, becomes meaningful and absolute; hence in faith there is a true ethic. And in faith existence in society becomes meaningful, too, as existence in true charity.

The faith is not what today is so often glibly called a "mystical experience"—something that can apparently be induced by the proper breathing exercises or by prolonged exposure to Bach. It can be attained only through despair, through suffering, through painful and

ceaseless struggle. It is not irrational, sentimental, emotional, or spontaneous. It comes as the result of serious thinking and learning, of rigid
discipline, of complete sobriety, of humbleness, and of the self's
subordination to a higher, an absolute will. The inner knowledge of
one's own unification in God—what St. Paul called hope and we call
saintliness—only a few can attain. But every man is capable of attaining faith. For every man knows despair.

Kierkegaard stands squarely in the great Western tradition of religious experience, the tradition of St. Augustine and St. Bonaventure,
of Luther, St. John of the Cross, and Pascal. What sets him apart, and
gives him this special urgency today, is his emphasis on the meaning
of life in time and society for the man of faith, the Christian.
Kierkegaard is "modern," not because he employs the modern vocabulary of psychology, aesthetics, and dialectics—the ephemeral traits
which the Kierkegaard boom ballyhoos—but because he concerns
himself with the specific disease of the modern West: the falling apart
of human existence, the denial of the simultaneity of life in the spirit
and life in the flesh, the denial of the meaningfulness of each for the
other.

Instead, we have today a complete divorce, the juxtaposition of
"Yogi" and "Commissar"—the terms are, of course, Arthur
Koestler's—as mutually exclusive possibilities: an either-or between
time and eternity, charity and faith, in which one pole of man's dual
existence is made the absolute. This amounts to a complete abdication
of faith: the "Commissar" gives up the entire realm of the spirit for the
sake of power and effectiveness; the "Yogi" assigns human existence
in time (that is, social life) to the devil and is willing to see millions
lose their lives and their souls if only his own "I" be saved. Both are
impossible positions for any religious man to take, but especially for a
Christian who must live in the spirit and yet must maintain that true
faith is effective in and through charity (i.e., in and through social
responsibility).

But at least both are honest positions, honestly admitting their
bankruptcy—in contrast to the attempt at evading the problem by way
of the various "Christian" political parties in Europe, Protestant and
Catholic, or the movement for "Social Christianity" still powerful in
this country. For these attempts substitute morality and good intentions
for faith and religious experience as mainsprings of action. While

sincere and earnest, while supported and sometimes led by good, even by saintly men, they must not only be as ineffectual in politics as the "Yogi" but must also fail, like the "Commissar," to give spiritual life; for they compromise both life in time and life in eternity. That Austrian cleric and Catholic party leader who, in the thirties, came out for Hitler with the argument, "At least he is opposed to mixed bathing," was a ghastly caricature of the Christian moralist in politics; but he caricatured something that is ever present where morality is confused with faith.

Kierkegaard offers no easy way out. Indeed it could be said of him, as of all religious thinkers who focus on experience rather than on reason and dogma, that he greatly overemphasizes life in the spirit, thus failing to integrate the two poles of human existence into one whole. But he not only saw the task; he also showed him his own life and in his works that there is no escape from the reality of human existence, which is one in tension. It is no accident that the only part of Kierkegaard's tremendous literary output that did not originally appear under a pseudonym but under his own name was the *Edifying Discourses*. Not that he wanted to conceal his authorship of the other works—the pseudonyms could not have fooled anybody; but the "edifying" books alone translate faith into social effectiveness and are thus truly religious and not just "Yogi." It is also not an accident that Kierkegaard's whole work, his twenty years of seclusion, of writing, thinking, praying, and suffering, were but the preparation for the violent political action to which he dedicated the last months of his life—a furious one-man war on the established church of Denmark and its high clergy for confusing morality and tradition with charity and faith.

Though Kierkegaard's faith cannot overcome the awful loneliness, the isolation and dissonance of human existence, it can make it bearable by making it meaningful. The philosophy of the totalitarian creeds enables man to die. It is dangerous to underestimate the strength of such a philosophy; for, in a time of sorrow and suffering, of catastrophe and horror (that is, in our time), it is a great thing to be able to die. Yet it is not enough. Kierkegaard's faith, too, enables man to die; but it also enables him to live.

Afterword:
Reflections of a Social Ecologist

When asked what I do, I say: "I write." Technically this is correct. Since I was twenty, writing has been the foundation of everything else I have been doing, such as teaching and consulting. And among my twenty-five books there are two novels and one quasi-autobiographical volume of stories. Yet I certainly am not a "literary person" and have never been mistaken for one.

But when people then ask: "What are you writing about?", I become evasive. I have written quite a bit about economics but surely am not an economist. I have written quite a bit about history but surely am not a historian. I have written quite a bit about government and politics; but though I started out as a political scientist I long ago moved out of the field. And I am also not a sociologist as the term is now defined. I myself, however, know very well—and have known for many years—what I am trying to do. I consider myself a "social ecologist" concerned with man's man-made environment the way the natural ecologist studies the biological environment. Even my two novels, while pure fiction, are social ecology. The central character in one is European society before the First World War; the central character in the other is an American Catholic university around the year 1980.

The term "social ecology" is of my own coinage. But the discipline itself—and I consider it a discipline—boasts old and distinguished lineage. Its greatest document is Alexis de Tocqueville's *Democracy*

Written for this essay volume in 1992.

in America. I count among its leading practitioners another French-
man, Bertrand de Jouvenel, two Germans, Ferdinand Toennies and
Georg Simmel, and three Americans, Henry Adams, John R. Commons (whose "Institutional Economics" was not too different from
what I call Social Ecology) and, above all, Thorstein Veblen.

But none of these is as close to me in temperament, concepts, and
approach as a mid-Victorian Englishman: Walter Bagehot. Living, (as
I have lived) in an age of great social change—he died, aged fifty-one,
in 1877—Bagehot first saw the emergence of new institutions: civil
service and cabinet government as the cores of a functioning democracy, and banking as the center of a functioning economy. Similarly I
was the first, a hundred years later, to identify management as the new
social institution of the emerging society of organizations, and, a little
later, to spot the emergence of knowledge as the new central resource
and of knowledge workers as the new ruling class of a society that is
not only "postindustrial" but postsocialist and, increasingly, post-
capitalist. Like Bagehot I see as central to society and to civilization
the tension between the need for continuity (Bagehot called it "the
cake of custom," I call it civilization) and the need for innovation and
change. Thus, I know what Bagehot meant when he said that he saw
himself sometimes as a liberal Conservative and sometimes as a
conservative Liberal but never as a "conservative Conservative" or a
"liberal Liberal."

The Themes

It was with the tension between continuity and change that my own
work began.

I was barely twenty at the time, in early 1930. I was then living in
Frankfurt and working as foreign and financial writer for the city's
main afternoon paper. But I also attended law school at the university
where I studied international relations and international law, political
theory, and the history of legal institutions. The Nazis were not yet in
power. Indeed reasonable people were quite sure that they would never
get into power—I knew even at that time that they were likely to be
wrong. All around me society, economy, and government—indeed
civilization—were collapsing. There was a total lack of continuity.

And this drew my attention to that remarkable trio of German thinkers who in a similar period of social collapse, a little over a hundred years earlier, had created stability by inventing what came to be known as *Der Rechtsstaat*—the best translation of that difficult word may be the Justice State. They were a remarkable trio, both because of the breadth of interests and activities of each of them, but also because they were respectively an agnostic Protestant, a romantic Catholic, and a converted Jew. The first of them, Wilhelm von Humboldt (1767-1835) was the last great figure of the European Enlightenment, a leading statesman during the Napoleonic Wars, the founder of the first modern university, the University of Berlin in 1809, and later the founder of scientific linguistics. The second, Joseph von Radowitz (1797-1853) was a professional soldier and the King's confidant and first minister, but also a crusading magazine editor and the progenitor of all Catholic parties in Europe—in Germany, in France, in Italy, in Holland, in Belgium, in Austria.

The third and last, Friedrich Julius Stahl (1802-1861), was Hegel's successor as professor of philosophy at the University of Berlin. A legal philosopher he also revived the moribund theology of Lutheran Protestantism. And he was the most brilliant parliamentarian, in fact the only brilliant parliamentarian in German history.

The three do not enjoy a good press. They are suspect precisely because they tried to balance continuity and change, that is, because they were neither unabashed liberals nor unabashed reactionaries. They tried to create a stable society and a stable polity that would preserve the traditions of the past and yet make possible change, and indeed very rapid change. And they succeeded brilliantly. They created the only political theory that originated on the continent of Europe in modern times—at least until Karl Marx fifty years later. But they also created a political structure that survived for almost a hundred years, until it came crashing down with World War I.

Its failure, the inability to get the military under civilian control—the major cause of the collapse of nineteenth-century Europe—the *Rechtsstaat* shared with all continental regimes, whether democracies, like the France of the Second Republic (as was demonstrated only too clearly in the Dreyfus Affair of 1896), or the absolute monarchy of the Russian Tsar—or of course with the constitutional monarchy of

Meiji Japan as well. During the entire nineteenth century only English-speaking countries succeeded in establishing civilian control of the military and with it civilian control of foreign policy.

Of course, Humboldt, Radowitz, and Stahl did not realize that what they were trying to do had actually been accomplished in the United States. They did not realize that the United States Constitution first and so far practically alone among written constitutions, contains explicit provisions how to be changed. This probably explains more than anything else why, alone of all written constitutions, the American Constitution is still in force and a living document. Even less did they realize the importance of the Supreme Court as the institution which basically represents both conservation and continuity, and innovation and change and balances the two.

But then nobody else in Europe has ever seen this. Tocqueville did not see this. Nor did Bagehot, otherwise a shrewd student of America, nor did Bryce in his monumental work on the American Common-wealth. To this day, by the way, few people in Europe seem to understand this. At least I have never been able to explain to any European the particular role of the Supreme Court—nor by the way was Mr. Justice Holmes able to explain it to his great English friend, Sir Frederick Pollock, as witness their correspondence.

To a European it is simply axiomatic that an independent judiciary and a political role are incompatible and are in fact a contradiction in terms.

And I too, it should be said, had no inkling in 1930 that what the three Germans in the early years of the nineteenth century had tried to accomplish had already been done, and far more successfully, by the Founding Fathers and by Chief Justice Marshall in the infant United States.

What Humboldt did was to balance two conserving institutions: a professional and university-trained civil service and a professional army, with two innovating institutions: the research university with complete freedom of research, publishing, and teaching, and the free-market economy on Adam Smith's prescription. The Monarch—a strong executive, very similar to the way Humboldt saw George Washington in far away America—would preside above the four and would serve as the balancing wheel. And, to repeat, this worked for a hundred years. While far from a ''liberal democracy''—in fact there was not even a parliament in Humboldt's original formula—it put king, civil service, and, at least in theory, the military under the Law; and at the same time it provided the foundation for the explosive rise to world leadership of the German university and of the Germany economy.

The book on the three and on the *Rechtsstaat* became the first of many, many books I should have written but did not write, or at least did not finish. The only part of it that was published was a monograph on Stahl—all of thirty-two printed pages. It was written during 1932 as an anti-Nazi manifesto—for Stahl the great conservative had been a Jew. It was meant to make it impossible for the Nazis ever to have any use for me and for me ever to have any use for them. And it was meant also, in the event of a Nazi victory to be among the first books they would banish. It succeeded in its objectives. It was accepted by Germany's leading social science publisher, Mohr in Tuebingen, and published as Number 100 in the prestigious series on law and government—a singular honor for a totally unknown, twenty-three year old. It came out—a pure coincidence, but still very much to my delight—two weeks after Hitler and seized power in 1933—and it was promptly banned by the Nazis.

What made me abandon the work on the *Rechsstaat* was, of course, the coming to power of the Nazis in Germany, and with it the failure to maintain any continuity. Out of this came my first major book *The End of Economic Man* (finished in 1937, published in early 1939). It chronicles the collapse of a society that had lost all continuity and all belief, a society plunged into abject fear and despair. This was then followed by the *Future of Industrial Man* in 1942 in which I tried to develop social theory and social structure for an industrial society which can both conserve and innovate. This lead directly to my work on the analysis of the institution through which industrial society gives status and function and integrates individual efforts into common achievement, *Concept of the Corporation* published in 1946.

But over the years I began to realize that change too has to be managed. In fact, I came to realize that the only way in which an institution, whether a government, a university, a business, a labor union, an army, can maintain continuity is by building systematic, organized, innovation into its very structure. This finally lead to my

1986 book *Innovation and Entrepreneurship* which tries to develop a discipline of innovation as a systematic activity.

My concern with the tension between continuity and change as a central polarity in society also led to a growing interest in technology. But even though I got to serve as president of the Society for the History of Technology, my interest has never been in technology as technology. I became interested because I found that technology has never been integrated into the study of society. The technologists look upon technology as having to do with tools. Historians, economists, philosophers—excepting only Karl Marx and Joseph Schumpeter—see technology as a demonic force outside their universe and perpetually threatening it. I see technology as a human activity in society. I see technology in fact as Alfred Russell Wallace, a theorist of evolution and a contemporary of Charles Darwin saw it: "Man" Wallace said, "is the only animal capable of conscious evolution; he invents tools."

And it soon became obvious to me that *work* is a central factor in shaping and molding society, social order, and community. In fact to me it became more and more clear that society is held in tension between two poles, the pole of great ideas, especially of course great religious ideas, and the pole of how man works. To me therefore technology deals with how man works rather than with tools per se. And so I began to think about a book tentatively entitled *"A History of Work."* It too is one of the many books I never wrote. But the products of this interest, a number of short papers on technology's history and its characteristics, are included in this volume.

Out of the concern with the tension between continuity and change came finally my interest in organizations. It became clear to me, during the early days of World War II, that we had moved or were moving into a society of organizations in which major social tasks are being performed in and through managed institutions. The first one to attract my attention was the business enterprise—for the simple reason that it was the institution through which the tasks of war-time America were being discharged, the institution which enabled America to win the war.

I have already mentioned my book *Concept of the Corporation* (1946) which was the first book to attempt the study of a major enterprise—in this case the General Motors Corporation, then the world' premier manufacturing business and its most successful one—from the inside, as a social organization, and as one that organizes power, authority, responsibility, that is the tasks which always had been seen as tasks of governance if not of government. This book was then followed in 1949 by *Landmarks of Tomorrow*—and ten years later by *The New Society*. In *The New Society* I first talked of a society of organizations but also of knowledge work and knowledge worker as becoming the new social and economic centers. Fifteen years later I then commented on another major change in the society of organization in a 1976 book (reprinted in the Transaction series under the title *The Pension Fund Revolution*); the original title was *The Unseen Revolution*.

When I began to study organizations in the 1940s everybody, including myself, saw only two: government, the old one, and business enterprise, the new one. By the late 1940s I myself had already begun to work actively with nonprofit institutions, for example, hospitals. But I did so as something ordinary decency demanded of a citizen. I did it as an act of charity in other words. It slowly dawned on me—I must admit very slowly—that these institutions were of importance in their own right and particularly in America. It gradually dawned on me that this "third sector" is fundamentally distinct from both government and business. Government commands; it tries to obtain compliance. Business supplies; it tries to get paid. The nonprofit institutions however are human-change agents. Their "product" is neither compliance nor a sale. It is a patient who leaves the hospital cured. It is a student who has learned something. It is a churchgoer whose life is being changed.

In addition, as I learned, gradually these institutions discharge a second and equally important task in American society: they provide effective citizenship. In modern society direct citizenship is no longer possible. All we can do is vote and pay taxes. As volunteers in the nonprofit institutions we are again citizens. We again have an impact on social order, social values, social behavior, social vision. We have social results. It is thus the nonprofit institution which increasingly creates citizenship and community. And in the United States every other adult by 1990 had come to work as a "volunteer" in a nonprofit institution, on average three hours a week. And so increasingly I began to work with the nonprofit sector as a major constituent of society—the end result was my 1990 book *Managing the Non-Profit Organization.*

But long before that, of course, I had begun to study the management of organizations—at first of the business enterprise and then of organizations altogether—in my various management books, beginning with my 1954 *Practice of Management* leading to the first book on what is now called strategy: *Managing for Results* (1964), then the *Effective Executive* (1966), an attempt to project what had been known from Plato to Machiavelli as "the education of the ruler" into "the education of the executive in the organization." In 1973 I then put together twenty years of management work in a big book *Management, Tasks, Responsibilities, Practices.*

The focus of my early work, the work from the 1930s until the early 1960s, was on the new social phenomenon, the organization, its structure, its constitution, its management, its functions. But I knew myself from the early days to be out of sympathy with the basic trend of my times, the trend towards centralization, towards monolithic society and towards government omnipotence.

This trend had its origin in World War I which was as much a triumph of the civilian bureaucracy as it was an abject failure of military leadership. The Western world came out of World War I convinced of its ability to predict the future and to make it through planning. It came out of it with the belief that there was no limit to government's ability to tax and therefore to government's ability to spend—of all the lessons of World War I this was probably the most deleterious one. And it came out with the conviction that government could do and in fact should do everything. Everybody believes that democracies and totalitarian regimes had nothing in common in the seventy years between 1920 and the collapse of communism. This is half-truth only. Both shared the belief in centralization and government, a belief that probably reached its peak during the Kennedy years in the United States. Indeed the belief was so prevalent that Stalinism, that most traditional of oriental despotisms, could wrap itself in the cloak of being "progressive" and indeed "revolutionary" by waving the banner of planning, of state control, of total centralization.

Of course, there was always opposition. But the opposition did not as a rule question the competence and ability of government to do everything and to be everything. It questioned the wisdom of doing so. When Friedrich von Hayek (1899–1991) in 1944 published *The Road to Serfdom*—and thereby started what we now call neoconservatism— he pointed out that socialist government must be a tyranny. But he did not question its ability to perform politically or economically (he did this only more than forty years later in his 1988 book *The Fatal Conceit* by which time the total incompetence, political, social, moral, economic, of socialist regimes had become clear to everybody).

In other words for five decades we asked what *should* government do? Very few people, if any, asked what *can* government do.

To me this question became, however, a central question fairly early. I probably owe this also to my interest in those German liberal conservatives (or conservative liberals), Humboldt, Radowitz, and Stahl. For Humboldt, while a very young man, barely twenty-five, and living as a student in the Paris of the French Revolution, had written a little book called *The Limits of the Effectiveness of Government* (*Die Grenzen der Wirksamkeit des Staates*)—which however was not published during his lifetime, being much too unfashionable even then. I began to ask the same question: what are the limits of government's effectiveness? in the early years after World War II and began to ask it with increasing urgency as we went into the Eisenhower administration. I first raised it obliquely in my 1959 book *Landmarks of Tomorrow* and then head-on, in my 1969 book *The Age of Discontinuity*. It is a central theme of my 1989 book The New Realities and is being discussed in considerable depths in a chapter entitled "the Megastate and Its Failure" in a book I am presently (1992) working on, ten- tatively entitled *The Post-Capitalist Society*.

And thus my work began to embrace both the political and the social institutions of modern society.

By the late 1950s another major theme had begun to appear in my work: the emergence of knowledge as central resource, and of the knowledge society (a word I coined in the late 1950s). The characteris- tics of knowledge—and it is totally different from any other resource in society and economy—the responsibilities of knowledge; the place and function of the knowledge worker; and the productivity of knowledge work, are themes discussed in my writing, and in a good many of the articles in this volume.

Finally there is one continuing theme, from my earliest to my latest book: the freedom, the dignity, the status of the person in modern society, the role and function of organization as instrument of human achievement, human growth and human fulfillment, and the need of the individual for both, society and community.

The Work of the Social Ecologist

If social ecology is a discipline, it not only has its own subject matter. It also has its own work—at least for me. But it is easier to say what the work is not than to be specific about what it consists of.

I am often called a "futurist." But if there is one thing I am not—one thing a social ecologist must not be—it is a "futurist." In the first place it is futile to try to foresee the future. This is not given to mortal man. And the idea that ignorance and uncertainty become vision by being put into a computer is not a particularly intelligent one. One problem is that the things the most brilliant and most successful predictor never predicts are always the things that are more important than the things he does predict. Futurists, always measure their batting average by how many of things they have predicted came true. They never count how many of the important things that came true they did not predict.

A good example is the most successful futurist in recorded history, the French science fiction writer Jules Verne (1828-1905). Most of the technologies he predicted have come true. But he—probably quite unconsciously—assumed that society and economy would remain what they were around 1870—and the changes in society and economy have, of course, been at least as important as the new inventions.

But also, and more important, the work of the social ecologist is to identify the changes that have already happened. The important challenge in society, economy, politics is to exploit the changes that have already occurred and to use them as opportunities. The important thing is to identify the "Future That Has Already Happened"—the tentative title of another book which I did not write, a book which was intended to develop the methodology for perceiving and analyzing these changes which had happened—and irreversibly so—but which had not yet had a impact, and were indeed not yet generally seen.

One example is my identifying around 1960 the social and industrial turnaround which had already occurred in Japan and which, within another ten years, propelled Japan into the first rank of economic powers. It was not hard to do so; one only had to look. But nobody else looked at that time. (While I did not write the "The Future That Has Already Happened," I incorporated a good deal of the material into my 1986 book *Innovation and Entrepreneurship* which shows how one systematically looks to the changes in society, in demographics, in meaning, in science and technology, as opportunities to make the future.)

While I am wrongly acclaimed as a "futurist," I am equally wrongly criticized for not being a quantifier. Actually I am an old quantifier. In 1929, not yet twenty, I published one of the first econometric studies. Like so much of this work it proceeded from a self-evident assumption by means of impeccable mathematics to an asinine conclusion to whit that the New York Stock Market could go only one way, that is up. The way such things go, this paper appeared in a highly prestigious economic journal in August or September of 1929, just a few weeks before the event which I had just proven to be "impossible," actually occurred—that is a few weeks before the October 1929 Stock Exchange Crash (fortunately, there is no copy of the journal left, at least not in my possession). It was the last prediction, by the way, I allowed myself to indulge in. But I also taught statistics at one time, and helped organize the first operations research departments in American industry (in GE and the Bell Telephone System), and thus can hardly be accused of being unfamiliar with quantitative methods.

But quantification is scaffolding rather than the building itself. And one removes the scaffolding when the building is finished. More important, quantification for most of the phenomena in a social ecology is misleading or at best useless. The data are much too poor—as the great mathematician and methodologist of quantification, Oskar von Morgenstern (1902-1977) showed in his 1950 book *On the Accuracy of Economic Observation*. Moreover, most of the socioeconomic figures we use in the United States (gross national product for instance, balance of payments, occupational classifications in the census, and many more) are based on aggregates developed in the 1920s, and especially in the U.S. Department of Commerce while Herbert Hoover was its secretary from 1924-1928. And it is a rule in social and economic statistics that aggregates if more than thirty or forty years old are highly suspect; they must be assumed to be outdated.

But the most important reason why I am not a quantifier is that in social and affairs, the event that matters cannot be quantified. It is the unique event that changes the statistical "universe" and with it what is "normal distribution."

One example is Henry Ford's ignorance in 1900 or 1903 of the prevailing economic wisdom according to which the way to maximize profit is to be a monopolist, that is to keep production low and prices high. Instead of which he assumed—foolishly according to his partners and to the other people who had invested in his business—that the way to make money was to keep prices low and production high. This, the invention of "mass production," totally changed industrial economics—though a good many economists still have not heard of it. It would have been impossible, however, to quantify the impact even as late as 1918 or 1920, that is a full ten years after Ford's success had made him the richest industrialist in the United States (and probably in the world), had revolutionized industrial production, the automobile industry and the economy altogether, and had, above all, totally changed our perception of industry.

The unique event that changes the universe is an event "at the margin." By the time it becomes statistically significant, it is no longer "future" it is indeed no longer even "present." It is already "past."

To quantify social events that make a difference we would need a mathematics that was first called for by the seventeenth-century philosopher and mathematician Gottfried Leibnitz—the co-inventor of the calculus. He spoke of a "calculus of relevance," that is of a calculus of qualitative change. But nobody even responded to the challenge for 250 years until John Maynard Keynes in his first major work. *A Treatise on Probability* (1911) tried to develop the statistics of the unique event—totally without success as he himself later admitted. New mathematical theories now promise to define the *probability* of the unique event. But so far there is not even the slightest sign of a quantitative method to identify and to define the unique event, that is of a quantitative method to show changes in meaning.

And until such a quantitative method is developed—if it ever will—

the social ecologist has to use qualitative means to find and assess qualitative change. This, however, is not "guesswork." It is not "hunch." It has to be a rigorous method of looking, of identifying, of testing—what such a method has to be I tried to develop, at least in rough outline, in my 1986 book *Innovation and Entrepreneurship*.

Now as to what the work of the social ecologist is or should be, at least for me.

First of all, it means looking at society and community with the question: What changes have already happened that do not fit "what everybody knows;" what are the "paradigm changes," to use the now popular term? The next question: Is there any evidence that this is a *change* and not a fad? The best test: Are there results of this change? Does it make a difference, in other words? And, finally, one then asks: If this change is relevant and meaningful, what opportunities does it offer?

A simple example would be the emergence of knowledge as a key resource. For me, the decisive change, the event that alerted me that something was happening, was the GI Bill of Rights in the United States after World War II. This bill gave every returning veteran the right to attend college with the government paying the bill. It was a totally unprecedented development. In retrospect it is easy to see that it was the GI Bills of Rights that ushered in the knowledge society. But nobody saw it in this light in 1947 or 1948. I, however, asked myself a simple question: "If there had been such a Bill of Rights after World War I, would it have had an impact? Would many people have taken advantage of it?" It did not of course occur to anybody in 1918—less than thirty years before the post-World War II bill—to offer such a veteran's benefit, though 1918 America was as generous to its veterans as 1946 America, if not more so. But if such a bill had been offered in 1919 and had passed the Congress—a most unlikely assumption— practically nobody would have taken advantage of it. Then going to college was not seen as a meaningful opportunity for anyone but the children of the wealthy who did not need a GI Bill of Rights, or for a few who were so talented that their poverty did not matter. Universal *high school* education became meaningful after World War I; universal *college education* would have been meaningless then.

This then led to the question, "What impact does this have on expectations, on values, on social structure, on employment, and so on?'' And once this question was asked—and I first asked it in the late 1940s—it became at once clear that knowledge had attained a position in society and as a productive resource which it had never had before in human history. It became very clear that we were on the threshold of a major change.

Ten years later, that is by the mid-fifties, this had become so clear that one could with confidence talk of a "knowledge society," of "knowledge work" as the new center of the economy, and of the "knowledge worker" as the new work force and the one in the ascendant.

The next thing to say about the work of social ecologist is that he must aim at impact. His goal is not knowledge; it is right action. In that sense social ecology is a practice—as is medicine, or law, or for that matter the ecology of the physical universe. Its aim is to maintain the balance between continuity and conservation on the one hand, and change and innovation on the other. Its aim is to create a society in dynamic disequilibrium. Only such a society has stability and indeed has cohesion.

This implies, however, that the social ecologist has a responsibility to make his work easily accessible. It rules out being "erudite." In fact, in social ecology being "erudite" is incompatible with, and the foe of, being "learned." The conceit that science is not "science," is in fact not "respectable" unless it is unaccessible, is obscurantism. Before World War II the first-rate scholars in America, and especially scholars in the social and political sciences, wrote for the educated layman—as did the first rate historians of that day, as did Reinhold and Richard Niebuhr both, Margaret Mead and Ruth Benedict, and even the economists of that day, and the psychologists. Max Weber wrote simply, clearly, and accessibly and published in magazines for the laity and even in the daily press—and so did Thorstein Veblen.

In 1927 a French philosopher, Julien Benda (1867-1956), published *La Trahison de Clerks* (*The Treason of the Intellectuals*), a blistering attack on the intellectuals of his day who for fashion's sake betrayed their duty and embraced racism and demagoguery, whether Nazism or Communism. The following decades then amply proved Benda's criticism—by the willingness of intellectuals all over Europe—and not just in Germany—to support Hitler, and equally by their willingness to support if not to idolize, Stalin.

I consider the obscurantism of today's intellectuals equally to be betrayal and treason. In large part they bear the blame for the debasement of culture, especially in the United States. The intellectuals themselves plead that the laity has lost receptivity to knowledge, to science, to discourse and to reason. But this is simply not true. Whenever a scholar deigns to write decent prose he or she immediately finds a wide audience. I myself am an example. But so was Barbara Tuchman, among historians, Rachel Carson or Loren Eisele among ecologists of the physical universe, Irving Louis Horowitz among sociologists, and a good many others. The receptivity is there, and so is the need. What today passes for scholarship is nothing but arrogance.

This arrogance is least justifiable for a social ecologist. His job is not to create knowledge. It is to create vision. He has to be an educator.

Finally, the work of the social ecologist requires, I am convinced, respect for language.

I would have respected language regardless of my field. The Vienna in which I was born in 1909 was extremely language-conscious. With his 1899 book *Zur Kritik der Sprache* (*The Critique of Language*), an Austrian, Fritz Mautner (1849-1923), founded what we now call the philosophy of language. His book was in the library of every educated Viennese—as it was on the bookshelf of my home. Mautner first pointed out that language is not "message." It is not "medium." It is meaning as well. In fact, Ludwig Wittgenstein is Mautner's heir, and, so to speak, Mautner's testamentary executor—most of what is in Wittgenstein Mautner had groped for forty and fifty years earlier. But then, the Vienna in which I grew up, was also the home of Karl Kraus (1874-1936), arguably the greatest master of the German language in this century. And for Kraus language was morality. Language was integrity. To corrupt language was to corrupt society and individual alike.

I thus grew up in an atmosphere in which language was taken seriously. And then at nineteen and a trainee in an export firm in

Hamburg, I encountered Kierkegaard—then still almost unknown outside his native Denmark, and largely untranslated. The preface to my essay on Kierkegaard in this volume recounts the impact of this experience on me, my work, and my life. And Kierkegaard preached the sanctity of language. For Kierkegaard, language is aesthetics and aesthetics is morality. Long before George Orwell I therefore knew that the corruption of language is the tool of the tyrant. It is both a sin and a crime.

For the social ecologist, language is however doubly important. For language is in itself social ecology. For the social ecologist language is not "communication." It is not just "message." It is substance. It is the cement that holds humanity together. It creates community and communion. Thus I always thought that the social ecologist has a responsibility to language. Social ecologists need not be "great" writers; but they have to be respectful writers, caring writers.

The Discipline

I called social ecology a "discipline." I did not call it a "science." Indeed I would be quite uncomfortable if anyone spoke of social ecology as a science. It is as different from any of the social sciences as the ecology of the physical universe is different from any of the physical sciences.

"Zum Sehen Geboren; Zum Schauen Bestellt"

(Born to See; Meant to Look)

sings the look-out in Goethe's *Faust*. This, I think, is the motto of the social ecologist and of social ecology as a discipline. It is based on looking rather than on analysis. It is based on perception. This, I submit, distinguishes it from what is normally meant by a science. It is not only that it can not be reductionist. By definition it deals with configurations. They may not be greater than the sums of their parts. But they are fundamentally different.

But also social ecology as a discipline deals with action. Knowledge is a tool to action rather than an end in itself. Social Ecology, as I said before, is a "practice."

Finally, social ecology is not "value free." If it is a science at all, it is a "moral science"—to use an old term that has been out of fashion for 200 years. The physical ecologist believes, must believe, in the sanctity of natural creation. The social ecologist believes, must believe, in the sanctity of spiritual creation.

There is a great deal of talk today about "empowering people." This is a term I have never used and never will use. Fundamental to the discipline of social ecology, as I see it, is not a belief in power. It is the belief in responsibility, in authority grounded in competence, and in compassion.

Index